ANALYSIS AND DESIGN OF MOSFETs

MODELING, SIMULATION, AND PARAMETER EXTRACTION

ANALYSIS AND DESIGN OF MOSFETs

Modeling, Simulation and Parameter Extraction

by

J. J. Liou
Department of Electrical and Computer Engineering
University of Central Florida
Orlando, Florida USA

and

A. Ortiz-Conde and F. Garcia-Sanchez
Electronics Engineering Department
University of Simon Bolivar
Caracas, Venezuela

Springer Science+Business Media, LLC

ISBN 978-1-4613-7473-2 ISBN 978-1-4615-5415-8 (eBook)
DOI 10.1007/978-1-4615-5415-8

Library of Congress Cataloging-in-Publication Data

A C.I.P. Catalogue record for this book is available
from the Library of Congress.

Originally published by Kluwer Academic Publishers in 1998
Softcover reprint of the hardcover 1st edition 1998

Printed on acid-free paper.

Contents

Appendices

Preface

Since the early 1980's, the metal-oxide-semiconductor field-effect transistor (MOSFET) has become the most widely used semiconductor device in very large scale integrated-circuits. This is due manly to the fact that the MOSFET has a simpler structure, costs less to fabricate, and consumes less power compared to its bipolar transistor counterpart. A main disadvantage of the MOSFET is its limited transconductance, and thus limited switching speed, although such a drawback has been reduced by the continuously decreasing of the MOSFET size due to the advance of photolithography and other processing technologies.

This book covers important and up-to-date knowledge and information about the physical mechanisms governing the MOSFET behavior, the insight and approach of MOSFET modeling and simulation, as well as issues related to extraction of MOSFET device parameters. It is intended to serve as a text for a one-semester course for first-year graduate students and qualified seniors in electrical engineering, and as a reference for engineers and researches involving with design, analysis, and fabrication of MOSFET devices and circuits. The readers should have acquired a basic understanding of semiconductor device processing and operations through a fundamental semiconductor device course normally offered in the sophomore or junior level. A unique feature of the book is its integration of MOSFET device physics, modeling, simulation, and extraction methods for the main device parameters such as the effective channel length, threshold voltage, and drain and source series resistances. In particular, the usefulness and applications of device simulation to device parameter extraction are clearly demonstrated.

The book is organized into six chapters. In Chapter 1, the physics and analytical modeling of MOSFET are treated. First, a brief background on the evolution of MOSFET and its integrated circuits is provided. Then the fundamentals of the MOS system, properties and physical insight of the surface region of MOSFET, such as the potential, electric field, and free-carrier charges, are discussed. Various MOSFET models developed in the past thirty years are then derived. This is followed by the discussions of the MOSFET behavior in weak and moderate inversion conditions. The increasingly important short-channel, narrow-channel, and hot-carrier effects are considered, and the quantum-mechanical effects prominent in modern MOS devices is also addressed. Finally, MOSFETs with advanced structures, such as the silicon-on-insulator

(SOI) and light-doped drain (LDD) are presented. The LDD structure reduces the electric field in the channel and thus reduces the hot-carrier effect, whereas the SOI structure provides the advantages of reduced latch-up and power consumption.

Chapter 2 first introduces the existing device simulators and then focuses on a two-dimensional device simulator called MEDICI. The basic equations used in MEDICI are introduced, and the physical mechanisms, such as the free-carrier mobility and recombination statistics, implemented in MEDICI are discussed. Also addressed are the numerical methods available in MEDICI, and the grid specifications for defining device structure meshes in MEDICI. Examples of MOSFETs and LDD MOSFETs simulations under various bias conditions are then presented to illustrate the capability of MEDICI. A brief discussion on the three-dimensional device simulator is also given at the end of the chapter.

Chapter 3 covers the topic of extracting the threshold voltage of MOSFETs. An overview is first provided to discuss and compare the advantages and disadvantages of various existing extraction methods for the threshold voltage. A new and improved extraction method is then developed and verified in the environment of circuit simulation, device simulation, and measurements. The physical mechanisms underlying a unique phenomenon called the threshold voltage reversal occurred in short-channel MOSFETs is also investigated in this chapter. The last subject covered in the chapter is the determination of threshold voltage shift due to quantum mechanical effects associated with the thin oxide and heavily doped substrate in modern MOS devices.

Chapter 4 is concerned with the various methods for extracting the effective channel length, probably the most important device parameter of MOSFET. They include a method based on metallurgical junctions, current-voltage method, capacitance-voltage method, shift and ratio method, and method based on device simulation. The procedures and developments of these methods are discussed in details, and their accuracy, advantages, and disadvantages are compared.

Chapter 5 deals with the extraction of the drain and source resistances of MOSFET, which are important device parameters in characterizing the voltage drops in the drain and source regions of MOSFET. Models and procedures for extracting both the total drain and source resistance and the difference between the drain and source resistances are developed, which can then be used to determine the individual values of the drain and source resistances. The results extracted from measurements as well as from device simulations are presented and discussed.

Chapter 6 treats the parameter extraction of LDD MOSFETs. The validity of the widely used Terada-Muta method, developed intended for conventional MOSFETs, for extracting the effective channel of LDD MOSFET is first investigated. Three other methods for extracting the effective channel length of LDD MOSFET are then discussed: gate-voltage-dependent effective channel length method, constant effective channel length method, and method based on device simulation. A capacitance-based method to extract the metallurgical channel length of LDD MOSFET is also presented. This is followed by the discussions of the extraction of the drain and source series resistances, as well as their dependence on the gate-oxide thickness of LDD MOSFET.

This is the first book devoted entirely to the MOSFET covering all aspects of modeling, simulation, and parameter extraction. All the chapters contain useful and simple figures to illustrate the trends and characteristics of the MOSFET, and some computer files for device simulation are also included in Chapter 2. Fairly extensive references have also been given as an aid to the reader who wishes to carry out an in-depth study of a particular topic.

It is our sincere hope that this book will be useful to engineers and researchers who are dealing with MOSFET projects or are interested in the topic.

Acknowledgments

The authors are indebted to the anonymous reviewers for their comments and suggestions. Special thanks are due to the students taking semiconductor device courses at University of Central Florida, Orlando, Florida and Universidad Simón Bolívar, Caracas, Venezuela, where we were able to test and improve this material. We are also grateful to the following present and former graduate and undergraduate students for their invaluable contribution to our MOSFET research: R. Narayanan, M. García Núñez, Z. Latif, Md. Rofiqul Hassan, E. Gouveia Fernandes, O. Montilla Castillo, A. Parthasarathy, J. Rodríguez, Y. Yue and M. Lei.

We further acknowledge Drs. R. L. Anderson at University of Vermont, W. Wong at Intel, P. Schmidt at Florida International University, J. Andrian at Florida International University and G. Mercato at Universidad Simón Bolívar for their research collaboration which has contributed greatly in inspiring the concept of this book.

Finally, one of the authors (JJL) is grateful to Prof. Y. T. Yeow at the University of Queensland, Australia, for sharing his expertise and books on MOSFETs modeling and simulation during the time the author was on sabbatical leave at the National University of Singapore.

J. J. Liou, Orlando, Florida, USA
A. Ortiz-Conde, Caracas, Venezuela
F .J. García Sánchez, Caracas, Venezuela

Chapter 1

MOSFET physics and modeling

Since the early 1980s, the metal-oxide-semiconductor field-effect transistor (MOSFET) has become the most widely used semiconductor device in very large scale integrated circuits. This is due mainly to the fact that the MOSFET has a simpler structure, costs less to fabricate, and consumes less power than its bipolar transistor counterpart. In this chapter, we will first present an overview of the MOSFET evolution, the so-called Moore's law, and the progress of microprocessors based on MOSFETs. Then the fundamentals of semiconductor and MOS system will be introduced. This will be followed by the physics and modeling of MOSFETs, including devices with a conventional structure (i.e., conventional MOSFET), silicon-on-insulator structure (i.e., SOI MOSFET), and lightly-doped drain structure (i.e., LDD MOSFET). Results obtained from device simulation will be included to aid the understanding of the MOSFET behavior and physical insight. The increasingly important short-channel, narrow-channel, hot-carrier, and quantum-mechanical effects on the MOSFET performance will also be addressed.

1.1 MOSFET evolution and its integrated circuits

1.1.1 The origin of the MOSFET

To understand the present status and importance of MOSFETs, we will first discuss the evolution of electronics since the invention of the first transistor until the latest microprocessor Pentium II.

The first transistor, nowadays known as a metal-semiconductor field-effect transistor (MESFET), was proposed in a patent by Lilienfeld in 1926 [1] who, two years later, presented the idea of the depletion-mode MOSFET [2]. The notion of the inversion-mode MOSFET was proposed, seven years later, by Heil [3]. Nevertheless, the first MOSFET was not fabricated until 1960 by Kahng and Atalla [4]. The immense delay between the idea and the fabrication was due to the technical difficulties to obtain a good oxide and to the lack of fundamental and basic semiconductor notions. For example, the idea of a "hole," developed by Wilson in 1931 [5], was used in 1941 by Ohl [6] to explain the very interesting photoelectric characteristic of a device he fabricated using a junction of a "purified silicon" and a "commercial silicon", now known as the p/n junction. Additional details about the origins of the p/n junction can be found in a recent review article by Riordan and Hoddeson [7].

After World War II, researchers at Bell Labs were trying unsuccessfully to make field effect devices and apparently they were not aware of the previous work by Lilienfeld and Heil. Ironically, this unsuccessful research on field effect devices led to the birth of the bipolar transistor. Brattain wrote [8] many years later the following:

> *"Bardeen and I were simply trying to make a good Field Effect device and as a result we were put in a position to observe, for the first time, a phenomenon now called the Transistor Effect."*

According to Shockley [9], the first bipolar transistor, which was called a "point contact transistor," was fabricated and used on 16 December 1947 by Bardeen and Brattain [10-11] and presented to Bell Labs executives on 23 December 1947. This device was built by connecting very closely two gold wires (emitter and collector) to the top of a semiconductor and the third connection was made to the bottom (base) of the semiconductor [12]. The semiconductor was n-type germanium, the top connections were designed to produce rectifying metal-semiconductor contacts, and the bottom connection was an ohmic contact. This device is basically an n/p/n transistor in which the p/n junctions have been replaced by Schottky diodes.

Using the ideas of the "point contact transistor", Shockley completed the conception of the bipolar junction transistor on 23 January 1948 and filed it for a patent on 26 June 1948 [13]. Shockley described in [9] the invention of the transistor as a "creative-failure methodology":

> *"The point contact transistor became a creative failure by setting up challenging scientific problems. My response to this challenge was what finally led me to the conception of the junction transistor"*

and

> *"What I say about myself -and I am sure most creative people would say the same thing- is that, when we look at how long it took us to get certain ideas, we are impressed with how dumb we were- on how long it took us, and how stupid we were. But we have learned to live with this stupidity, and to find from it what relationships we should have seen in the first place."*

In a confidential conference [9] at Bell Labs held on 18 February 1948, Shive reported the fabrication of the first bipolar junction transistor. Shockley and his two Bell Labs colleagues, Brattain and Bardeen, received the Nobel prize in

Physics in 1956 for the "invention of the transistor." A better denomination for this Nobel prize would have been "invention of the bipolar transistor" because the first transistor was proposed by Lilienfeld in 1926 [14-15]. It is interesting that, in contrast to the case of the MOSFET, the theories of the bipolar devices were developed after the devices were fabricated [9].

The interest in the MOSFET was resuscitated in 1960 by Kahng and Atalla [4] by presenting the first successful silicon inversion-channel MOSFET using thermally grown oxide for the gate insulator. The MOSFET dramatically increased its importance three years later when Wanlass and Sah invented the CMOS (complementary Metal Oxide Semiconductor) circuit [16-17]. The low power dissipation of CMOS in VLSI (very large scale integrated) circuits has made the MOSFET the most widely used semiconductor device since the 1980s [14].

1.1.2 The integrated circuit trend

It is understandable that the evolution of integrated circuits has always followed but has been somewhat behind that of semiconductor devices. In 1959, Kilby fabricated a flip-flop from a single chip of germanium by using gold wires for intra connections [14]. This is called today a hybrid integrated circuit or the predecessor of the monolithic integrated circuit. Kilby presented in 1976 [18] his personal account about his previous invention of the integrated circuit at Texas Instruments.

The development of the first planar transistor by Fairchild in the late 1950s [19] allowed Noyce to invent in 1959 [20] the monolithic silicon integrated circuit. Moore, a cofounder of both Fairchild and of Intel, predicted in 1965 [21] that the number of devices per integrated circuit would increase exponentially with the time "as far as the eye could see." This astonishing prediction, which has been named Moore's Law [22-23] and is still valid today, was based on only five data points available from Fairchild. Recently Yu [24] augmented Moore's Law by predicting that the cost of fabrication increases exponentially as the sophistication of chips increases. To verify the validity of Moore's Law and to understand the evolution of electronics, we will look into the case of the Intel microprocessors.

Table 1.1 summarizes the general information about the Intel's microprocessors. Figure 1.1 presents the number of transistors per die versus the year, which shows that Moore's law is still valid to-date. In this figure, the dots are the experimental data and the solid line is the curve fitting using the following algorithm of Marquardt-Levenberg [25]:

Processor	Introduction Date	Number of Transistors	Area (mm^2)	Clock Speed (MHz)	Technology, Voltage, Power, Bits and Pins
4004	Nov. 1971	2.3×10^3	12	0.108	10 μm, pMOS, 12V, 0.3 W, 4 Bits and 18 Pins
8008	April 1972	3.5×10^3	N.A.*	0.2	10 μm, pMOS, 12V, 8 Bits and 18 Pins
8080	April 1974	6×10^3	25	2	6 μm, nMOS, 12V and 5V, 8 Bits and 40 Pins
8085	March 1976	6.5×10^3	N.A.*	5	3 μm, nMOS, 5V, 8 Bits and 40 Pins
8086	June 1978	29×10^3	31	5	3 μm, nMOS, 5V, 16 Bits and 40 Pins
80286	Feb. 1982	134×10^3	45	6	1.5 μm, nMOS, 5V, 16 Bits and 68 Pins
80386	Oct. 1985	275×10^3	100	16	1.5 μm, CMOS, 5V, 32 Bits and 132 Pins two-metal-layer
80486	April 1989	1.2×10^6	173	25	1 μm, CMOS, two-metal-layer, 5V, 32 Bits and 168 Pins
Pentium (P5)	March 1993	3.1×10^6	294	66	0.8 μm, BiCMOS, three-metal-layer, 3.5V, 15 W, 32 Bits and 273 Pins
Pentium Pro (P6)	Nov. 1995	5.5×10^6	306	200	0.6 μm, BiCMOS, four-metal-layer, 2.9V, 23 W, 64 Bits and 387 Pins
Pentium MMX (P55C)	Jan. 1997	4.5×10^6	140	200	0.35 μm, CMOS, four-metal-layer, 2.9V, 32 Bits, and 273 Pins
Pentium II (Klamath)	May 1997	7.5×10^6	203	300	0.35 μm, CMOS, four-metal-layer, 2.8V, 64 Bits

* Not Available

Table 1.1 : The evolution of Intel's microprocessors

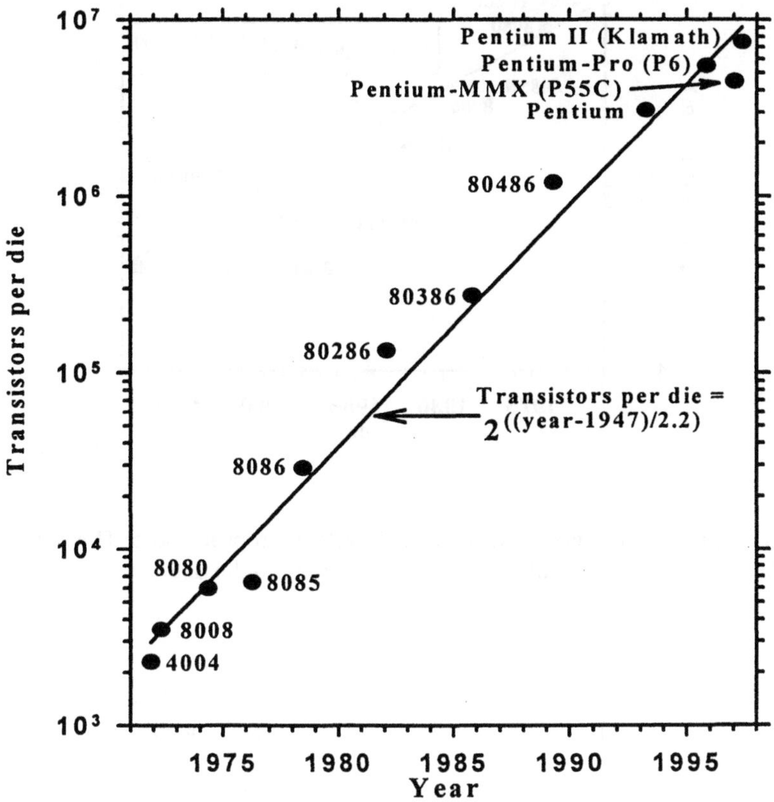

Figure 1.1 : Transistors per die versus the year for Intel's microprocessors. The dots are the experimental data and the solid line is the curve fitting to the data.

$$Transistors\ per\ die = 2^{\frac{(year - 1947)}{2.2}} \tag{1.1}$$

This implies that every 2.2 years the number of transistors per die duplicates, a trend made possible by the decrease in the line width illustrated in Fig. 1.2, and increase in the die area shown in Fig. 1.3. The last four points of Fig. 1.3 seem to indicate that the die area has reached its maximum and is saturating. In contrast, the line width is still decreasing and can be approximated by:

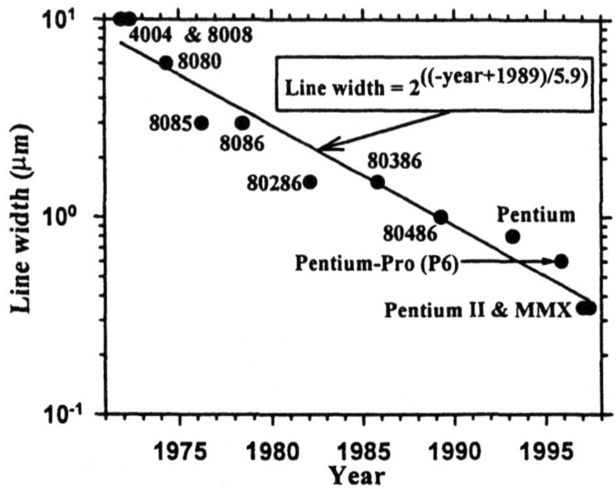

Figure 1.2 : Line width versus the year for Intel's microprocessors. The dots are the experimental data and the solid line is the curve fitting to the data.

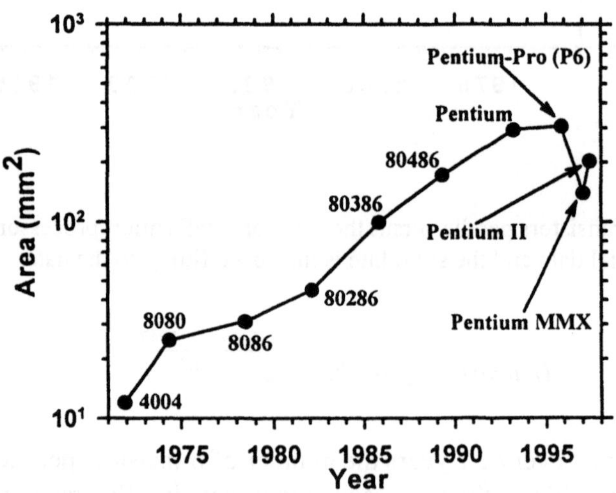

Figure 1.3 : Die area versus the year for Intel's microprocessors.

$$line\ width = 2^{\frac{(1989 - year)}{5.9}} \tag{1.2}$$

It is shown in Table 1.1 that, from 1971 to 1997, the power consumption of the microprocessor has increased from 0.3 W to more than 20 W, the number of pins has increased from 18 to 387, the speed has increased from 108 KHz to 300 MHZ, and the technology is now a four-layer metallization. In order to decrease the power, the supply voltage has been decreased from 12 V to 2.8 V, and the pMOS and nMOS technologies have been replaced by CMOS.

1.2 MOS fundamentals

1.2.1 Energy band theory of semiconductors

We begin the topic of MOS fundamentals by discussing the energy band theory of semiconductors, which was obtained from the quantum mechanisms and the solution of Schrodinger wave equation [26]. Figure 1.4 shows the *E-k* relation (also called the energy band diagram) of silicon, germanium, and gallium arsenide. The most important features of the band structures are represented by the minima E_C of the lowest conduction band and by the maxima E_V of the highest valence band. These minima and maxima are the places where the free electrons and holes are most likely to be found. The valence band in each of the materials peaks at the zone center ($k = 0$) and is actually composed of three subbands. Two of the bands have the same allowed energy at $k = 0$, while the third band maximizes at a slightly smaller energy. In silicon, the two upper bands are indistinguishable, and the maximum of the third band is barely 0.044 eV below the maximum of the other two bands. The band with a smaller curvature around $k = 0$ is called the heavy-hole band and the band with a larger curvature around $k = 0$ is called the light-hole band. The conduction bands in each case are also composed of a number of subbands. In silicon, the conduction band minimum occurs at $k = 0.8(2\pi/a)$ from the zone center along the <100> direction. Other minima in the silicon conduction band structure occur at considerably higher energies and are typically ignored. As a result, there is a main conduction band with the conduction band edge E_C and there are two main valence bands (heavy and light valence bands) with the valence band edge E_V. The forbidden bandgap E_G is defined as $E_C - E_V$ ($E_G = 1.1$ eV for Si and $E_G = 1.4$ eV for GaAs at room temperature).

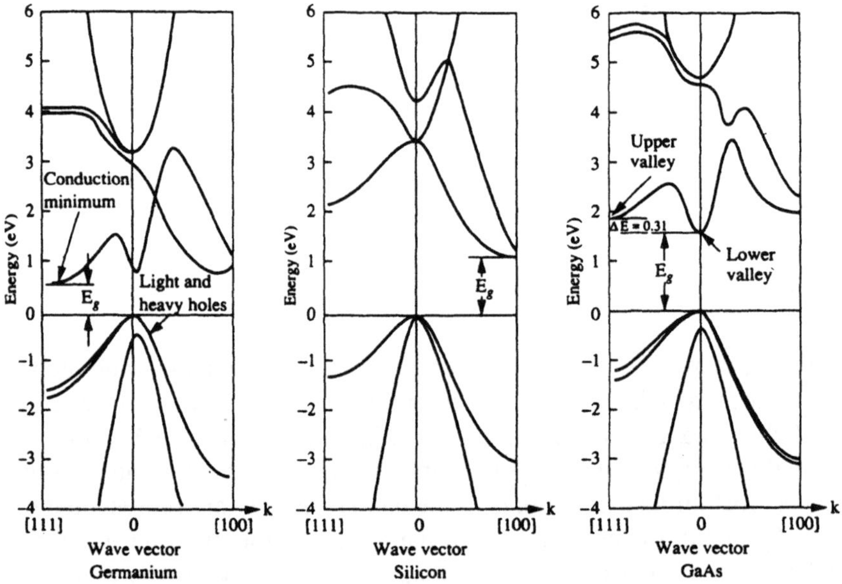

Figure 1.4 : Energy band structure of Si, Ge, and GaAs.

To discuss the free-carrier generation concept, we first consider an intrinsic silicon and use a simplified energy band diagram consisting of only the two band edges E_C and E_V (Fig. 1.5(a)). Electrons in the valence band are not free electrons because they are bonded, to a lesser extent, to the atom nucleus by Coulomb force and, to a larger extent, to the covalence bonding formed by the two neighboring electrons. If an electron in the valence band gains energy (such as that from thermal excitation) that is greater than or equal to E_G, then the electron can break the covalence bonding and become a free electron. In the energy band picture, this represents an electron going from the valence band to the conduction band (see Fig. 1.5(a)). While this process, called the intrinsic generation process, yields an electron (free electron) in the conduction band, it also results in a vacancy in the valence band. Since this vacancy can be filled easily by another electron in the valence band, it is said that the vacancy can move freely and that it is a free hole. Thus, the numbers of free electrons and free holes in the "intrinsic" semiconductor are always equal.

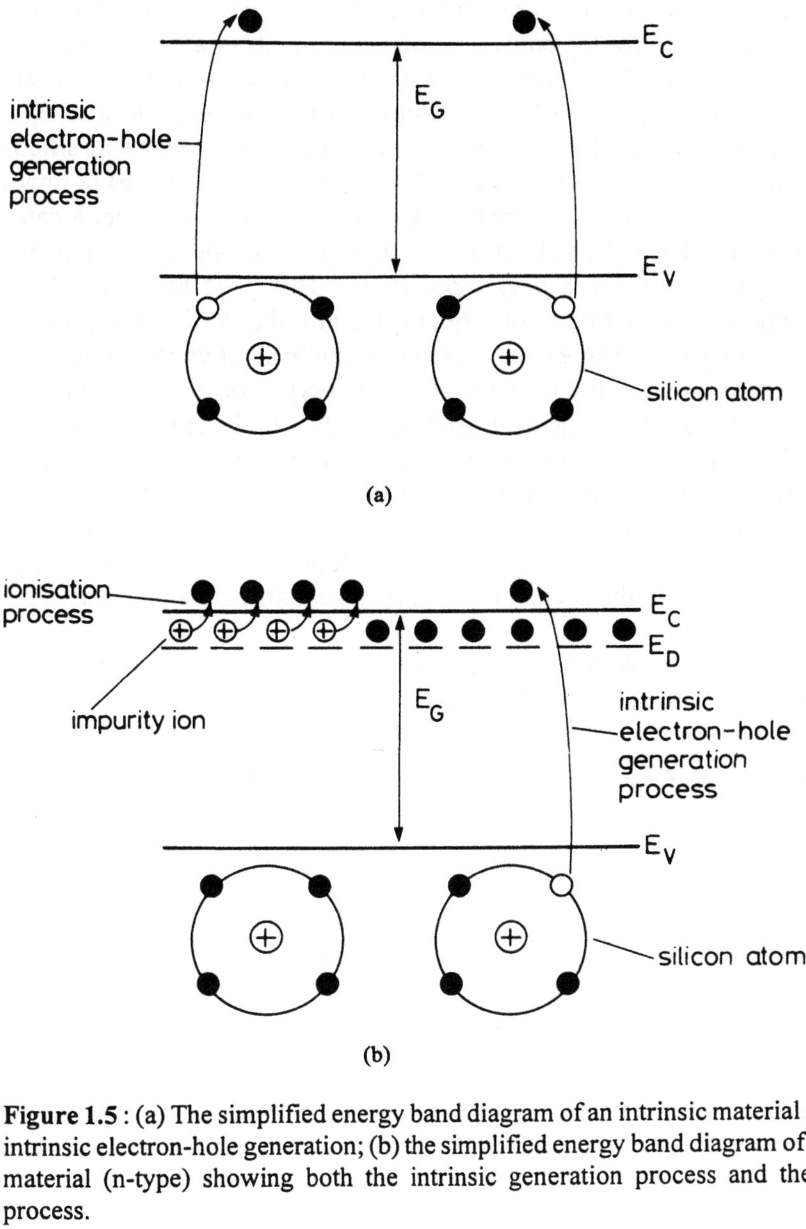

Figure 1.5 : (a) The simplified energy band diagram of an intrinsic material showing the intrinsic electron-hole generation; (b) the simplified energy band diagram of an extrinsic material (n-type) showing both the intrinsic generation process and the ionization process.

The numbers of electrons and holes in silicon can be varied by introducing impurity atoms into the intrinsic silicon (i.e., extrinsic silicon). Introducing group V atoms, such as phosphorus, can increase the electron density, whereas introducing group III atoms, such as boron, can increase the hole density. The statistics of electrons and holes in the semiconductor will be discussed in the next section. It is also commonly assumed that the energy band structure (i.e., E_G) is not altered if the doping concentration is not much smaller compared to the silicon atom density. The impurity dopant will introduce an energy state very close to E_C if group V dopant (donor dopant) is used, and very close to E_V if group III dopant (acceptor dopant) is used. Let us focus on the extrinsic silicon doped with donor dopants (n-type silicon). As shown in Fig. 1.5(b), a donor energy state E_D is present in the energy bandgap, and each donor atom located at E_D (represented by a dashed line) can easily donate one electron to the conduction band, since E_D is very close to E_C. This process is called the ionization process. Generation of electrons from the ionization process, however, will not generate holes; rather, it will result in positively charged ions. At the same time, the electron-hole generation process that occurs in the intrinsic material (i.e., intrinsic generation process) also occurs in the extrinsic material (see Fig. 1.5(b)). Except at very high temperatures, the electron concentration in an extrinsic material is governed by the ionization process and thus by the number of the donor concentration, leading to an electron concentration much larger than the hole concentration in the donor-doped extrinsic material. The same concept applies to the acceptor-doped (p-type) extrinsic material as well.

For the sake of brevity, we will hereafter refer to free electrons and free holes as electrons and holes, respectively.

1.2.2 Statistics of free carriers in semiconductors

1.2.2.1 Fermi-Dirac statistics

There are three basic principles governing free carrier distributions in a semiconductor: (1) conservation of particles; (2) conservation of energy; and (3) the Pauli exclusion principle, which states that no more than one electron with the same spin is allowed to occupy a quantum state. These principles are expressed in mathematics as [27]

$$\sum_{j=1}^{q} n_j = n \tag{1.3}$$

$$\sum_{j=1}^{q} n_j E_j = E \tag{1.4}$$

$$n_j \leq g_j \tag{1.5}$$

where n_j is the electron density (#/cm^3) at a particular energy level E_j, n is the total electron density among q energy states, E is the total energy, and g_j is the degeneracy of the quantum states at a particular energy level. From these, the electron Fermi-Dirac distribution function $f_e(E_j) \equiv n_j/g_j$ can be derived [27]:

$$f_e(E_j) = \frac{1}{1 + \exp\left(\frac{E_j - E_{fn}}{kT}\right)} \tag{1.6}$$

Here, E_{fn} is the electrochemical energy or quasi-Fermi energy for electrons in the conduction band. A similar approach yields the hole Fermi-Dirac distribution function f_h at E_j:

$$\begin{aligned} f_h(E_j) &= 1 - \frac{1}{1 + \exp\left(\frac{E_j - E_{fp}}{kT}\right)} \\ &= \frac{1}{1 + \exp\left(\frac{E_{fp} - E_j}{kT}\right)} \end{aligned} \tag{1.7}$$

where E_{fp} is the quasi-Fermi energy for holes in the valence band.

1.2.2.2 Maxwell-Boltzmann statistics

When the number of free carriers is small in a semiconductor, it is sometimes adequate to treat them as particles that are distinguishable and that are not interacting with each other. Under this condition (i.e., nondegenerate case), one may relax the constraint of the Pauli exclusion principle used in deriving the Fermi-Dirac statistics. Using only particle and energy conservation, the electron Maxwell-Boltzmann distribution function can be expressed by [27]

$$f_e(E_j) = A \exp(-\beta E_j) \tag{1.8}$$

where constants A and β can be determined from the distribution of gas molecules in an ideal gas, which obeys the Maxwell-Boltzmann statistics:

$$A = n\left(\frac{m_e^*}{2\pi k T}\right)^{1.5} \quad and \quad \beta = \frac{1}{kT} \quad . \tag{1.9}$$

Thus, the electron and hole Maxwell-Boltzmann distribution functions are

$$f_e(E_j) = n\left(\frac{m_e^*}{2\pi k T}\right)^{1.5} \exp\left(-\frac{E_j}{kT}\right) \tag{1.10}$$

$$f_h(E_j) = 1 - f_e(E_j) \tag{1.11}$$

where m_e^* is the effective electron mass.

1.2.2.3 Free-carrier concentration in semiconductors

Under the thermal equilibrium condition (no external excitation such as optical or electrical excitation), $E_{fn} = E_{fp} \equiv E_f$. Thus, the electron Fermi-Dirac distribution function becomes

$$f^0{}_e(E) = \frac{1}{1 + \exp\left(\dfrac{E - E_f}{kT}\right)} \tag{1.12}$$

and $f^0{}_h(E) = 1 - f^0{}_e(E)$, where superscript 0 denotes equilibrium. The equilibrium electron and hole concentrations (n_0 and p_0) in the conduction and valence bands can be expressed as

$$n_0 = \int_{E_C}^{\infty} f^0{}_e(E)\, g_C(E)\, dE \tag{1.13}$$

and

$$p_0 = \int_{-\infty}^{E_V} f^0{}_h(E)\, g_V(E)\, dE \tag{1.14}$$

where g_c and g_v are the degeneracy of quantum states in the conduction and valence bands, respectively. Using a parabolic band model for g_c and g_v which assumes that the energy versus the wave number relation is quadratic, n_0 and p_0 can be written as

$$n_0 = N_C\, \mathscr{F}_{\frac{1}{2}}\left(\frac{E_f - E_C}{k\,T}\right) \tag{1.15}$$

and

$$p_0 = N_V\, \mathscr{F}_{\frac{1}{2}}\left(\frac{E_V - E_f}{k\,T}\right) \tag{1.16}$$

where N_C and N_V are the effective density of states in the conduction and valence bands, respectively:

$$N_C = 2\left(\frac{2\,\pi\, m_e^*\, k\,T}{h^2}\right)^{1.5}$$

and (1.17)

$$N_V = 2\left(\frac{2\,\pi\, m_h^*\, k\,T}{h^2}\right)^{1.5}$$

and $\mathcal{F}_{1/2}(\eta) = (2/\sqrt{\pi})\int_0^\infty \delta^{1/2} d\delta/[1 + e^{(\delta - \eta)}]$ is the Fermi-Dirac integral of order of ½. Similarly, the nonequilibrium electron and hole concentrations (n and p) are

$$n = N_C \mathcal{F}_{\frac{1}{2}}\left(\frac{E_{fn} - E_C}{k T}\right) \tag{1.18}$$

and

$$p = N_V \mathcal{F}_{\frac{1}{2}}\left(\frac{E_V - E_{fp}}{k T}\right) \tag{1.19}$$

where $n = n_0 + \Delta n$ and $p = p_0 + \Delta p$. Here Δn and Δp are the excess electron and hole concentrations resulting from excitations other than thermal excitation.

When the semiconductor is nondegenerate, the simpler Maxwell-Boltzmann statistics are applicable. Using such a statistics in the above analysis yields

$$n_o = N_C \exp\left(\frac{E_f - E_C}{k T}\right) \tag{1.20}$$

and

$$p_o = N_V \exp\left(\frac{E_V - E_f}{k T}\right) \tag{1.21}$$

for equilibrium, and

$$n = N_C \exp\left(\frac{E_{fn} - E_C}{k T}\right) \tag{1.22}$$

and

$$p = N_V \exp\left(\frac{E_V - E_{fp}}{k\,T} \right) \tag{1.23}$$

for nonequilibrium.

A parameter called the intrinsic free-carrier concentration n_i is frequently used in device modeling, which is related to n_0 and p_0 by the mass action law:

$$n_i = (\, n_0 \, p_0 \,)^{1/2} = (\, N_C \, N_V \,)^{1/2} \exp\left(\frac{-\,E_G}{k\,T} \right) \tag{1.24}$$

Since the parameters in (1.24), N_C, N_V, E_G, and T, are all temperature dependent, n_i is a strong function of temperature [28]. Figure 1.6 shows the intrinsic free-carrier concentration versus temperature for Ge, Si, and GaAs.

1.2.3 MOS system

The concept of a metal-oxide-semiconductor system, called MOS capacitor shown in Fig. 1.7, needs to be discussed before the theory and operation of MOSFET can be understood. Such a system consisting of a gate metal, oxide layer, and n-type silicon. Figure 1.8 shows the individual energy band diagrams for the metal, insulator, and semiconductor. In the figure, ϕ_m and ϕ_S are the workfunctions of the metal and Si, respectively. Assuming ϕ_m and ϕ_S are equal to each other and no charge in the oxide and silicon-oxide interface (i.e., ideal MOS capacitor), then the MOS energy band diagram under thermal equilibrium is the one shown in Fig. 1.9, where the Fermi energies of metal and Si are located at the same level. The use of the ideal MOS capacitor implies that the flatband voltage for the MOS structure is zero. Such a voltage will be addressed in more detail later.

When a voltage V_G is applied to the gate terminal, the Fermi energy of Si will rise above or drop below the Fermi energy of metal depending on the polarity of the voltage. For example, when a positive voltage is applied, a sheet of positive charges appears on the gate metal. This then creates an electric field in the direction from the metal into semiconductor and thus induces a sheet of equal but negative charge in the semiconductor (see Fig. 1.10(a)).

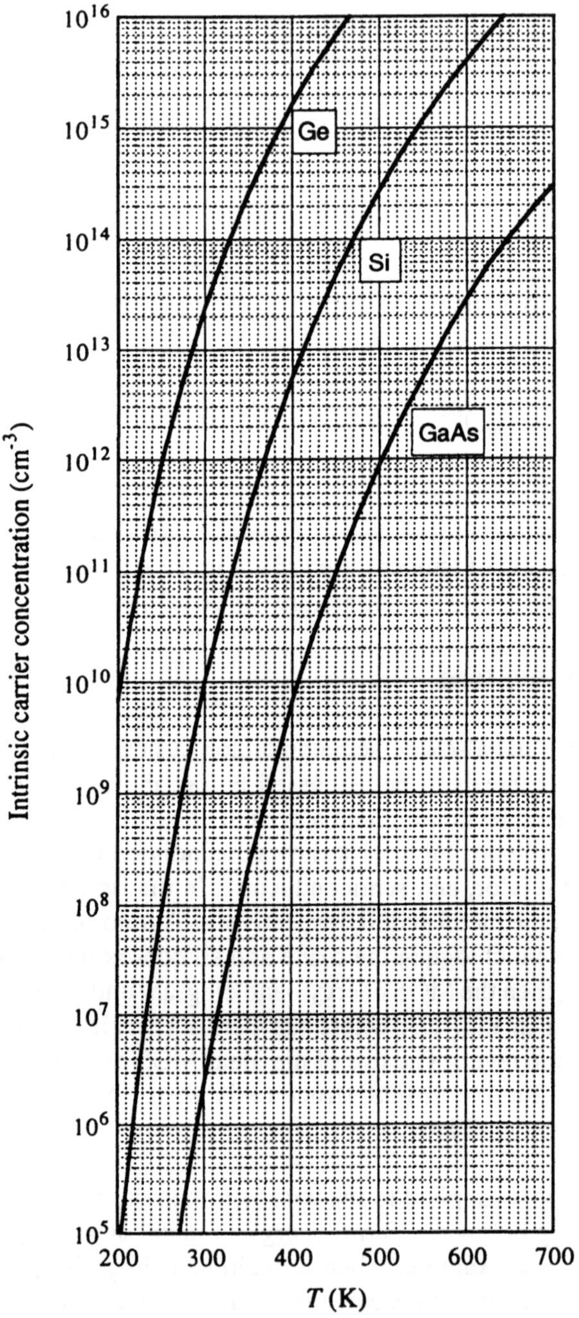

Figure 1.6 : Intrinsic free-carrier concentration in Ge, Si, and GaAs as a function of temperature (after Pierret [28]).

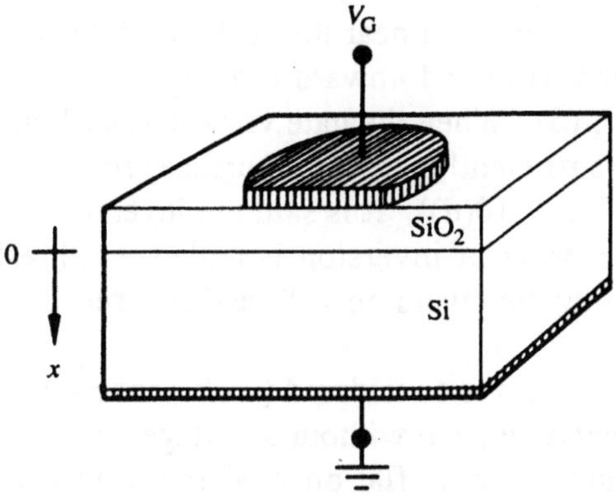

Figure 1.7 : Schematic of an MOS capacitor (after Pierret [28]).

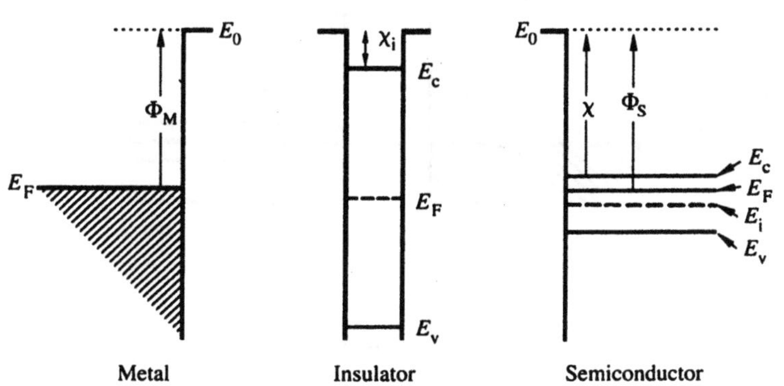

Figure 1.8 : Individual energy band diagrams for the metal, oxide, and semiconductor (after Pierret [28]).

The same physics can be illustrated in the energy band diagram by raising the Fermi energy in semiconductor above that in metal by an amount of qV_G, which consequently raises the conduction and valence band edges and results in a downward band bending. This is called the accumulation operation because more electrons are accumulated near the surface of the n-type silicon. On the other hand, the bands will bend upward if a negative V_G is applied, as can be seen in Figs. 1.10(b)-(d). When the gate voltage is sufficiently large, and thus the band bending is sufficiently large, the surface region is inverted from the n-type to p-type (see Fig. 1.10(d)). It is said the inversion has occurred, and the gate voltage for the onset of inversion is called the threshold voltage. The conditions for defining the inversion will be discussed later.

If a non-ideal MOS system is considered (with nonzero flatband voltage), then there will be band bending even without a voltage applying to the system (i.e., at $V_G = 0$). The bands become flat only when a gate voltage equal to V_{FB} is applied.

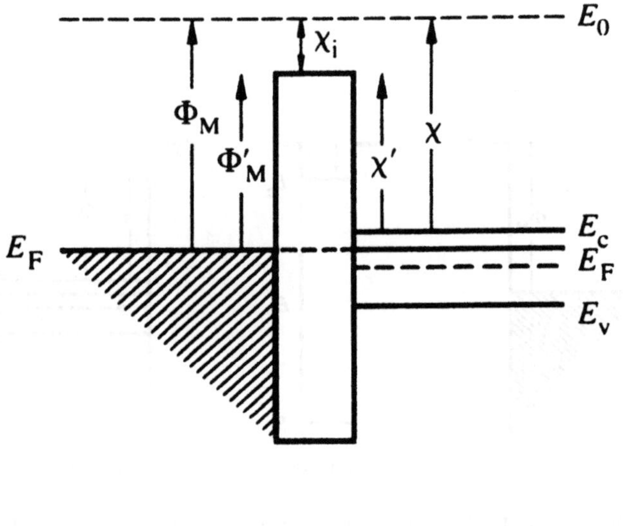

Figure 1.9 : Energy band diagram of an ideal MOS system under equilibrium condition (after Pierret [28]).

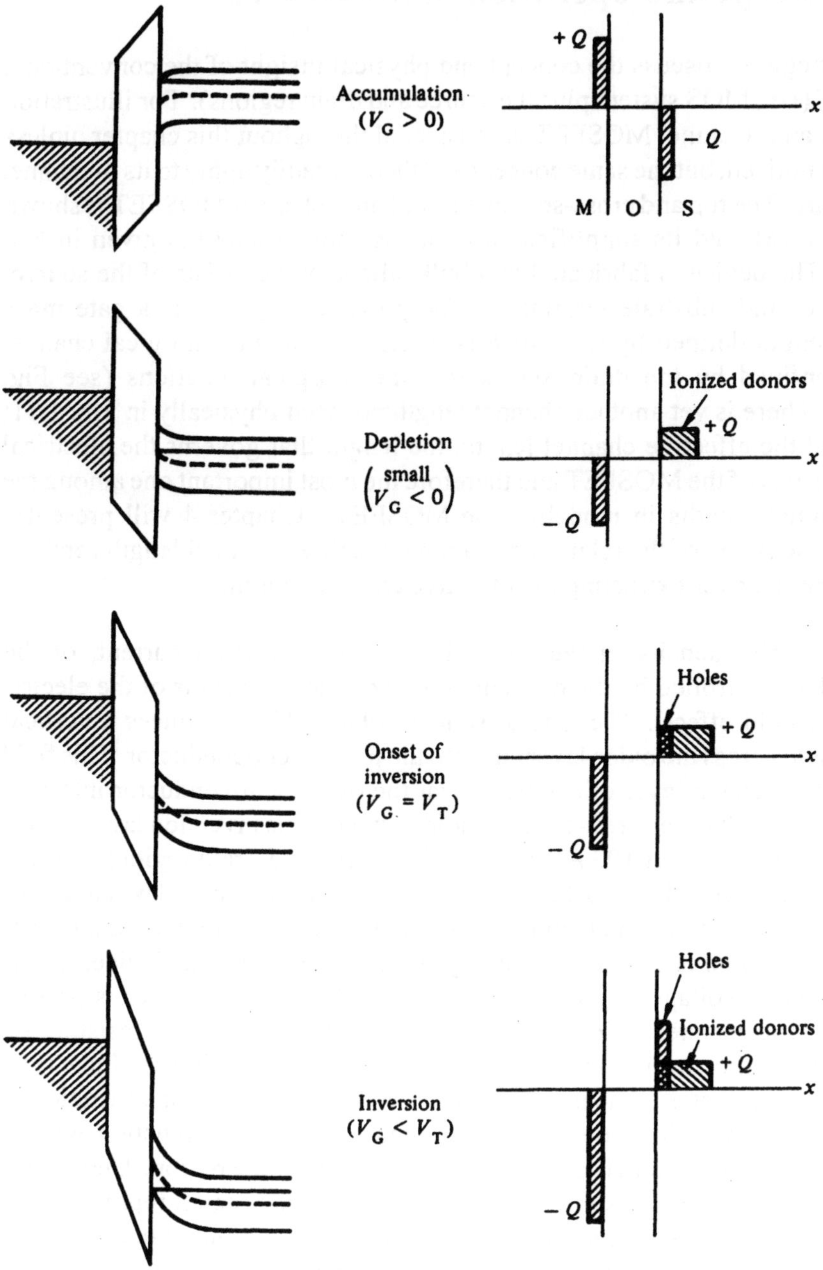

Figure 1.10 : Energy band diagrams and corresponding block charge diagrams for MOS system under various bias conditions (after Pierret [28]).

1.3 Concept and operation of MOSFETs

Here we begin to discuss the concept and physical insight of the conventional MOSFET (i.e., MOS system plus the source and drain regions). For illustration purposes, an n-channel MOSFET is considered throughout this chapter, unless otherwise noticed, but the same concept and theory readily apply to its p-channel counterpart. The top and cross-section view of an n-channel MOSFET is shown in Fig. 1.11(a), and its simplified two-dimensional structure is given in Fig. 1.11(b). The device is fabricated in a bulk silicon and consists of the source, drain, gate, and substrate terminals. The polysilicon gate has a gate mask channel length defined by L_m, which is larger than the metallurgical channel length confined by the drain and source metallurgical junctions (see Fig. 1.11(b)). There is yet another channel length not seen physically in Fig. 1.11; it is called the effective channel length, the length that governs the electrical characteristics of the MOSFET and therefore the most important one among the three channel lengths in modeling the MOSFET. Chapter 4 will present a detailed discussion of the relationship among the three channel lengths and the various methods for extracting the effective channel length.

The free-carrier density in the channel, and thus the drain current, of the MOSFET is controlled by the gate and drain voltages by means of the electric field (i.e., field effect). The gate to source voltage, V_{GS}, produces a vertical electric field across the oxide layer and into the p-type semiconductor. This field creates an inversion layer of electrons near the oxide-semiconductor interface when the gate voltage exceeds the threshold voltage. This inversion layer, which has a thickness of about 100 Å, connects the source and drain regions. When applying a positive voltage to the drain terminal, the drain terminal will drain the electrons (holes if p-channel device) out of the device and, to maintain charge neutrality, the source region will supply the electrons into the device. If the drain-to-source voltage V_{DS} is relatively small, the inversion layer is present throughout the channel region, and the MOSFET is analogous to a resistor. As a result, the drain current varies linearly with respect to the drain voltage (i.e., the MOSFET operates in the linear region). On the other hand, for a sufficiently large V_{DS}, the inversion layer near the reverse-biased drain junction will be depleted (i.e., channel pinch-off) due to a high field near the region. Under such a condition, the device is similar to an $n^+/n/i/n^+$, where *i* region represents the pinch-off (or depleted) region. Since the current in the *i* region is predominantly drift and since the free-carrier velocity saturates in this region with a very high field, the drain current is nearly constant with respect to the drain voltage (i.e., the MOSFET operates in the saturation region).

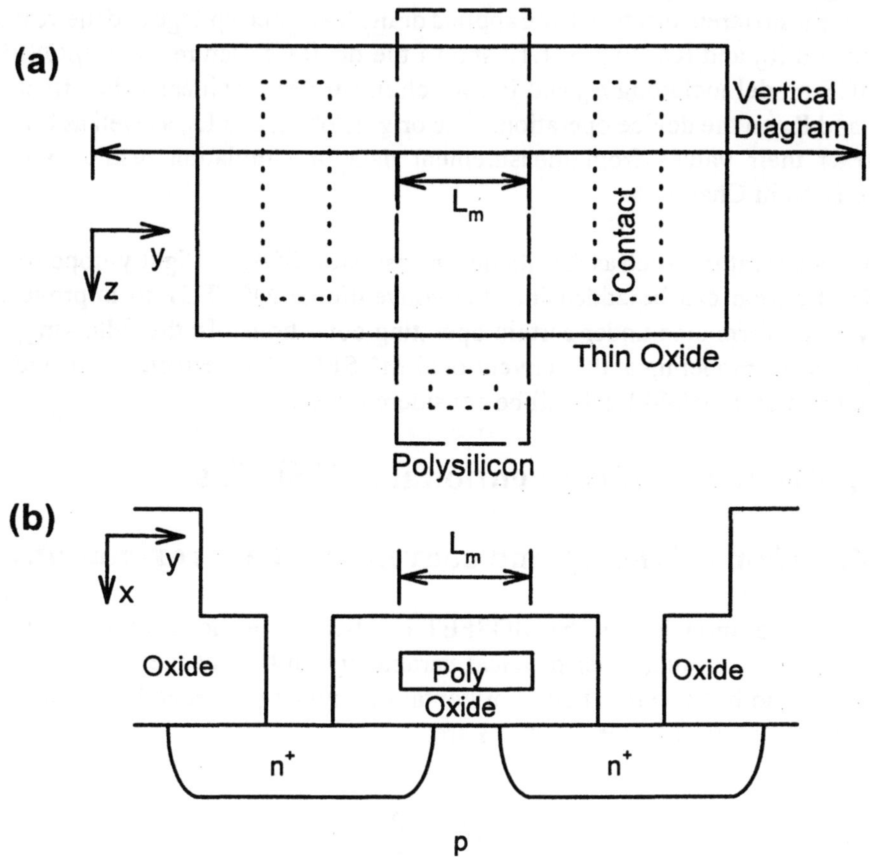

Figure 1.11 : (a) Top and cross-section view of a conventional MOSFET, and (b) the simplified two-dimensional MOSFET structure.

Since there are series resistances R_D and R_S associated with the drain and source regions, respectively, the drain-to-source voltage V_{DS} mentioned above and used throughout this chapter is the intrinsic drain-to-source voltage. In other words, V_{DS} is the difference between the applied drain-to-source voltage and the voltage drops on R_D and R_S. Figure 1.12 shows the device structure and a MOSFET SPICE model including R_D and R_S, which illustrates graphically the effects of R_D and R_S on the device operation. The origins of R_D and R_S, as well as how to extract their values from measurement data or simulation results will be discussed in Chapter 5.

Advanced features such as the silicon-on-insulator (SOI) and lightly-doped drain (LDD) region can be added into the conventional MOSFET to improve the device performance under certain operating conditions. In the following, the physics and modeling of the conventional MOSFET will be treated first, and the LDD and SOI MOSFETS will be considered later.

1.4 Modeling of conventional MOSFETs

1.4.1 General theory and conventional approximations

Clearly, the modeling of the MOSFET is a two-dimensional problem, as the electric field, among other physical parameters, in the channel of MOSFETs varies in both the vertical and horizontal directions (i.e., x and y directions). The basic semiconductor equations are:

$$\nabla \xi = \frac{\rho}{\epsilon_s} \tag{1.25}$$

$$\xi \equiv -\nabla \phi \tag{1.26}$$

$$\nabla^2 \phi = -\frac{\rho}{\epsilon_s} \tag{1.27}$$

$$\rho = -q\,(p - n + N_D^+ - N_A^-) \tag{1.28}$$

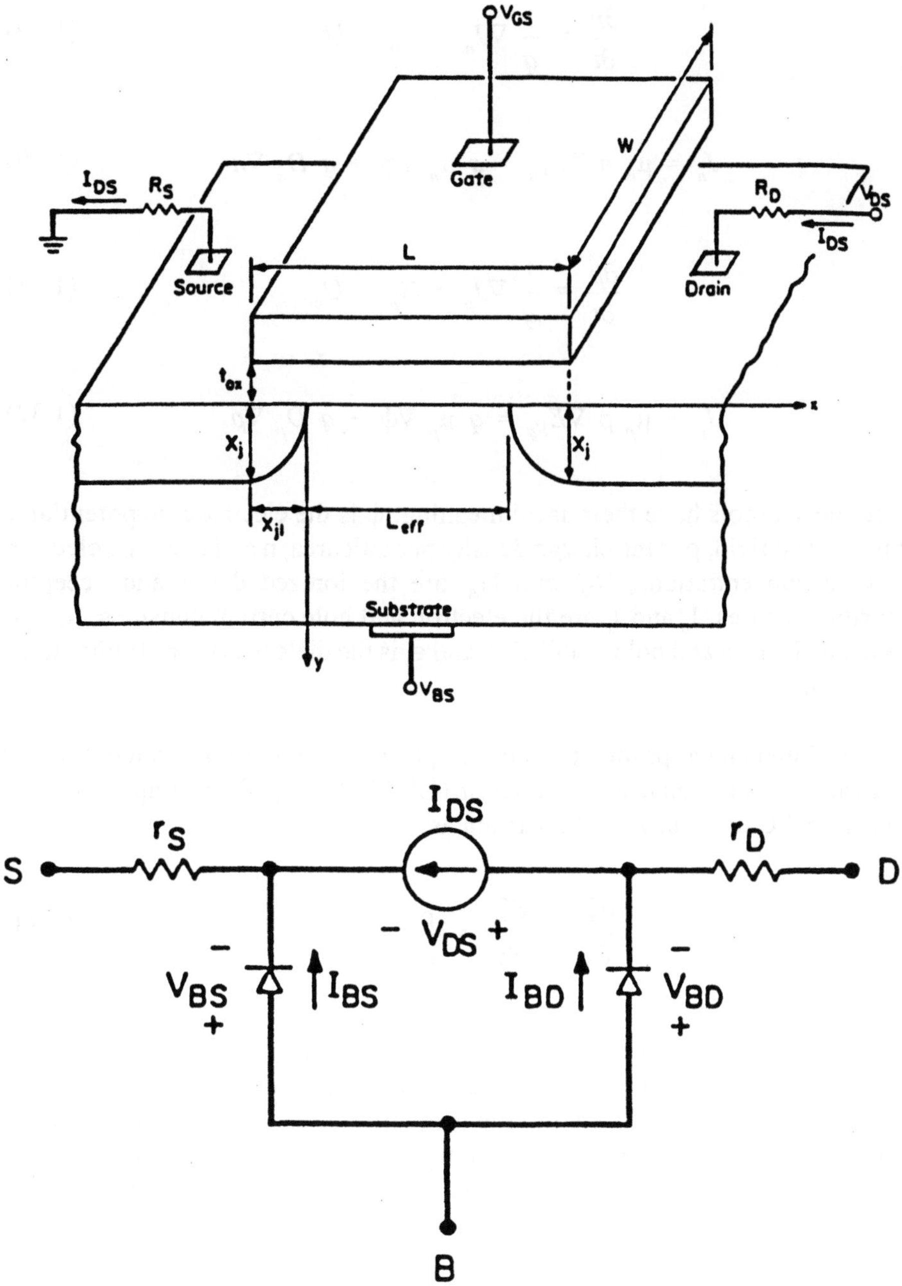

Figure 1.12 : MOSFET structure and SPICE model including the effects of the drain and source series resistances (R_D and R_S).

$$\frac{\partial n}{\partial t} = \frac{1}{q} \nabla J_n + G_n - U_n \qquad (1.29)$$

$$J_n = \mu_n n \nabla E_{Fn} = q \mu_n \nabla\phi + q D_n \nabla n \qquad (1.30)$$

$$\frac{\partial p}{\partial t} = \frac{1}{q} \nabla J_p + G_p + U_p \qquad (1.31)$$

$$J_p = \mu_p p \nabla E_{Fp} = q \mu_p \nabla\phi - q D_p \nabla p \qquad (1.32)$$

where the symbols have their usual meaning: ϕ is the electrostatic potential, ξ is the electric field, ρ is the charge density per unit area, n and p are the electron and hole concentrations, N_D^+ and N_A^- are the ionized donor and acceptor impurity densities, J_n and J_p are the electron and hole current densities, μ_n and μ_p are the electron and hole mobilities, and ϵ_s is the dielectric permittivity in the semiconductor.

The two-dimensional problem can be simplified to a one-dimensional problem using the gradual channel approximation (GCA) [29-30]. For example, the two-dimensional Gauss's law can be written as

$$\frac{\partial \xi_x}{\partial x} + \frac{\partial \xi_y}{\partial y} = \frac{\rho}{\epsilon_s} \qquad (1.33)$$

where ξ_x and ξ_y are the vertical and lateral components of the electric field. The graduate channel approximation states that the inequality of the x- and y-direction field can be translated to the inequality of their derivatives. In other words, if $\xi_y << \xi_x$, which holds for long-channel MOSFET, then it can be assumed $\partial\xi_y / \partial y << \xi_x / \partial x$. Therefore, based on such an approximation, (1.33) reduces to

$$\frac{\partial \xi_x}{\partial x} \approx \frac{\rho}{\epsilon_s} \qquad (1.34)$$

which is the one-dimensional Gauss's law. As will be shown later, the GCA is reasonably accurate for long-channel MOSFETs. For short-channel devices, however, the GCA becomes questionable due to the strong and inseparable interaction between the x- and y-direction physical mechanisms, such as the electric field, in the channel region of such devices.

Another approximation used frequently to simplify MOSFET modeling is the quasi-equilibrium approximation [31], which assumes that electron and hole quasi-fermi levels are nearly flat in the vertical direction due to the fact that the electron and hole currents in the vertical direction are very small. This approximation reduces the problem of solving three simultaneous equations (Poisson's equation, plus the electron and hole continuity equations) into the problem of solving only one equation (Poisson's equation).

In the following, we will present two-dimensional numerical simulation results of a long-channel device in order to explain and verify the validity of the GCA and quasi-equilibrium approximation frequently used to develop MOSFET's models. A detailed discussion on the device simulation and its application will be presented in Chapter 2.

We have simulated an n-channel conventional MOSFETs using a device simulator called MICROTEC [32] run on a PC platform. The device, shown in Fig. 1.11(b), has a substrate doping concentration of 10^{17} cm^{-3}, and the same source and drain make-up as follows: a heavily doped n^+-type Gaussian profile with a peak doping concentration of 10^{20} cm^{-3}, a junction depth of 0.12 μm with a lateral extent of 0.09 μm (0.75 of its vertical extent), and a contact width of 1 μm. The separation between the source- and drain-contact to the gate is 1 μm. The x- and y-directions denote the vertical and lateral coordinates, respectively, and the point x = y = 0 has been defined at the upper left-hand corner of the device. The mask channel length L_m is 10 μm and it extends from y = 2 μm to y = 12 μm. The metallurgical junctions of source- and drain-substrate are located at y = 2.09 μm and y = 11.91 μm, respectively. The gate oxide thickness is 25 nm (i.e., the oxide capacitance is 1.38×10^{-7} F/cm^2).

The Energy band diagram gives good physical insight and is an indispensable aid in understanding the operation of the device. Figure 1.13 presents the simulated energy band diagrams in the vertical direction with a range of $0 < x < 0.3$ μm, taken at the source and drain, and for $V_{GS} = 3$ V (strong inversion), $V_{DS} = 0.1$ V (linear region), $V_{BS} = 0$ (no body effects). To avoid the two-dimensional effects of source- and drain-substrate junctions, the "source" and "drain" regions were defined at y = 3 μm and y = 11 μm, respectively. The use of the quasi-equilibrium approximation [31,33-35] in the MOSFET modeling is supported by the results in Fig. 1.13; that is, the quasi-Fermi energies for electrons and holes, E_{Fn} and E_{Fp}, are nearly flat in the vertical direction. The separation of quasi-Fermi energies at the drain, for the range of $0 < x < 0.3$ μm, is approximately equal to $V_{DS} = 0.1$ V. Figure 1.14 shows the same calculations presented in Fig. 1.13, but for an extended range of $0 < x < 40$ μm. At the drain, we find that E_{Fn} approaches E_{Fp} at a vertical distance of about 10 μm. Although there is a significant gradient of E_{Fn} in this region, the electron current is small because the electron concentration is small. Also, it can be seen that two regions exist in the semiconductor: the surface region, $x < X_d$ (where the bends bending occurs), where X_d is the depletion region thickness, and the bulk region, $x > X_d$ (where the bends are flat). For this particular simulation, X_d is in the order of 0.1 μm.

Figure 1.15 shows the electron concentration, n, versus the vertical distance, taken at the source and drain, for the same bias conditions used in the previous figures. The results indicate that the thickness x_i of the inversion layer (i.e., the region in which the electron concentration is nearly constant) is about 0.01 μm, which is in the order of the electron de Broglie wavelength. Therefore, while the classical treatment has been used frequently to model the inversion layer charge, such a quantity should be more precisely described quantum-mechanically [36]. The results show that the electron concentrations at the source and drain are almost identical. This is due to the fact that a small drain voltage has been used, thus resulting a nearly uniform inversion layer in the lateral direction. The inversion carrier density near the drain will become smaller than that near the source, however, if the drain voltage is increased and therefore the electric field near the reverse-biased drain junction is increased. The inversion layer charge density Q_n (Coulomb per unit area) can be expressed in term of the inversion electron concentration as

$$Q_n = -q \int_0^{x_i} (n - n_o)\, dx \tag{1.35}$$

where n_o is the electron density under thermal equilibrium condition.

The lateral and vertical drain current densities at the source and drain versus vertical distance are presented in Fig. 1.16 for the same bias conditions used in Fig. 1.15. We see that the vertical current is almost zero and the lateral current is high only inside the inversion layer thickness (i.e., $x < 0.1$ μm).

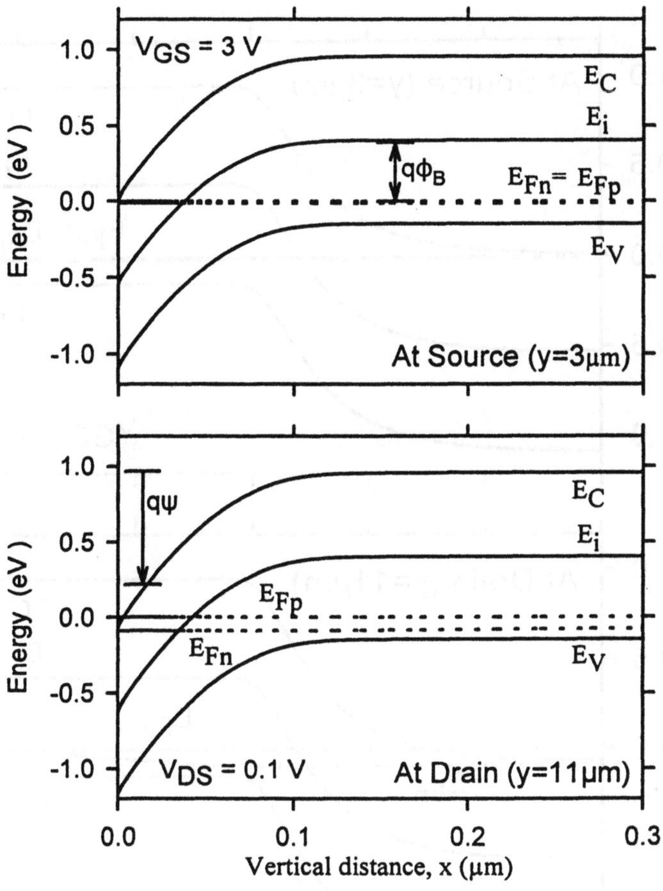

Figure 1.13 : Energy band diagram in the vertical direction (i.e., x direction) at the source and drain junctions for a long-channel MOSFET simulated using two-dimensional device simulator MICROTEC. The vertical distance illustrated is up to 0.3 μm. The bias conditions used are V_{GS} = 3 V (strong inversion), V_{DS} = 0.1 V (linear region), and V_{BS} = 0 (no body effect). The same device make-up and bias conditions will be used in Figs. 1.14 to 1.20.

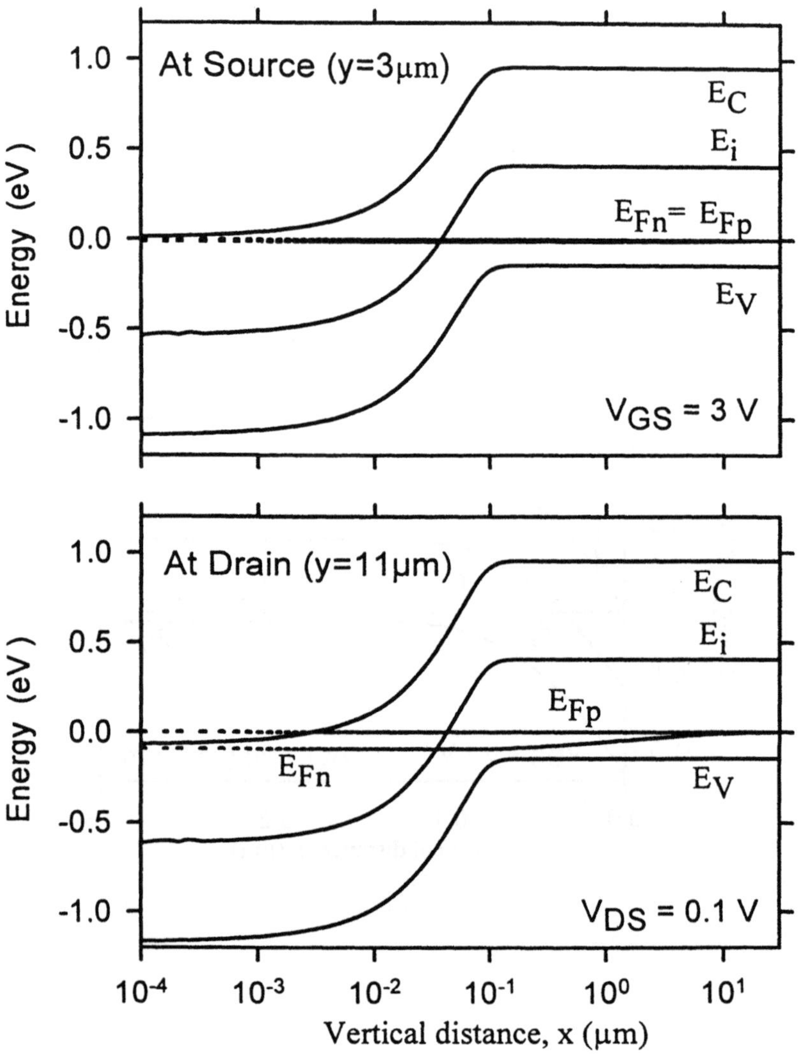

Figure 1.14 : The same energy band diagram presented in Fig. 1.6 but for a vertical distance up to 40 μm using a logarithmic scale.

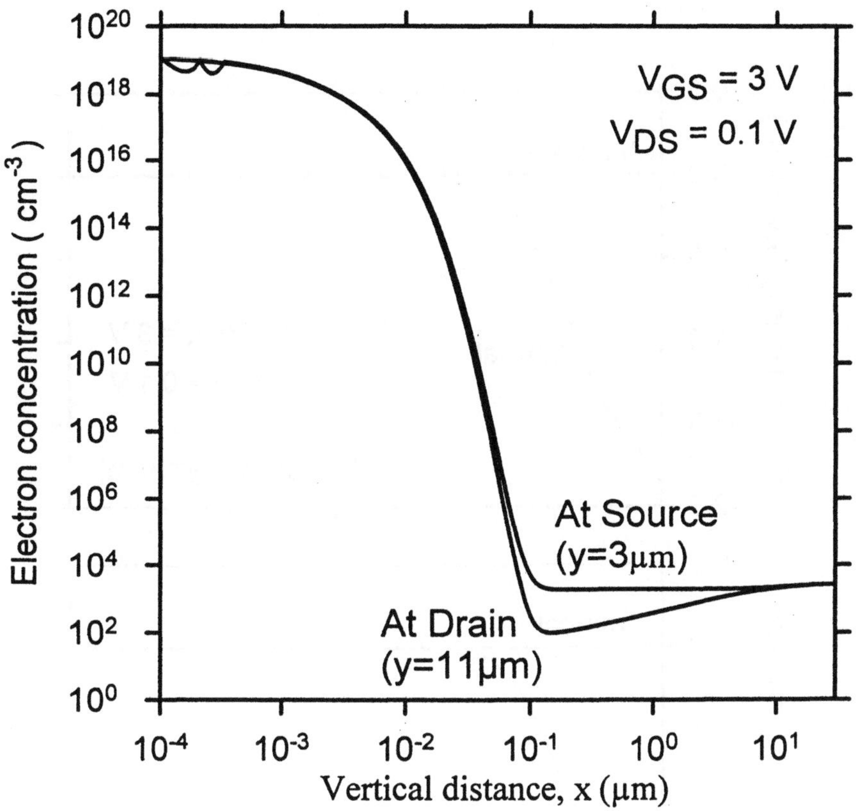

Figure 1.15 : Electron concentration versus the vertical distance taken at source and drain junctions.

The lateral and vertical electric fields versus the vertical distance, taken at the source and drain, are shown in Fig. 1.17. The vertical electric field is high only inside the inversion layer. Furthermore, for this case, it can be concluded that the gradual channel approximation (GCA) [30,34] is valid because the lateral electric field is much smaller than the vertical electric field and thus the two fields can be assumed not interacting with each other and be treated separately. The GCA will allow the development of analytic models for MOSFET, a subject to be treated in the next section.

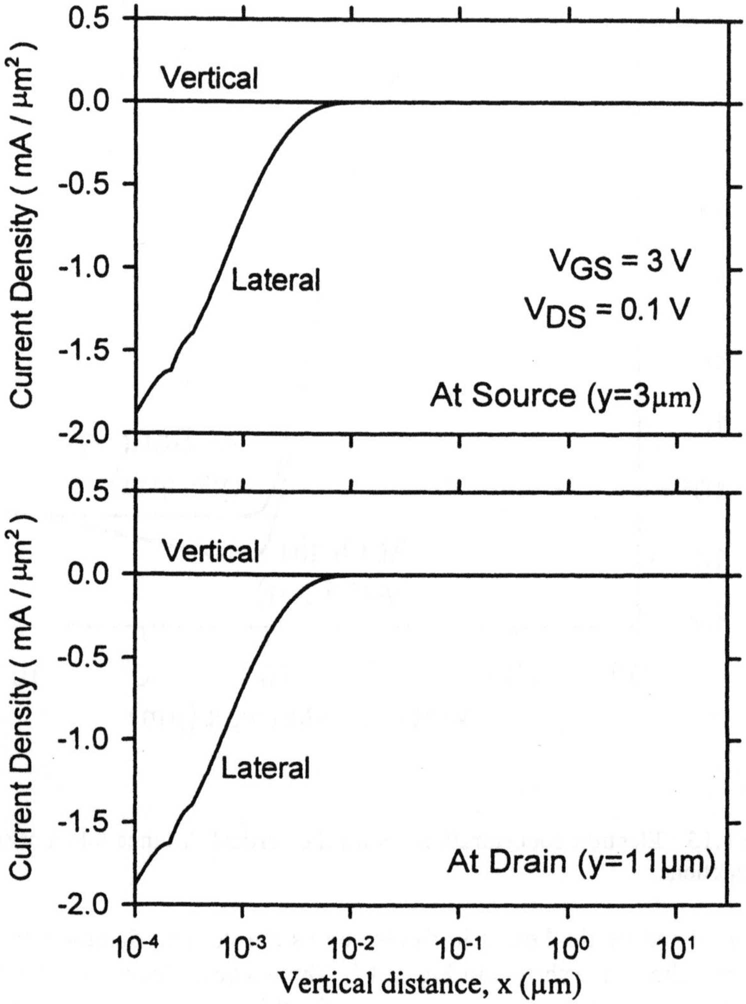

Figure 1.16 : Lateral and vertical drain current densities at the source and drain versus the vertical distance.

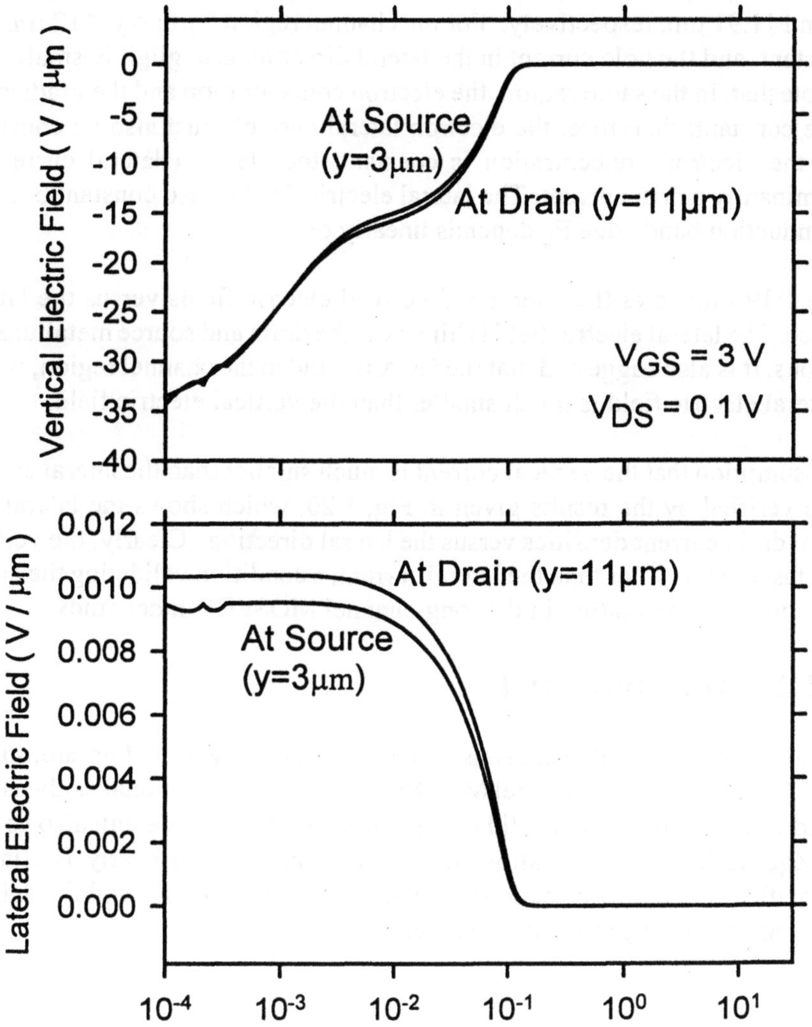

Figure 1.17 : Lateral and vertical electric fields at the source and drain versus the vertical distance.

Figure 1.18 presents the energy band diagram, the electron concentration and the doping density versus the lateral distance. From the plot of the doping density, the locations of the source and drain metallurgical junctions can be estimated as 2.09 and 11.91 μm, respectively. For the channel region 2 μm < y < 12 μm, E_{Fp} is constant, and the hole current in the lateral direction is negligibly small. We also note that, in the same region, the electron concentration and the gradient of E_{Fn} are constant; therefore, the electron lateral current must also be constant. Since the electron concentration is constant, the electron lateral current is predominantly a drift current. The lateral electric field is also constant because the conduction band edge E_C depends linearly on y.

Figure 1.19 illustrates the lateral and vertical electric fields versus the lateral distance. The lateral electric field is high near the drain and source metallurgical junctions. It is also suggested that the GCA is valid in the channel region, where the lateral electric field is much smaller than the vertical electric field.

The assumption that the vertical current is much smaller than the lateral current can be verified by the results given in Fig, 1.20, which shows the lateral and vertical drain current densities versus the lateral direction. Clearly, the vertical current is much smaller than the lateral current, a condition validating the quasi-equilibrium approximation in the long-channel MOSFET under study.

1.4.2 Surface potential

We now begin our detailed analysis of the vertical energy band diagram, shown in Fig. 1.13, for the long-channel MOSFET. The one-dimensional analysis will be used because the GCA is valid in the vertical direction. We will also use the quasi-equilibrium approximation; that is, E_{Fp} and $E_{Fn,}$ are nearly flat in the vertical direction (i.e., x direction). The separation between E_{Fp} and E_{Fn} defines the voltage drop V(y) along the channel,

$$V(y) = \frac{E_{Fp} - E_{Fn}}{q} \quad , \tag{1.36}$$

which varies only on the lateral direction (i.e., y direction).

The electrostatic potential $\phi(x)$ in the surface region, using the intrinsic Fermi energy E_i as the reference, is defined as

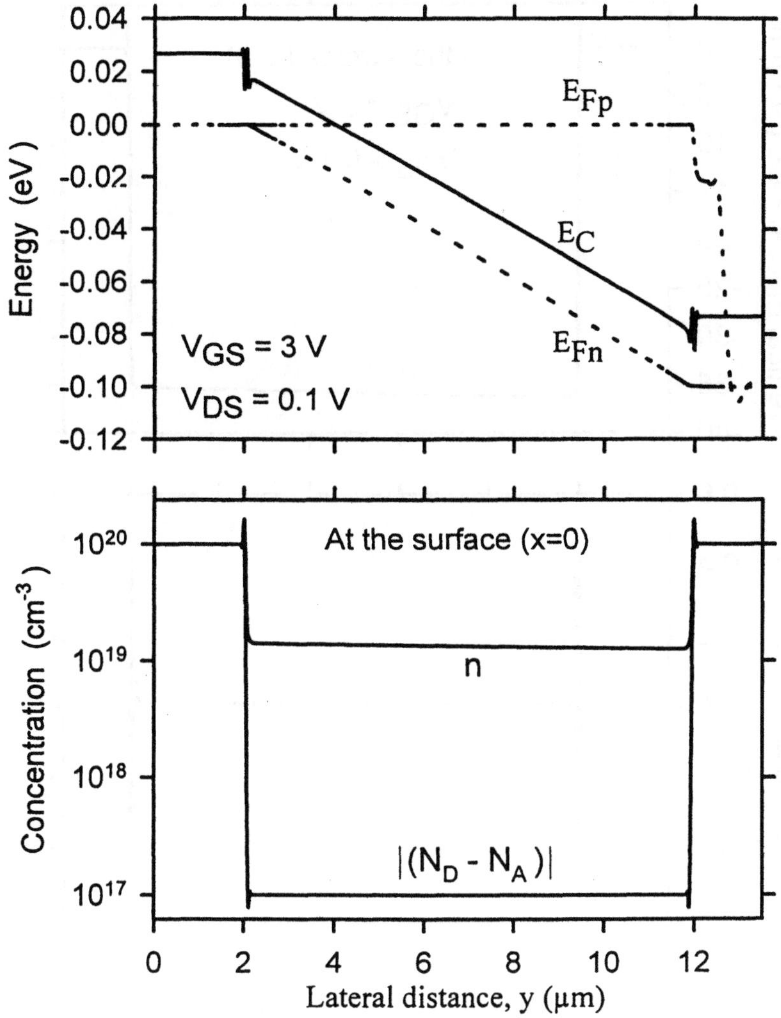

Figure 1.18 : Energy band diagram, the electron concentration, and the doping density versus the lateral distance (i.e., y direction).

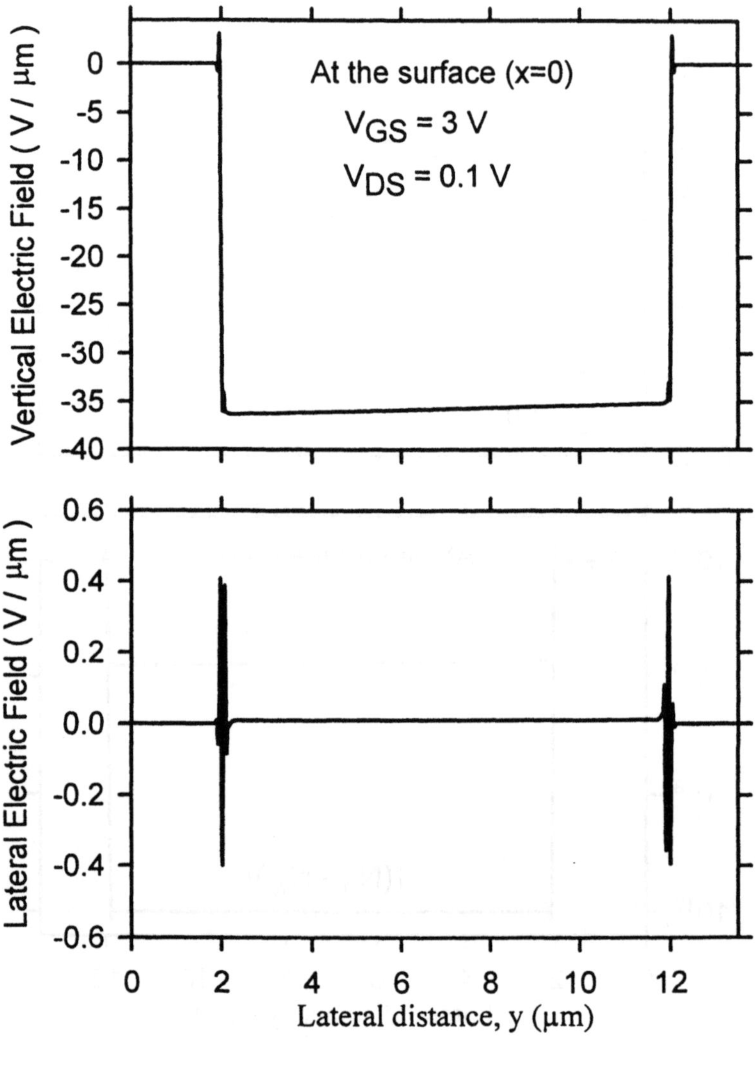

Figure 1.19 : Lateral and vertical electric field versus the lateral distance. The two impulses are the large electric fields at the drain and source metallurgical junctions.

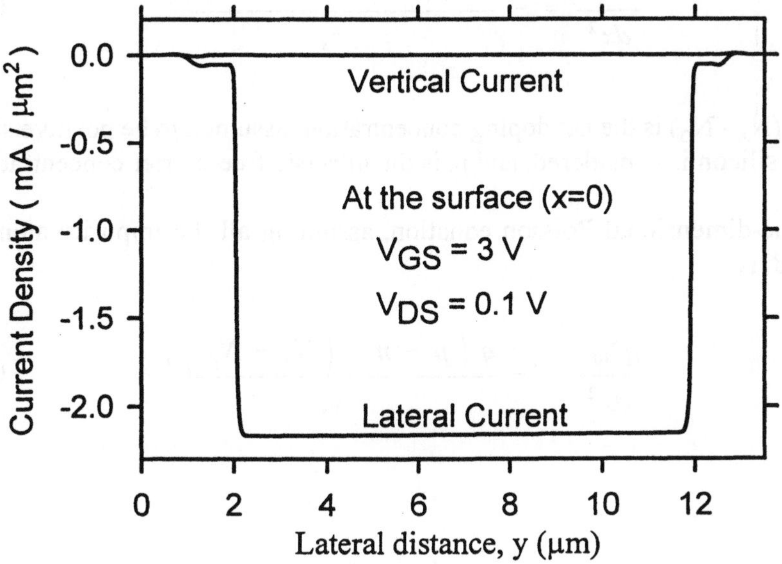

Figure 1.20 : Lateral and vertical drain current densities versus the lateral distance.

$$\phi(x) \equiv \frac{E_i(x) - E_{Fp}(x=\infty)}{q} \tag{1.37}$$

where $\phi(x = 0) \equiv -\phi_S$ is the surface potential, and $\phi(x = \infty) \equiv \phi_B$ is the bulk potential. Both ϕ_S and ϕ_B are positive quantities. The band bending $\psi(x)$ in the surface region, using the conduction band edge E_C in the bulk as the reference, is related to $\phi(x)$ by

$$\psi(x) \equiv \phi_B + \phi(x) \tag{1.38}$$

and $\psi(x = 0) = \phi_B + \phi_S \equiv \psi_S$ and $\psi(x = \infty) = 0$. Note that $\xi = -d\psi/dx = -d\phi/dx$, and ϕ and ψ are positive if the bands bend downward (depletion or inversion) and are negative if the bands bend upward (accumulation). Also, the bulk potential is related to the bulk doping density by

$$\frac{d^2\psi}{dx^2} \approx -\frac{q\left(p - n - \left(N_A - N_D\right)\right)}{\epsilon_s} \tag{1.39}$$

where $(N_A - N_D)$ is the net doping concentration, assumed to be positive since a p-type silicon is considered, and n_i is the intrinsic free-carrier concentration.

The one-dimensional Poisson equation, assuming all the impurity atoms are ionized, is

$$\frac{d^2\psi}{dx^2} \approx -\frac{q\left(p - n - \left(N_A - N_D\right)\right)}{\epsilon_s} \tag{1.40}$$

where

$$p = n_i\, e^{\left(\frac{E_i - E_{Fp}}{k\,T}\right)} \tag{1.41}$$

is the hole concentration in the semiconductor, and

$$n = n_i\, e^{\left(\frac{E_{Fn} - E_i}{k\,T}\right)} \tag{1.42}$$

is the electron concentration. Equations (1.41) and (1.42) are valid for the nondegenerate case. Using the quasi-equilibrium approximation; i.e., E_{Fp} and E_{Fn} are nearly flat in the vertical direction, p and n can be rewritten as:

$$p = n_i\, e^{\left(-\frac{q\,\phi}{k\,T}\right)} = p_o\, e^{-\beta\psi} \tag{1.43}$$

$$n = n_i\, e^{\left(\frac{q\,(\phi\, -\, V(y))}{k\,T}\right)} = n_o\, e^{\beta(\psi\, -\, V(y))} \tag{1.44}$$

where p_o and n_o are the equilibrium hole and electron densities and $\beta = q/kT$ is the inverse of the thermal voltage.

The charge neutrality condition in the bulk region gives

$$\left(N_D - N_A \right) = - \left(p_o - n_o \right) . \tag{1.45}$$

Substituting (1.43)-(1.45) into (1.40) yields

$$\rho = q\, p_o \left((e^{-\beta\psi} - 1) - \frac{n_o}{p_o} \left(e^{\beta(\psi - V(y))} - 1\right) \right) . \tag{1.46}$$

The space-charge density ρ on the left-hand side of (1.46) can be related to ξ and ψ using the following expressions:

$$\begin{aligned} -\frac{\rho}{\epsilon_s} &= \frac{d^2\psi}{dx^2} = -\frac{d\xi}{dx} = \\ -\frac{d\xi}{d\psi}\frac{d\psi}{dx} &= \xi\frac{d\xi}{d\psi} = \frac{1}{2}\frac{d\xi^2}{d\psi} \end{aligned} \tag{1.47}$$

Equation (1.47) can be rewritten in the form

$$d\xi^2 = \frac{-2\,\rho}{\epsilon_s}\, d\psi \quad . \tag{1.48}$$

Now, combining (1.46) and (1.48), and integrating the resulting equation from the interface (at $x = 0$, where $\psi = \psi_S$, $\xi = \xi_S$, and ξ_S is the electric field at the surface) to any point inside the semiconductor (x, ψ and ξ), we obtain:

$$\xi_S^2 - F^2(\psi_S, V) = \xi^2 - F^2(\psi, V) \equiv \alpha \tag{1.49}$$

where α is a parameter that quantifies the charge between the front- and back-gates. For the conventional MOSFET with only the front-gate, $\alpha = 0$. The term $F^2(\psi,V)$ is the Kingston function defined by [37]:

$$F^2(\psi,V) \equiv \int \frac{-2\rho}{\epsilon_s}\, d\psi$$
$$= \frac{2}{\beta^2 L_D^2}\left((e^{-\beta\psi} + \beta\psi - 1) - \frac{n_o}{p_o}\left(e^{-\beta V}(e^{\beta\psi} - 1) - \beta\psi\right)\right) \tag{1.50}$$

Here, L_D is the extrinsic Debye length given by

$$L_D = \left(\frac{\epsilon_s}{q\,\beta\, p_o}\right)^{1/2} . \tag{1.51}$$

The boundary condition at the oxide-semiconductor interface is [26,38]:

$$V_{GS} - V_{FB} = \psi_S + \frac{\epsilon_s\, \xi_S}{C_o} \tag{1.52}$$

where C_o is the oxide capacitance and V_{FB} is the flatband voltage, which is defined as the gate voltage at which the energy bands in the semiconductor are flat.

A few words are needed to elaborate the relationship between the flatband voltage and the capacitance of the MOSFET. In general, the flatband voltage is a function of the work function difference W_{ms} between the gate and silicon, the interface trap charge, and the oxide charge. For the purpose of illustration, let us neglect the effect of interface trap charge and focus on the oxide charge Q_{ox}, then $V_{FB} = W_{ms} - Q_{ox}/C_o$ [29]. Figure 1.21 shows the cross section of an MOS device having a p-type substrate and having positive charge in the oxide. Also shown in the figure is that the total capacitance C of the MOS device is the series combination of the oxide capacitance C_o and the capacitance C_D in semiconductor, which consists of the capacitance associated with the free carriers in the channel and capacitance associated with the depletion charge in the channel. Figure 1.22 illustrates the measured total capacitance versus the gate voltage characteristics with frequency as parameter [29]. Note that at low frequencies, the total capacitance approaches to the oxide capacitance when the device is in the accumulation region (V_G < -32 V) or inversion region (V_G > -20 V). This is because C_D is much larger than C_o in both cases, due to the large number of free carriers in the channel, thus resulting in $C \approx C_o$. In the inversion region under high frequencies, however, the inverted free-carrier cannot response to the quick change of signal, and C_D becomes the capacitance associated with the depletion region in silicon. This leads to a constant total capacitance in both the depletion and inversion regions at high frequencies, as shown in Fig. 1.22. The specific effect of the oxide charge on the total capacitance is shown qualitatively in Fig. 1.23. For example, for a positive oxide charge in the MOS device having a p-type substrate, the presence of the oxide charge will shift the total capacitance to the left from its deal case with an amount of V_{FB} on the gate-voltage axis (Fig. 1.23(a)). The trend is reversed if the oxide charge is negative.

Combining (1.49) and (1.52), we get

$$V_{GS} - V_{FB} = \psi_S \pm \frac{\epsilon_s}{C_o}\left(\alpha + F^2(\psi_S, V) \right)^{1/2} \tag{1.53}$$

where - is used when the bands bend up, and + is used when the bands bend down. For the case of conventional MOSFET (or called the bulk MOSFET because it is fabricated in a bulk silicon), evaluating the middle term of (1.49) in the bulk ($x = \infty$, $\psi = 0$, and $\xi = 0$) gives $\alpha = 0$. It will be shown later that the parameter α is not zero for SOI MOSFETs due to the presence of the Si-SiO_2 back interface. For a given V_{GS}, ψ_S can be calculated numerically from (1.53).

Alternatively, the plot of ψ_S versus V_{GS} can be analytically obtained by calculating V_{GS} for a given ψ_S.

Figure 1.24 shows the ψ_S versus V_{GS} characteristics (line) obtained from MICROTEC simulation. Also included in the figure (open circles) are ψ_S versus V_{GS} calculated using (1.53) and V_{FB} = -0.73 V. It can be concluded that (1.53) is valid for long-channel devices because it gives the same results as the two-dimensional simulations. Three different asymptotic behaviors can be found in Fig. 1.24: a) $-\psi_S \propto \log(-V_{GS})$ for strong accumulation (i.e., $\psi_S < 0$); b) $\psi_S \propto V_{GS}$ for depletion (i.e., 0.1 V $< \psi_S <$ 0.8 V) ; and c) $\psi_S \propto \log(V_{GS})$ for strong inversion (i.e., $\psi_S >$ 0.84 V). The small and not well defined area between ψ_S = 0.8 and 0.85 V is the weak or moderate inversion region. It can be seen in Fig. 1.24 that the onset of strong inversion region occurs at $\psi_S \approx$ 0.84 V, which is close to $\psi_S \approx 2\phi_B$ = 0.816 V. Also note that ψ_S is insensitive to V_{GS} in the strong inversion region.

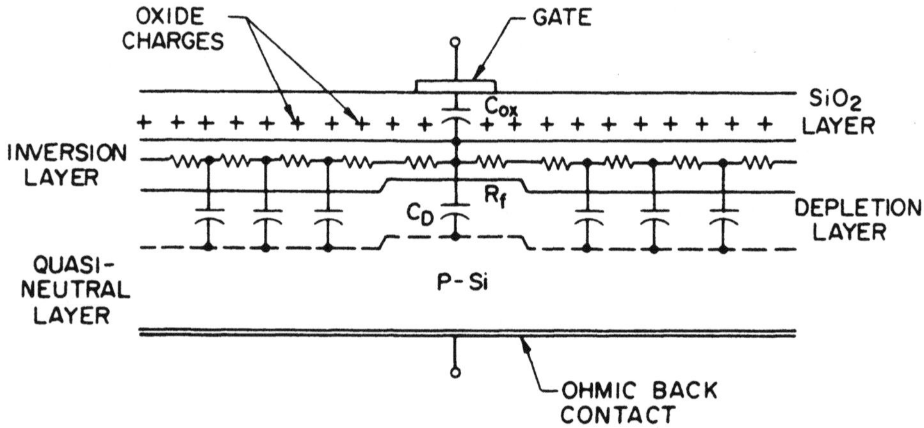

Figure 1.21 : Cross section of an MOS device showing the oxide charges, oxide capacitance, and capacitance in the semiconductor (after Nicollian and Brews [29]).

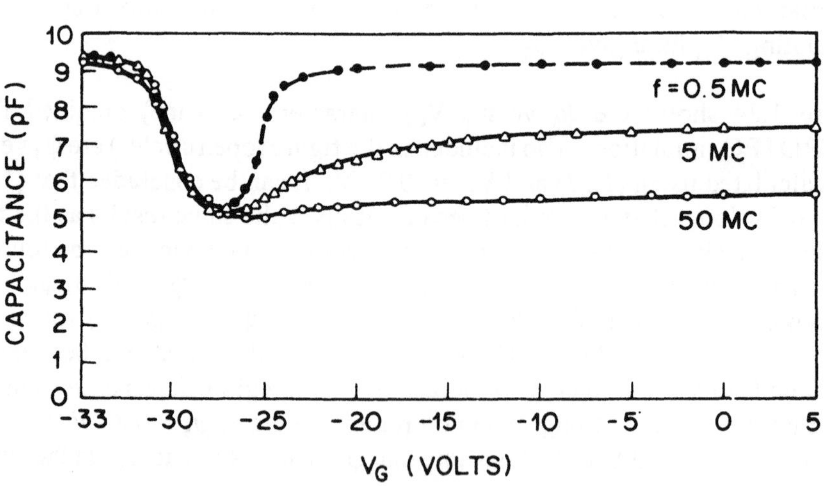

Figure 1.22 : Measured total capacitance versus the gate voltage as a function of frequency (after Nicollian and Brews [29]).

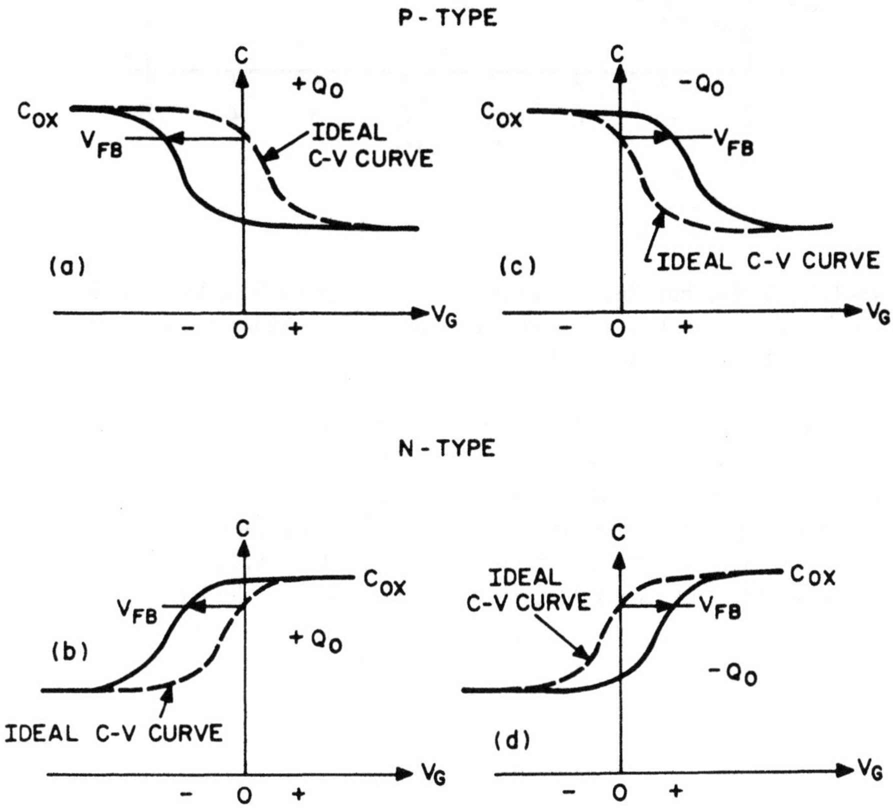

Figure 1.23 : Qualitative illustration of high frequency capacitance as a function of gate voltage with and without oxide charge (i.e., ideal C-V curve) for (a) positive oxide charge, p-type substrate, (b) positive oxide charge, n-type substrate, (c) negative oxide charge, p-type substrate, and (d) negative oxide charge, n-type substrate (after Nicollian and Brews [29]).

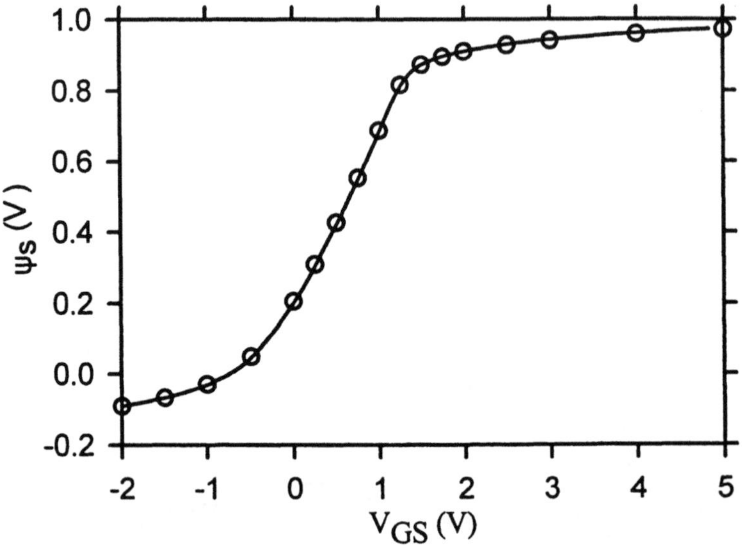

Figure 1.24 : Surface band bending versus gate bias simulated from MICROTEC (line). Also included in the figure (open circles) are values of surface band bending calculated using equation (1.53) and V_{FB} = -0.73V.

The surface band bending at the onset of inversion can also be derived using a more rigorous approach. First, the origins of each term in (1.50) are: a) the term exp(-$\beta\psi$) is due to hole accumulation charge; b) the term ($\beta\psi$ - 1) is due to depletion charge; and c) the term (n_o/p_o){exp(-βV)[exp($\beta\psi$) - 1] - $\beta\psi$} is due to inversion charge. The inversion charge term at the source (V = 0), which is approximated by (n_o/p_o)exp($\beta\psi$), is much larger than the depletion term for the case of $\beta\psi > 1$. Putting $p_o = (N_A - N_D)$ into (1.39), together with the assumption $\beta\psi >> 1$, and solving the resulting transcendental equation, we have

$$\frac{n_o}{p_o} e^{\beta\psi_S} = e^{\beta(\psi_S - 2\phi_B)} \approx \beta\psi_S \quad , \tag{1.54}$$

From (1.54), we obtain ψ_S = 0.9 V for $2\phi_B$ = 0.816 V. Rewriting (1.54),

$$\beta\psi_S - 2\beta\phi_B - \ln(\beta\psi_S) = 0 \quad , \tag{1.55}$$

and noting that the solution is around $\psi_S \approx 2\phi_B$ and that the logarithm changes very little in this range compared to a linear function, we can approximate $\ln(\beta\psi_S) \approx \ln(\beta 2\phi_B)$ to obtain:

$$\beta\psi_S \approx 2\beta\phi_B + \ln(2\beta\phi_B) \quad . \tag{1.56}$$

The solution of ψ_S from (1.56), called the improved definition, is in close agreement with the exact solution obtained from the transcendental equation in (1.54) or (1.55). On the other hand, solving the simplified equation $(n_o/p_o)\exp(\beta\psi_S) = 1$ would yield $\psi_S = 2\phi_B$, which is called the classical definition [39-41]. Figure 1.25 presents the ψ_S versus $2\phi_B$ characteristics obtained from the classical definition $\psi_S = 2\phi_B$, from the transcendental equation (1.54), and from the improved definition in (1.56). It is shown that the improved definition is in good agreement with the results calculated from the transcendental equation, but the classical definition underestimates the surface band bending at the onset of strong inversion.

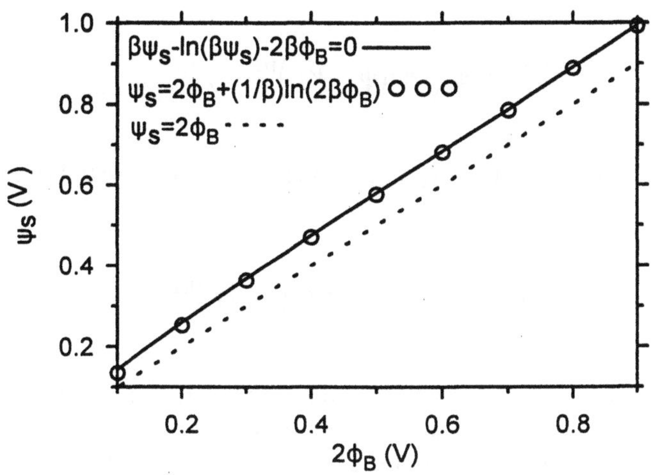

Figure 1.25 : Surface band bending at the onset of strong inversion versus the bulk potential obtained from various definitions. The dot line is the classical definition $\psi_S = 2\phi_B$; the solid line is the solution of the transcendental equation (1.54); and the symbols is the improved definition $\beta\psi_S = 2\beta\phi_B + \ln(2\beta\phi_B)$.

1.4.3 Threshold voltage

The threshold voltage V_T is defined as the gate voltage for the onset of strong inversion in the channel. According to the analysis given in Sec. 1.4.2, such a condition occurs when $\psi_S = 2\phi_B$ at the source. Using (1.54), together with α = 0, we obtain

$$V_T = V_{FB} + 2\phi_B + \frac{\epsilon_s}{C_o}\left(F^2(\psi = 2\phi_B, V = 0) \right)^{1/2} . \tag{1.57}$$

The Kingston function can be related to the electric field. Using the condition $\alpha = 0$ and (1.49), the electric field ξ in the surface region is

$$\xi = \pm \left(F^2(\psi, V) \right)^{1/2} \tag{1.58}$$

where - is used when the bands bend up, + is used when the bands bend down. At the surface, (1.58) becomes

$$\xi_S = \pm \left(F^2(\psi_S, V) \right)^{1/2} . \tag{1.59}$$

Using the Gauss's law, the total charge Q_s in the semiconductor can be expressed as

$$Q_S = - \epsilon_s \xi_S = \mp \epsilon_s \left(F^2(\psi_S, V) \right)^{1/2} . \tag{1.60}$$

Figure 1.26 is a plot of (1.60), which can be used to verify the condition of $\psi_S = 2\phi_B$ at onset of strong inversion derived and used in the previous sections. In Fig. 1.26, the point of strong inversion can be clearly identified as the ψ_S at which Q_S starts to increase very quickly (i.e., proportional to $\exp(q\psi_S/2kT)$). This point is $\psi_S = 2\phi_B$, which agrees with the finding given in Sec. 1.4.2. Of equal importance to note is the point of $\psi_S = \phi_B$, where weak inversion starts to occur. The MOSFET drain current models for both the strong and weak inversions will be developed in the next section.

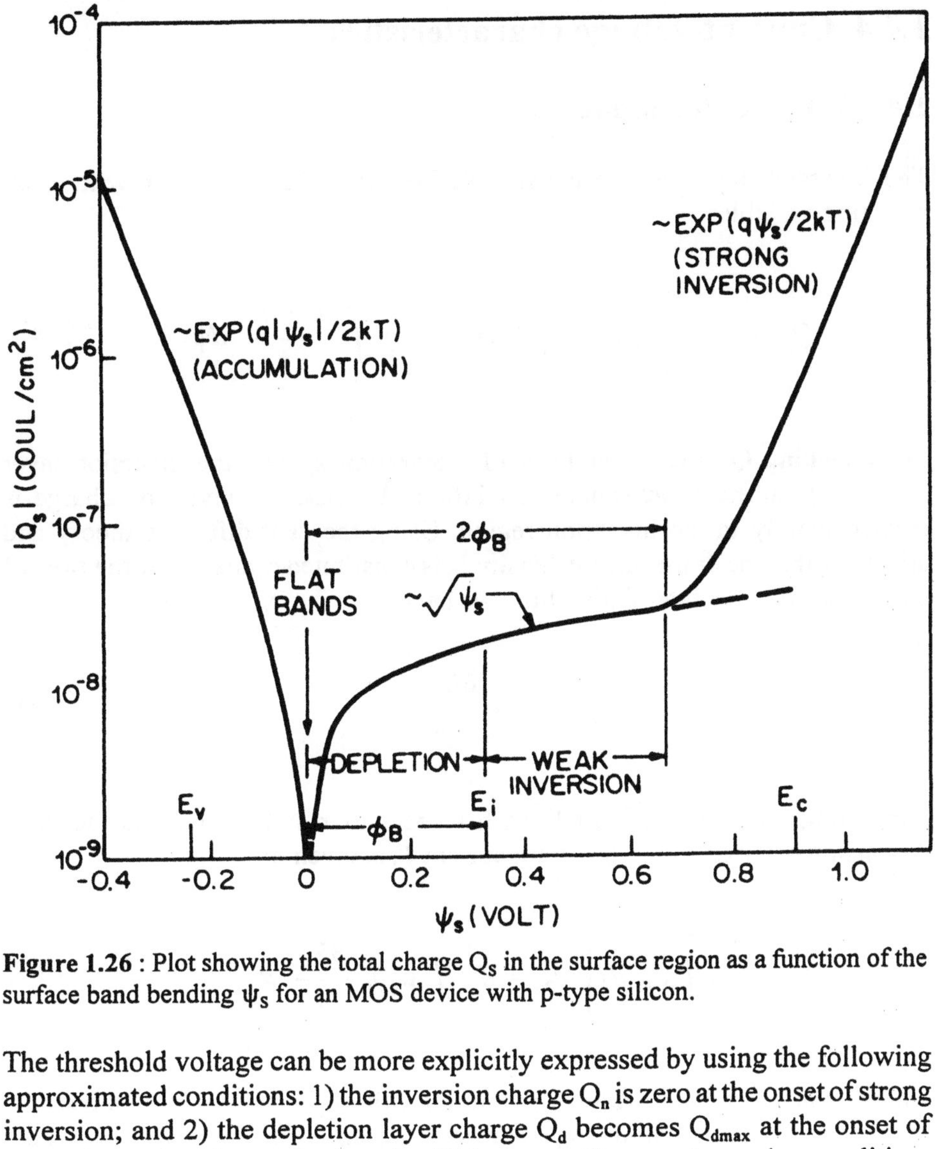

Figure 1.26 : Plot showing the total charge Q_S in the surface region as a function of the surface band bending ψ_S for an MOS device with p-type silicon.

The threshold voltage can be more explicitly expressed by using the following approximated conditions: 1) the inversion charge Q_n is zero at the onset of strong inversion; and 2) the depletion layer charge Q_d becomes Q_{dmax} at the onset of strong inversion due to the fact that X_d becomes X_{dmax} under such a condition. This yields

$$V_T = V_{FB} + 2\,\phi_B + \frac{q\,N_A\,X_{dmax}}{C_o} \tag{1.61}$$

where N_A is the p-type doping density in the substrate.

1.4.4 Current-voltage characteristics

1.4.4.1 Pao-Sah's model

The inversion layer charge density, defined in (1.35), can be rewritten by changing variables:

$$Q_n \equiv -q \int_0^{x_i} (n - n_o)\, dx = -q \int_0^{\psi_S} \frac{n - n_o}{\xi}\, d\psi \quad . \tag{1.62}$$

In calculating Q_n, the upper limit of integration x_i (i.e., the inversion layer thickness) can be approximated by infinity because the inversion charge is located mainly in the inversion region. Using the drift-diffusion theory and assuming that the drain current density J_D is constituted primarily by the flow of electrons along the lateral direction, we have

$$J_D = \mu_n\, n\, \frac{\partial E_{Fn}}{\partial y} \quad . \tag{1.63}$$

Integrating both sides of (1.63) with respect to x and z (i.e., z is the third dimension of MOSFET), the drain current I_D is given by

$$I_D = \int_0^W \left(\int_0^{x_i} J_D\, dx \right) dz = W \int_0^{x_i} \mu_n\, n\, \frac{\partial E_{Fn}}{\partial y}\, dx \tag{1.64}$$

where W is the channel width. The term $\partial E_{Fn}/\partial y$ depends weakly on x because of the quasi-equilibrium approximation; therefore, $\partial E_{Fn}/\partial y \approx dE_{Fn}/dy$. Furthermore, assuming μ_n does not depend on x, (1.64) can be simplified to

$$I_D = W \mu_n \frac{dE_{Fn}}{dy} \int_0^{x_i} n\, dx = -\frac{W}{q} \mu_n \frac{dE_{Fn}}{dy} Q_n \tag{1.65}$$

where Q_n is the inversion layer charge density defined in (1.35). Integrating (1.65) along the channel (i.e., in the lateral or y direction), we obtain

$$\int_{y_s}^{y_d} I_D\, dy = I_D\, L_{eff} = -\frac{W}{q}\int_{y_s}^{y_d} \mu_n \frac{dE_{Fn}}{dy} Q_n\, dy$$
$$= -\frac{W}{q}\int_{E_{Fn}(y_s)}^{E_{Fn}(y_d)} \mu_n\, Q_n\, dE_{Fn} = W\,\mu_n \int_0^{V_{DS}} Q_n\, dV \tag{1.66}$$

where y_s and y_d are the values of y at the source and drain ends of the channel, respectively, $L_{eff} \equiv (y_d - y_s)$ is the effective channel length, V is the voltage drop along the channel (i.e., the separation of the two quasi-Fermi energies E_{Fp} and E_{fn}). The boundary conditions of $V(y = 0) = 0$ and $V(y = L_{eff}) = V_{DS}$ have been used in obtaining (1.66). Rewriting (1.66) and combining it with (1.62), we obtain the so-called double-integral expression of Pao-Sah's model [42]:

$$I_D = \mu_n \frac{W}{L_{eff}} \int_0^{V_{DS}} Q_n\, dV =$$
$$-\,q\,\mu_n \frac{W}{L_{eff}} \int_0^{V_{DS}}\int_0^{\psi_S} \frac{n - n_o}{\xi}\, d\psi dV \tag{1.67}$$

Equations (1.62) and (1.67) can be used to calculate numerically the inversion layer charge Q_n and the drain current I_D for $V_D > 0$ under all inversion conditions, including weak inversion, moderate inversion, and strong inversion.

It is worth pointing out that the effective channel length L_{eff} is smaller than the mask channel length L_m but larger than the metallurgical channel length defined by the drain and source junctions (see Fig. 1.11). The determination of such a parameter, a topic to be addressed in Chapter 4, is critical to the accurate modeling of MOSFET because I_D depends strongly on L_{eff}, as indicated in (1.67).

1.4.4.2 Pierret-Shields's model

Pierret and Shields [43] transformed the double-integral expression of the Pao-Sah's model, without making any additional approximation, into the following completely equivalent single-integral equation:

$$I_D = \mu_n \frac{W}{L_{eff}} \left[C_o \left((V_{GS} - V_{FB})(\psi_{SL} - \psi_{So}) - \frac{(\psi_{SL}^2 - \psi_{So}^2)}{2} \right) \right. \tag{1.68}$$

$$\left. + \epsilon_s \int_0^{\psi_{So}} F(\psi, V=0)\, d\psi - \epsilon_s \int_0^{\psi_{SL}} F(\psi, V=V_{DS})\, d\psi \right]$$

where $\psi_S(y = y_s) \equiv \psi_{So}$ and $\psi_S(y = y_d) \equiv \psi_{SL}$. This model is also valid for long-channel MOSFETs under all inversion conditions.

1.4.4.3 Charge-sheet model

The current-voltage characteristics of MOSFETs can also be derived from the assumption that the inversion charge is an infinitesimally thick layer near the interface (i.e., x = 0) (i.e., the charge-sheet model) [44-46]. Based on this approach, the inversion layer charge can be expressed, using the Gauss's law, by

$$Q_n \approx - \epsilon_s (\xi_S - \xi_{Sd}) = Q_s - Q_d \tag{1.69}$$

where $\xi_S = \xi(x = 0^-)$, $\xi_{Sd} = \xi(x = 0^+)$, Q_s is the total charge in the semiconductor, and Q_d is the depletion charge. In other words, ξ_S and ξ_{Sd} are the surface electric fields on both sides of the inversion region. The term $Q_s = -\epsilon_s \xi_S$ was described in (1.60) and the depletion charge is given by:

$$Q_d = - \epsilon_s \xi_{Sd} \tag{1.70}$$

where ξ_{Sd} can be obtained from (1.59) and (1.50):

$$\xi_{Sd} = \left(F^2(\psi_S, V) \right)^{1/2} \approx \frac{2^{1/2}}{\beta L_D} \left((\beta\psi_S - 1) \right)^{1/2} \quad . \tag{1.71}$$

According to the conventional approach (i.e., without assuming an infinitesimally thick inversion layer), the electric field ξ_S at the surface can also be derived from the Poisson equation using the depletion approximation:

$$\frac{d^2\psi}{dx^2} = -\frac{\rho}{\epsilon_s} \approx \frac{q\,N_A}{\epsilon_s} , \tag{1.72}$$

This, together with the boundary conditions of $\psi(x=0)=\psi_S$, $\psi(x=X_d)=0$ and $\xi(x=X_d)=0$, yields:

$$\xi_S = \left(F^2(\psi_S, V) \right)^{1/2} \approx \frac{2^{1/2}}{\beta\, L_D} \left(\beta\psi_S \right)^{1/2} \tag{1.73}$$

The electric fields at the surface given in (1.71) and (1.73) differ by a "-1" term associated with $\beta\psi_s$, which can give rise to a large discrepancy between the conventional and charge-sheet models under certain bias conditions [45].

Brews, using (1.69)-(1.71) and some additional approximations, obtained the drain current [45-46]:

$$\begin{aligned} I_D = \mu_n \frac{W}{L_{eff}} \Big[C_o & \left((V_{GS} - V_{FB})(\psi_{SL} - \psi_{So}) - \frac{(\psi_{SL}^2 - \psi_{So}^2)}{2} \right) \\ & - \frac{q\,N_A\,L_D\,2^{3/2}}{3} \left((\beta\psi_{SL} - 1)^{3/2} - (\beta\psi_{So} - 1)^{3/2} \right) \\ & + q\,N_A\,L_D\,2^{1/2} \left((\beta\psi_{SL} - 1)^{1/2} - (\beta\psi_{So} - 1)^{1/2} \right) \Big] \end{aligned} \tag{1.74}$$

This model, which is also valid for long-channel MOSFETs under all inversion conditions, has an error of 5 % or less compared the Pao-Sah counterpart.

It should be pointed out that the Brews's model in (1.74) for the drain current can also be obtained from the Pierret's model using the following empirical approximation:

$$(F^2(\psi,V))^{1/2} \approx \frac{2^{1/2}}{\beta L_D} \left((\beta\psi - 1)^{1/2} - \frac{(\beta\psi - 1)^{-1/2}}{2} \right) \tag{1.75}$$

1.4.4.4 Strong inversion model

The drain current models discussed above can be simplified under the strong inversion condition. In this case, the surface band bending increases very little with increasing gate bias, as has been illustrated in Fig. 1.24. This allows one to assume the band bending is nearly independent of the gate bias under the strong inversion. Thus,

$$\psi_{So} \approx 2\phi_B \quad , \tag{1.76}$$

at the source, and

$$\psi_{SL} \approx 2\phi_B + V_{DS} \quad , \tag{1.77}$$

at the drain. Also, under the strong inversion, the inequality $\beta\psi >> 1$ is valid, and (1.50) can be approximated by

$$F^2(\psi,V) \approx \frac{2\psi}{\beta L_D^2} \tag{1.78}$$

Putting (1.76)-(1.78) into (1.68), and integrating the resulting equation yields the following analytic expression for the drain current:

$$I_D = \mu_n \frac{W}{L_{eff}} C_o \, [\, (V_{GS} - V_{FB} - 2\phi_B - \frac{V_{DS}}{2})V_{DS} - \frac{2 \, (2 \, \epsilon_s \, q \, N_A)^{1/2}}{3 \, C_o} \left((V_{DS} + 2\phi_B)^{3/2} - (2\phi_B)^{3/2} \right)] \quad . \tag{1.79}$$

Equation (1.79) can also be derived from (1.67) using the assumptions that the current is only due to drift and the device is in strong inversion [26].

It is important to mention that the model in (1.79) is valid only when the inversion layer is present in the entire channel, a case holds for a relatively small drain voltage (i.e., MOSFET operates in the linear region). For a sufficiently large drain voltage, the inversion layer will pinch-off near the drain junction due to a large lateral field in that region. Under such a condition, the drain current is near constant with respect to the drain voltage (i.e., MOSFET operates in the saturation region), and (1,79) needs to be modified to reflect this behavior [26].

1.4.4.5 Weak inversion model

Under the weak inversion condition, the surface is depleted or weakly inverted, and the current is very small. In this case, $F^2(\psi,V)$ can be approximated by

$$F^2(\psi,V) \approx \frac{2}{\beta^2 L_D^2}\left(\beta\psi - 1 - e^{\beta(\psi - V - 2\phi_B)}\right) \tag{1.80}$$

Note that the last term on the right-hand side of (1.80) is very small under weak inversion. Thus, the following approximation can also be used:

$$(\beta\psi - 1) \gg e^{\beta(\psi - V - 2\phi_B)} \tag{1.81}$$

It has been shown in Fig. 1.18 that the surface band bending depends approximately linearly on the gate bias for the weak inversion condition. This dependence can also be obtained from the combination of (1.80), (1.81) and (1.53), and using the condition $\alpha = 0$ and the Taylor expansion at $\psi = \phi_B$:

$$V_{GS} - V_{FB} = \psi_S \pm \frac{\epsilon_s}{C_o}\frac{2^{1/2}}{\beta L_D}(\beta\psi_S - 1)^{1/2} \tag{1.82}$$

Equation (1.82) indicates that the surface band bending is independent of the voltage drop along the channel. Thus,

$$\psi_{SL} \approx \psi_{So} \approx \psi_S \tag{1.83}$$

and the drift component of the current is negligibly small.

The drain current in the weak inversion is obtained by substituting (1.81) and (1.83) into (1.68):

$$I_D \approx \mu_n \frac{W}{L_{eff}} \epsilon_s \left(\int_0^{\psi_S} F(\psi, V{=}0)\, d\psi - \epsilon_s \int_0^{\psi_S} F(\psi, V{=}V_{DS})\, d\psi \right)$$

$$\approx \mu_n \frac{W}{L_{eff}} \left(2\, \epsilon_s\, p_o\, kT \right)^{1/2} \int_0^{\psi_S} \left[\left(\beta\psi - 1 - e^{\beta(\psi - 2\phi_B)} \right)^{1/2} \right.$$

$$\left. - \left(\beta\psi - 1 - e^{\beta(\psi - V_{DS} - 2\phi_B)} \right)^{1/2} \right] d\psi$$

$$\approx \mu_n \frac{W}{L_{eff}} \left(\frac{\epsilon_s\, p_o\, kT}{2} \right)^{1/2} \int_0^{\psi_S} \left(\frac{e^{\beta(\psi - 2\phi_B)} - e^{\beta(\psi - V_{DS} - 2\phi_B)}}{(\beta\psi - 1)^{1/2}} \right) d\psi \quad . \tag{1.84}$$

The approximation $(1 - x)^{0.5} \approx 1 - x/2$, valid for $x << 1$, was also used in obtaining (1.84). Since the term $(\beta\psi)^{0.5}$ depends much less on ψ than the term $\exp(\beta\psi)$, it can be assumed that $(\beta\psi)^{0.5}$ is constant with respect to ψ and that $\psi \approx \psi_S$. This results in (1.84) being simplified to

$$I_D \approx \mu_n \frac{W}{L_{eff}} \left(\frac{\epsilon_s\, p_o\, kT}{2} \right)^{1/2} \frac{e^{\beta(\psi_S - 2\phi_B)}}{\beta\, (\beta\psi_S - 1)^{1/2}} \left(1 - e^{-\beta V_{DS}} \right) \tag{1.85}$$

in which the approximation $\exp(\beta\psi_S) >> \beta\psi_S$ was also used. Equation (1.85) suggests that the drain current is independent of V_{DS}, for $\beta V_{DS} >> 1$, because the term $\exp(-\beta V_{DS})$ is negligibly small. Finally, noting that ψ_S depends approximately linearly on the gate bias and that I_D is proportional to $\exp(\beta\psi_S)$, we can conclude that the drain current depends exponentially on the gate voltage under the weak inversion.

1.4.4.6 SPICE model

For the purpose of SPICE circuit simulation, a further simplified model is needed. The simplest MOSFET SPICE model (i.e., level-1 model) can be obtained as follows [47]. Consider the case of strong inversion and assume

$$V_{DS} \ll 2\phi_B \tag{1.86}$$

Next, use the Taylor's series to approximate the terms having the power of 3/2 in (1.79), one obtain:

$$I_D = \mu_n \frac{W}{L_{eff}} C_o \left[V_{GS} - V_T - \frac{V_{DS}}{2} \right] V_{DS} \tag{1.87}$$

where

$$V_T \equiv V_{FB} + 2\phi_B + \frac{2 (\epsilon_s q N_A \phi_B)^{1/2}}{C_o} \tag{1.88}$$

is the threshold voltage. Equation (1.88) can be rewritten by noting that its last term is related to the maximum depletion charge Q_{dmax} (i.e., Q_d becomes Q_{dmax}) which occurs at the onset of inversion:

$$V_T \equiv V_{FB} + 2\phi_B - \frac{Q_{dmax}}{C_o} \quad . \tag{1.89}$$

It is important to point out that equation (1.88) is valid only when the substrate voltage V_{BS} is zero (i.e., no body effect). On the other hand, equation (1.89) is more general and is valid with the presence of body effect, provided the depletion charge accounts for the effect of V_{BS}.

Clearly, the threshold voltage is an important parameter for modeling the MOSFET. Beside modeling V_T, as was done in (1.88) and (1.89), such a parameter can be determined by extraction methods, which will be discussed in detail in Chapter 3.

The plot of I_D versus V_{DS} based on (1.87) is a parabola with a maximum at $V_{DS} = (V_{GS} - V_T)$. This value of V_{DS} is known as the drain saturation voltage. This is because for $V_{DS} > (V_{GS} - V_T)$, the drain current does not obey (1.87) and is nearly constant versus the drain voltage.

The simplest SPICE model described above can be improved by including the weak inversion characteristics, which results in a more accurate but complex SPICE model (i.e., level-10 model). The preceding analysis has shown that the drain current of a MOSFET possesses two different asymptotic behaviors: $I_D \propto V_{GS}$ for strong inversion ($V_{GS} >> V_T$), and $I_D \propto \exp(V_{GS})$ for weak inversion ($V_{GS} < V_T$). This problem is analogous to that of a diode with a current described by an ideal characteristic given by the saturation current I_s and voltage V (i.e., $I_s \exp(qV/kT)$) and an nonideal component of series resistor R [48-49]. A simple and approximated solution for this generic problem is of the type:

$$I_n \approx \ln\left(1 + e^{V_n}\right) \tag{1.90}$$

where I_n and V_n are the normalized current and voltage. Equation (1.90) has the characteristics of $I_n \approx V_n$ for $V_n >> 1$ and $I_n \approx \exp(V_n)$ for $V_n << 1$.

The approach of (1.90) has been implemented in SPICE level-10 model [49-50] to describe the inversion charge under both the strong and weak inversion conditions:

$$Q_n \approx 2\, Q_o \ln\left(1 + \frac{1}{2}\, e^{\beta(V_{GS} - V_T)}\right) \tag{1.91}$$

where Q_o is defined by

$$Q_o \equiv \frac{C_o\, \eta}{2\, \beta} \tag{1.92}$$

and η is the ideality factor in the subthreshold region.

The SPICE level-1 model uses two different equations for the triode region ($V_{DS} < V_{GS} - V_T$) and for the saturation region ($V_{DS} > V_{GS} - V_T$). This can give rise to a great difficulty for SPICE simulation of short-channel devices because the boundary between the saturation and triode region in such devices is unclear.

SPICE level-10 model evades this difficulty by describing the drain current in both the triode and saturation regions in a single expression:

$$I_D = \frac{g_{ch} V_{DS} (1 + \lambda V_{DS})}{\left(1 + (\frac{V_{DS}}{V_{sate}})^m \right)^{\frac{1}{m}}} \quad . \tag{1.93}$$

where λ is the channel-length modulation coefficient, m is an empirical parameter, V_{sate} is the effective voltage accounting for the effect of free-carrier drift velocity saturation in a short channel, and g_{ch} is the total channel conductance. Including the parasitic series resistances at the source and drain, the total channel conductance of the MOSFET is the series combination of the inverse of drain and source series resistance (i.e., $1/(R_D + R_S)$) and the intrinsic channel conductance g_{chi}:

$$\frac{1}{g_{ch}} = \frac{1}{g_{chi}} + (R_D + R_S) \tag{1.94}$$

where

$$g_{chi} = \frac{W}{L_{eff}} Q_n \mu_n \quad . \tag{1.95}$$

Models like (1.93) defined by a single equation are very useful for circuit simulation because they often eliminate the convergence problems related to the discontinuity of the derivatives [51]. Also, It can be seen that the accuracy of the model in (1.93) depends heavily on the accurate values of R_D and R_S, particularly for short-channel MOSFETs in which the importance of parasitic resistances increases. As will be shown in Chapter 5, such parameters can be extracted from measurements or simulation results.

1.5 Short-channel effects

Due largely to the relentless effort of increasing the number of transistors per chip and the advance in device processing technology, the MOSFET channel length continues to decrease, and the gradual channel approximation, which is the basis of all the long-channel MOSFET's models, becomes questionable. This is because the y-direction electric field is increased with decreasing channel length, and thus the strong interaction between the x- and y-direction fields in the short-channel device has invalidated the conventional approach of solving the two-dimensional problem in the individual x- and y-directions. Figures 1.27 and 1.28 show the electrostatic potential contours, simulated using MICROTEC, for a long- and a short-channel MOSFET, respectively. The long-channel device, with a mask channel length of 10 μm, is the same as that used and described in Section 1.4. The short-channel device has the same device make-up as the long-channel MOSFET, but the lateral dimension was scaled down by a factor of 20. Explicitly, the device make-up for the short-channel device is: a 0.5 μm mask channel length, a 0.05 μm contact width, and a 0.05 μm separation between the source- and drain-contact to the gate. It is clearly shown in Fig. 1.27 that, for the long-channel MOSFET, the effects of the source- and drain-body junctions on the electrostatic potential in the channel are insignificant. On the other hand, shown in Fig. 1.28 for the short-channel MOSFET, the effects of the source- and drain-body junctions are very important in the channel region, as the channel region and the drain/source junctions become indistinguishable.

We also find that when the channel length of the MOSFET is decreased, the relative importance of the drain/source lateral diffusions with respect to the channel length is increased. This, as will be shown later, can give rise to a reduction in the threshold voltage. Other physical insight of the short-channel MOSFET are illustrated the following figures. Figure 1.29 shows the energy band diagram, the electron concentration and the doping density versus the lateral distance. Figure 1.30 shows the lateral and vertical electric fields versus the lateral distance. Figure 1.31 shows the lateral and vertical drain current densities versus the lateral distance.

The effects of short channel on the threshold voltage and on the charge transport in the channel are discussed below.

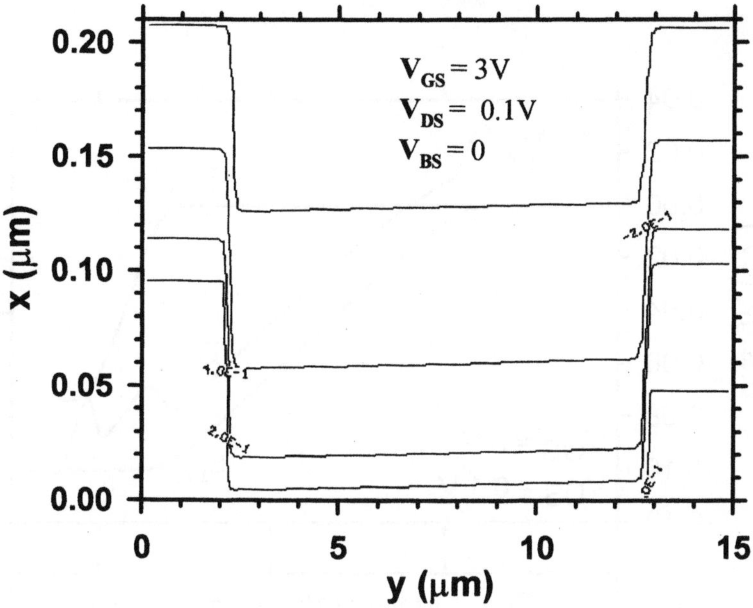

Figure 1.27 : Electrostatic potential contours for a MOSFET with a mask channel length of 10 μm.

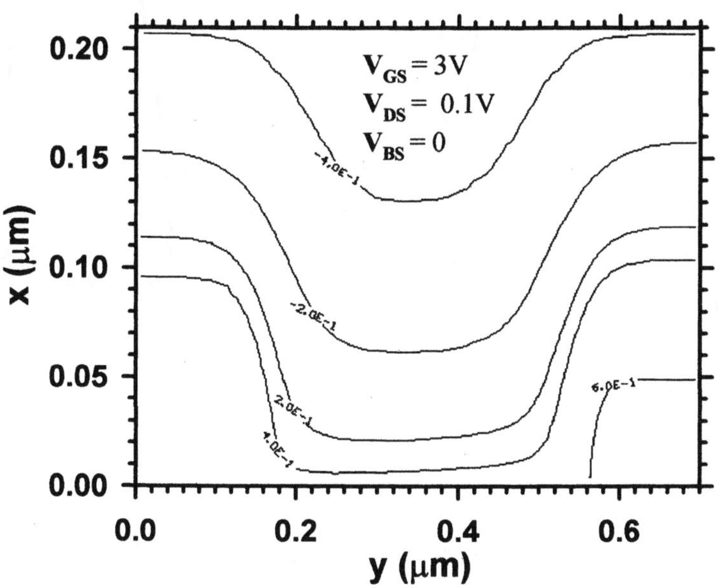

Figure 1.28 : Electrostatic potential contours for a MOSFET with a mask channel length of 0.5 μm.

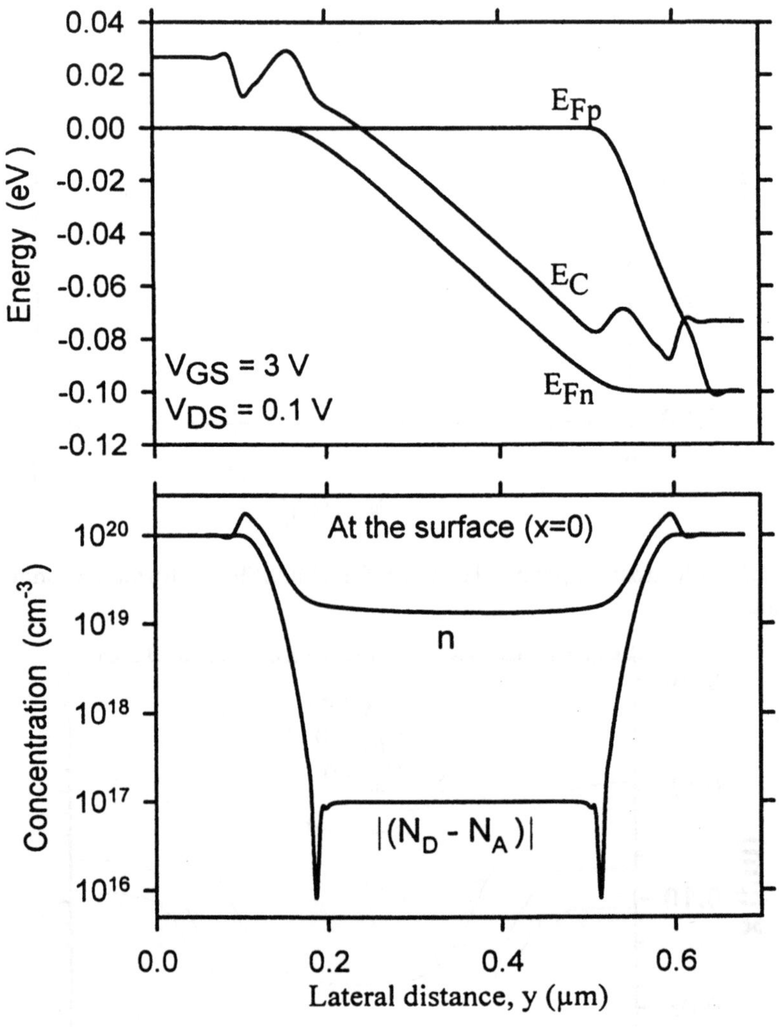

Figure 1.29 : Energy band diagram, electron concentration, and doping density versus the lateral distance simulated for a MOSFET with a mask channel length of 0.5 μm.

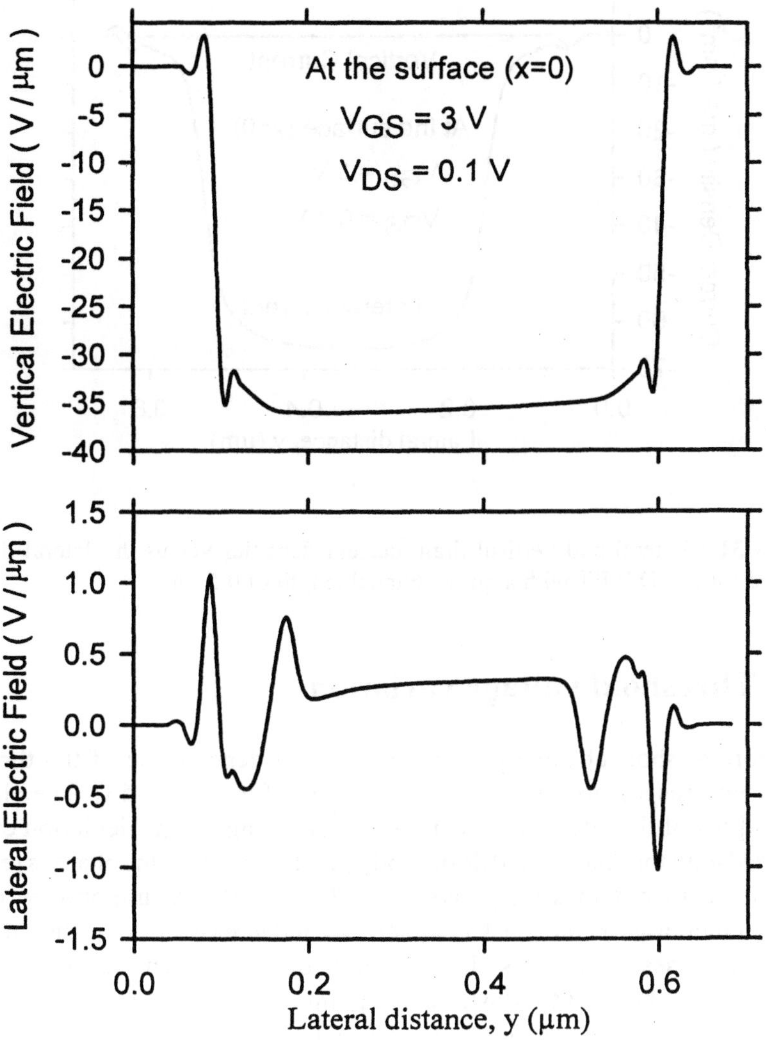

Figure 1.30 : Lateral and vertical electric fields versus the lateral distance simulated for a MOSFET with a mask channel length of 0.5 μm..

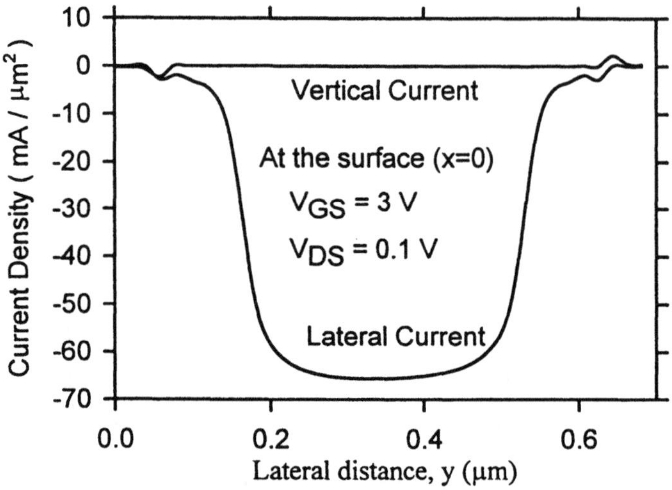

Figure 1.31 : Lateral and vertical drain current densities versus the lateral distance simulated for a MOSFET with a mask channel length of 0.5 μm.

1.5.1 Threshold voltage variation

An important short-channel phenomenon is the dependance of the threshold voltage with respect to the mask channel length [26, 43, 52-53]. The classical analysis [26] of this effect is based on the following: 1) the depletion charges associated with the source- and drain-body junctions reduce the charges required for the gate to obtain strong inversion; 2) the relative importance of this reduction increases as the mask channel length decreases; and 3) the threshold voltage decreases as the mask channel length decreases. The schematic shown in Fig. 1.32 can be used to illustrate this point. For a long-channel device, the depletion charge in the channel is $Q_{dmax}[(L_e + L)/2L] \approx Q_{dmax}$ (i.e., $L \approx L_e$), where L and L_e are the upper and lower channel lengths, respectively, as shown in Fig. 1.32. This is not true, however, for short-channel devices due to the fact that the lateral extent of the source and drain (defined by the radius r, see Fig. 1.32) plays a more important role and the ratio of L_e to L is decreased. The threshold voltage V'_T taking this into account can be expressed as

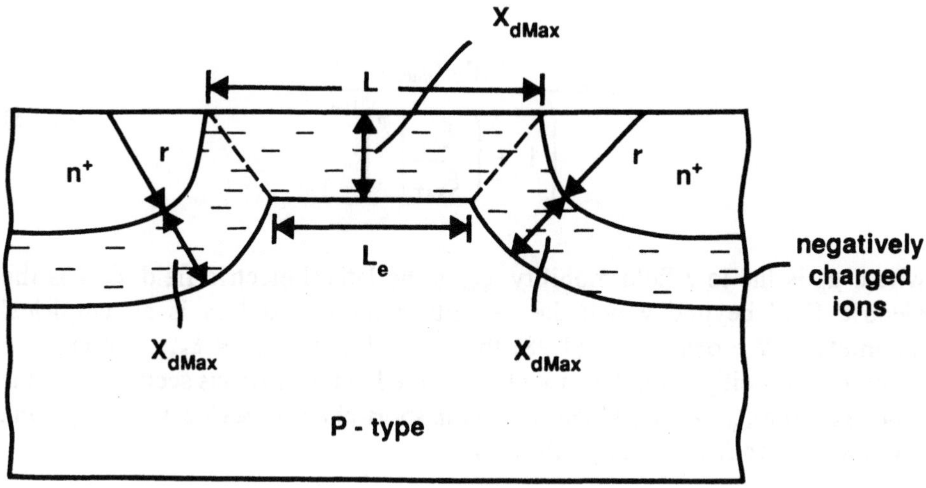

Figure 1.32 : Schematic of short-channel MOSFET showing the effect of drain/source lateral extent (defined by the radius r) on the depletion region charge.

$$V'_T = V_T - \Delta' V_T = V_{FB} + 2\,\phi_B - \frac{\dfrac{Q_{dmax}}{C_o}}{\dfrac{L_e + L}{2\,L}} \tag{1.96}$$

where $\Delta' V_T$ is the reduction in the threshold voltage due to the short-channel effect:

$$\Delta' V_T = 2\,\alpha_1 \frac{\epsilon_s}{\epsilon_{ox}} \frac{Tox}{L} (2\,\phi_B - V_B) \tag{1.97}$$

Here α_1 is a channel-length geometry factor involving the effect of lateral-extent radius r, and T_{ox} is the oxide layer thickness.

1.5.2 Velocity saturation

The lateral electric field in the channel region is inversely proportional to the channel length. Furthermore, the free-carrier drift velocity v saturates at a sufficiently large field. As a result, v is likely to saturate in the short-channel device in which the electric field is high. The drift velocity can be modeled by [49]

$$v = \frac{\mu_{lf}\, \xi_{lat}}{\left(1 + \left(\frac{\xi_{lat}}{\xi_{SAT}}\right)^{m}\right)^{1/m}} \tag{1.98}$$

where μ_{lf} is the low field mobility, ξ_{lat} is the lateral electric field, ξ_{SAT} is the electric field beyond which the velocity saturates, and m is an empirical parameter. We observe in (1.98) that $v \approx \mu_{lf}\xi_{lat}$ for $\xi_{lat} << \xi_{SAT}$, a condition assumed implicitly in all the models presented in the previous sections. For the other extreme $\xi_{lat} >> \xi_{SAT}$, which is likely in short-channel device, $v \approx \mu_{lf}\xi_{SAT}$, and the free-carrier drift velocity saturates.

Taking the drift-velocity saturation effect into account, the drain current I'_D in the linear region can be related to the long-channel drain current I_D by

$$I'_D = \frac{I_D}{1 + \frac{V_{DS}}{L\, \xi_{SAT}}} \tag{1.99}$$

Apparently, the drift velocity saturation will decrease the drain current.

1.5.3 Channel-length modulation

When the channel length of the MOSFET is decreased and the MOSFET is operated beyond channel pinch-off as shown in Fig. 1.33, the relative importance of the pinch-off length ΔL with respect to the physical channel length is increased. Such an effect, called the channel-length modulation, can be included into the saturation drain current I'_{Dsat} as

$$I'_{Dsat} = \frac{I_{Dsat}}{1 - \frac{\Delta L}{L}} \tag{1.100}$$

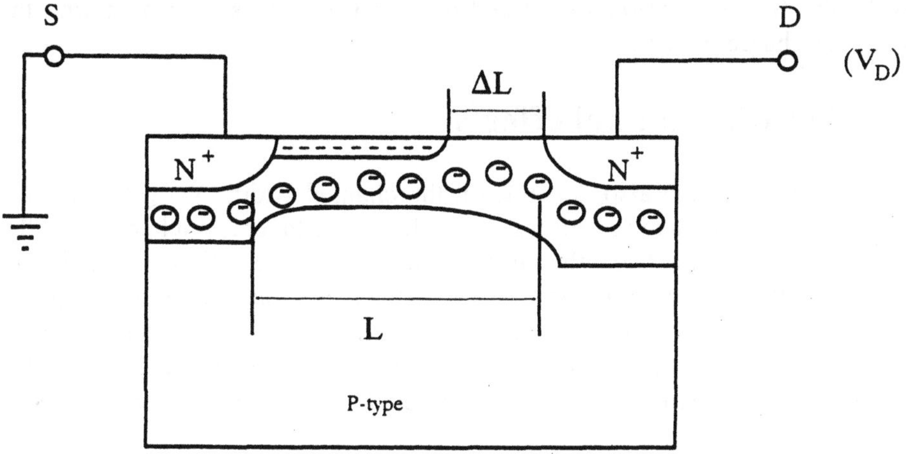

Figure 1.33 : A short-channel MOSFET in saturation, showing the length ΔL of the pinch-off region is significant comparing to the physical channel length L.

where I_{Dsat} is the saturation drain current without including the channel-length modulation effect and

$$\Delta L = \left(\frac{2\, \epsilon_s}{q\, N_A} \right)^{1/2} \left(\left(\phi_D + V_{DS} - V_{Dsat} \right)^{1/2} - \phi_D^{1/2} \right) \tag{1.101}$$

where $\phi_D = (\epsilon_s/qN_A)(V_{DS}/L)^2$. Equation (1.100) can also be written in the form of

$$I'_{Dsat} = I_{Dsat} \left(1 + \lambda\, V_{DS} \right) \tag{1.102}$$

Here, λ is called the channel-length modulation coefficient.

1.5.4 Ballistic transport

An implicit assumption used in all previous models is that the free carriers in the channel experience considerable scattering when they travel from the source to drain. In other words, it has been assumed that the effective channel length is much larger that the mean free path of free-carrier scattering process. When the effective channel length is less than the mean free path, as would be the case of short-channel MOSFET, many carriers will be subjected to very few scattering when traveling along the channel. This is called the ballistic transport, and it invalidates the conventional drift and diffusion model used in modeling the MOSFET characteristics.

1.6 Narrow-channel effects

The threshold voltage also depends on the channel width W [26,43]. This dependence is explained as follows: 1) the thin oxide layer is thicker toward the edges of the channel (i.e., the bird's beak); 2) the depletion region is not confined to the area under the thin oxide; 3) the depletion charge in the regions underneath the bird's beak is negligibly small compared to those in the channel for a large W but becomes more important as W is decreased; and 4) the threshold voltage, which is function of the depletion charge, is increased as W decreases.

Figure 1.34 illustrates the depletion charge in the channel (underneath the thin oxide layer) and in regions under the bird's beak (underneath the thick oxide layer). The threshold voltage V''_T taking into account the narrow channel effect is

$$V''_T = V_T - \Delta'' V_T$$

and (1.103)

$$\Delta V''_T = \alpha_2 \pi \frac{\epsilon_s}{\epsilon_{ox}} \frac{T_{ox}}{W} (2 \phi_B - V_B)$$

where $\Delta'' V_T$ is the threshold voltage increase due to the narrow channel effect and α_2 is the channel width geometry factor involving the effect of bird's beak shape, which is a function of the processing conditions (i.e., oxidation time, wet or dry oxidation, etc.).

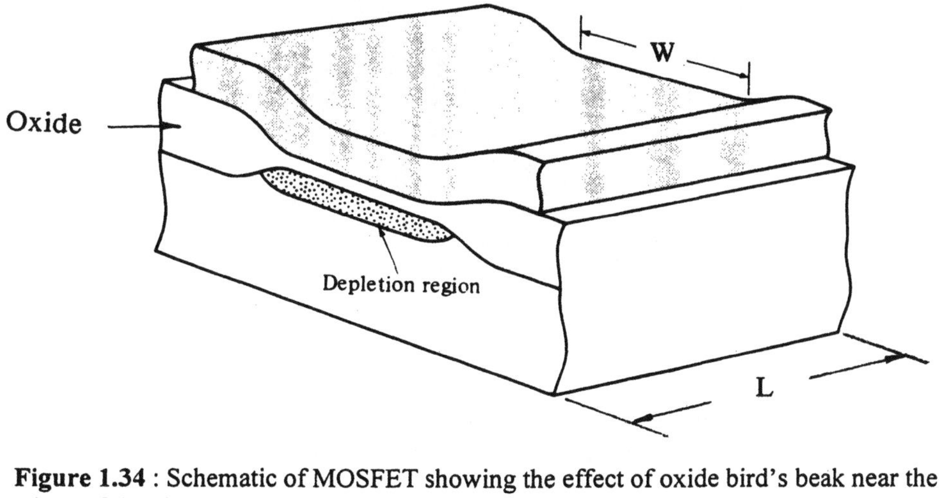

Figure 1.34 : Schematic of MOSFET showing the effect of oxide bird's beak near the edges of the channel width on the depletion region charge.

1.7 Hot-carrier effects

The scale-down in the channel length of advanced MOSFETs has produced a very high electric field near the drain junction. This high field heats up the electrons in the channel and thus causes the so-called hot-carrier effects. The hot electrons acquire enough energy to inject into the oxide, resulting in an increase in the gate current as well as a shifting in the threshold voltage. The high field near the drain junction also activates impact ionization in the region, which then increases the drain and substrate currents. Moreover, the field distribution near the drain junction in such devices is highly two-dimensional, a condition invalidates the conventional MOSFET model derived based on the simplified one-dimensional approach [54-56]. Figure 1.35 shows the two-dimensional electric field distribution in the channel of a 0.8-μm MOSFET obtained from a numerical study which couples drift-diffusion and Monte Carlo models [57]. The very high field and the multi-dimensional nature of the field near the drain region are clearly illustrated.

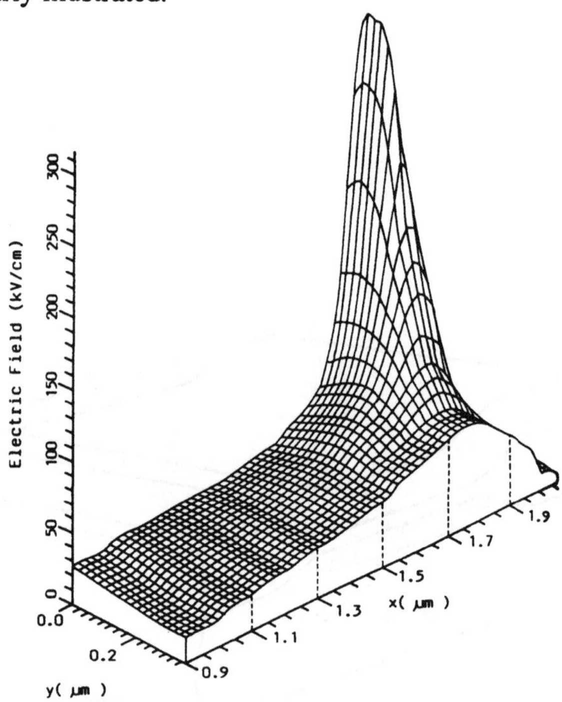

Figure 1.35 : Two-dimensional electric field distribution for an n-channel MOSFET with a channel length of 0.8 μm, gate oxide thickness of 100 Å, gate width of 20 μm, and doping concentration of 10^{16} cm^{-3}. The increasing x-direction is the channel direction toward the drain region. (Source: Huang et al. [57]. Reprinted with permission).

1.7.1 Drain current including hot-carrier effects

While the conventional drain current model for the linear region is still applicable (derived in the previous sections) because of the relatively low drain voltage, and thus the relatively low field in the channel, the conventional model for the saturation region needs to be modified to account for the two-dimensional effect associated with the hot-carrier phenomenon. A simple, analytical, quasi-two-dimensional model including hot-carrier effect has been proposed for this objective [58]. Consider an n-channel MOSFET under the saturation region. The channel can be divided into two regions, as shown in Fig. 1.36: the source region where the conventional gradual-channel approximation holds, extending from the source junction to the point of saturation ($y = 0$), and the drain region where the quasi-two-dimensional analysis is used, extending from the saturation point to the drain junction. The drain current I_D in the saturation region including the hot-carrier effect is given by

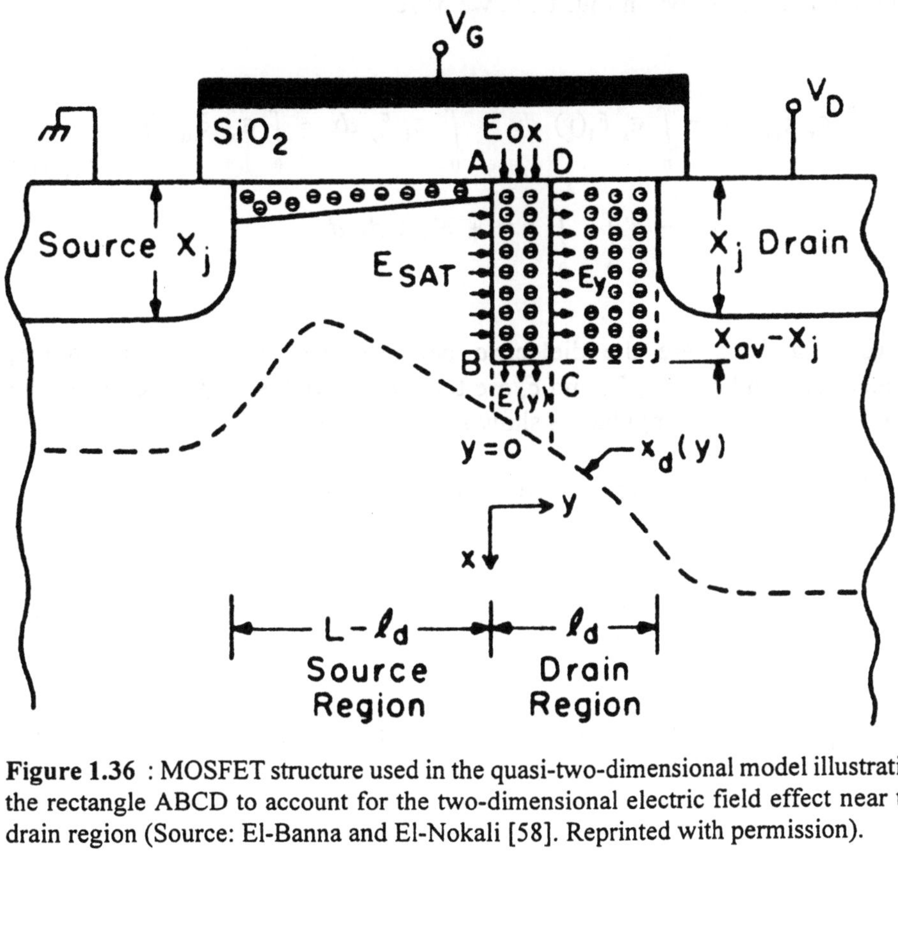

Figure 1.36 : MOSFET structure used in the quasi-two-dimensional model illustrating the rectangle ABCD to account for the two-dimensional electric field effect near the drain region (Source: El-Banna and El-Nokali [58]. Reprinted with permission).

$$I_D = \frac{I_{Dsat}}{1 - \dfrac{l_d}{L_{eq}}} \tag{1.104}$$

$$L_{eq} = L\left(1 + \frac{V_{Dsat}}{L\,\xi_C}\right) \tag{1.105}$$

where l_d is the length of the drain region (Fig. 1.36), L is the metallurgical channel length, and ξ_C is the critical field beyond which the velocity saturates. In the drain region ($0 < y < l_d$), the mobile electrons are assumed spread over an average width $X_{av} = (X_1 + X_j)/2$, where X_1 is the depletion region thickness at y = 0 and X_j is the drain junction depth. Applying Gauss's law to the sides of the rectangle ABCD shown in Fig. 1.36, we have

$$\begin{aligned} -\int_0^{X_{av}} \epsilon_s\,\xi_{sat}\,dx + \int_0^{y} \epsilon_s\,\xi_1(y)\,dy + \int_0^{X_{av}} \epsilon_s\,\xi_y\,dx - \int_0^{y} \epsilon_{ox}\,\xi_{ox}\,dy \\ = -q\int_0^{y}\int_0^{X_{av}} (n + N_A)\,dx\,dy \end{aligned} \tag{1.106}$$

where ϵ_s and ϵ_{ox} are the dielectric permittivities in silicon and oxide, respectively, and ξ_{sat}, ξ_{ox}, ξ_1, and ξ_y are the electric fields perpendicular to the boundaries of the rectangular, as shown in Fig. 1.36.

The electric field $\xi_1(y)$ can be solved from the one-dimensional Poisson equation subject to the following boundary conditions:

$$\psi = 0 \quad and \quad \frac{d\psi}{dx} = 0 \quad at \quad x = X_d(y) \tag{1.107}$$

where X_d is the x-direction depletion region thickness (see Fig. 1.36). This yields

$$\xi_1(y) = \frac{q\,N_A}{\epsilon_s}\left((X_2 - X_1)\frac{y}{l_d} + X_1 - X_{av} \right) \tag{1.108}$$

Here X_2 is the y-direction depletion region thickness at the drain junction ($y = l_d$).

Equation (1.106) can be solved using the following boundary conditions:

$$V_y(y{=}0) = V_{Dsat} \quad and \quad \xi_y = \xi_{sat} \tag{1.109}$$

The resulting equation is [58]

$$V_y = V_{Dsat} + \frac{C}{A^2\,[\cosh(A\,y) - 1]} + \frac{1}{A\,(\xi_{sat} + \xi)\sinh(A\,y)} - \xi y \tag{1.110}$$

Here $C = I_{Dsat}/(X_{av}\epsilon_s W v_{max})$, $A^2 = C_{ox}/X_{av}\epsilon_s$, and $\zeta = \mathrm{k}qN_A/C_{ox}$. One can then derive the following expression involving l_d from (1.110) at the drain end:

$$V_D = V_{Dsat} + \frac{C}{A^2\,[\cosh(A\,l_d) - 1]} + \frac{1}{A\,(\xi_{sat} + \xi)\sinh(A\,l_d)} - \xi\,l_d \tag{1.111}$$

The expressions for ξ_{sat} and V_{Dsat} can be derived using the condition that the conductance is continuous at the transition point ($y = 0$). This gives

$$\frac{\xi^3_{sat}}{\xi^2_{01}} + \xi_{sat} - \frac{\xi_{02}}{2} = 0 \tag{1.112}$$

$$a_0 \, V_{Dsat} \, \xi_{sat} \left(L + \frac{V_{Dsat}}{\xi_C} \right) + \left(V_G - V_T - a_0 \, V_{Dsat} \right) \left(V_{Dsat} - L \, \xi_{sat} \right) = 0 \tag{1.113}$$

$$\xi_{01} = \frac{\left(V_G - V_T - a_0 \, V_{Dsat} \right) V_{Dsat}}{L \left(a_0 + \left(V_G - V_T \right) L \, \xi_C \right)^{1/2}} \tag{1.114}$$

$$\xi_{02} = C \, L_{eq} \tag{1.115}$$

$$a_0 = \frac{1}{2} + \frac{f_{st} \left(2 \, \epsilon_s \, q \, N_A \right)^{1/2}}{4 \, C_{ox} \left(2 \, \phi_B + V_B \right)^{1/2}} \tag{1.116}$$

Here f_{st} is a parameter accounting the short-channel geometry effect [26].

Equations (1.112) and (1.113) can be solved numerically to obtain V_{Dsat} and ξ_{sat}, which are then used to find l_d from (1.111). The drain current can be calculated from (1.104) and (1.105) after l_d is found.

Figure 1.37 shows the comparison of the I_D-V_D characteristics calculated from the above model including hot-carrier effect, calculated from the conventional model, and obtained from measurements for an n-channel MOSFET with a channel length of 1.09 μm, channel width of 100 μm, doping concentration of 6.6×10^{15} cm^{-3}, and oxide thickness of 358 Å. Note that the hot-carrier effect is important only if the gate or drain voltage is large. Under such bias conditions, the conventional model can overestimate the drain current considerably, particularly if both the gate and drain voltages are large.

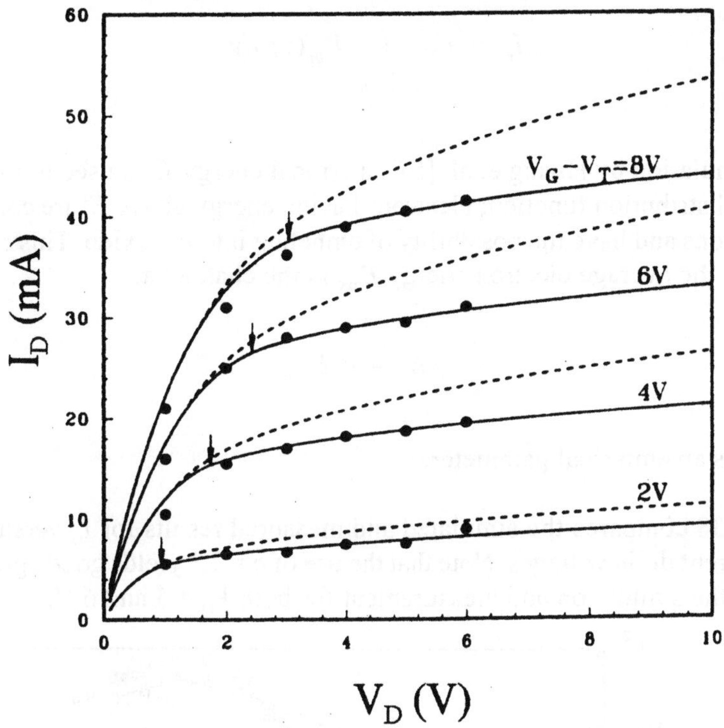

Figure 1.37 : Comparison of I_D-V_D characteristics obtained from the model with hot-carrier effect (solid lines), without hot-carrier effect (dashed lines), and measurements (closed circles) (Source: El-Banna and El-Nokali [58]. Reprinted with permission).

1.7.2 Modeling the gate current

The hot carriers, when injected into the gate oxide, can increase the gate current. The probability P_{inj} of electron injection into the gate is [57]

$$P_{inj}(y) = \int_0^{\infty} P_c(y)\ f(y,E)\ D(E)\ P_t(y,E)\ dE \tag{1.117}$$

where $D(E)$ is the electron density of state at a particular energy E, P_c is the collision probability, f is the electron distribution function, p_t is the oxide transmission probability. Integrating (1.117) along the channel, the resulting gate leakage current I_G is obtained:

$$I_G = I_D \int_{channel} P_{inj}(y)\, dy \qquad (1.118)$$

In the simulation by Huang et al. [57], a critical energy E' is used to model the electron distribution function; electrons having energy above E' are considered hot electrons and have the possibility of tunneling into the oxide. This energy is related to the average electron energy E_{avg} in the channel as

$$E' = c\, E_{avg} \qquad (1.119)$$

where c is an empirical parameter.

Figure 1.38 compares the simulated and measured results for I_G versus V_G for two different drain voltages. Note that the use of $c = 2.1$ yields good agreements between the simulation and measurement for both $V_D = 5$ and 6 V.

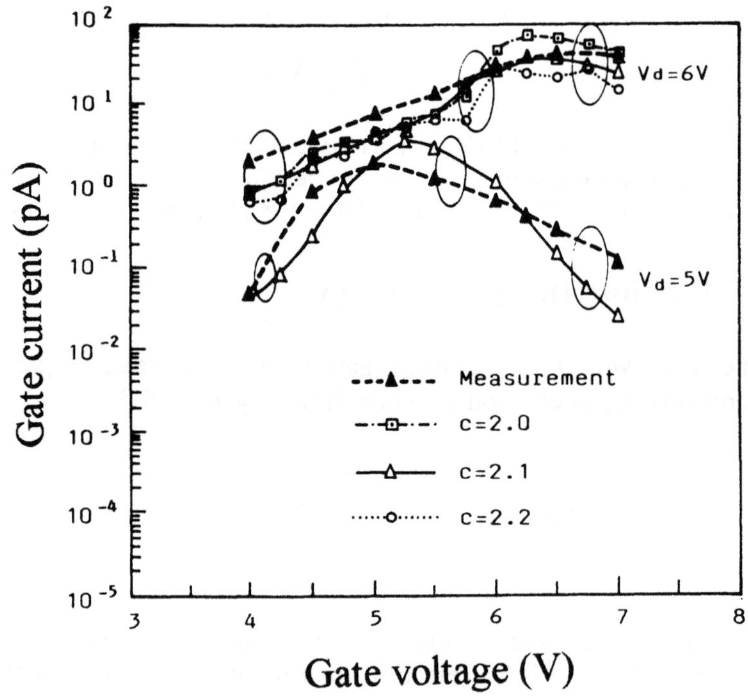

Figure 1.38 : Simulated and measured gate leakage current for the same device considered in Figure 1.35. The empirical parameter c is defined in the text (Source: Huang et al. [57]. Reprinted with permission).

1.7.3 Modeling the substrate current

The substrate current I_{sub} is a useful indicator for the hot-carrier effect and the lifetime of the MOSFET subject to hot-carrier stress. Such a current is due primary to the impact ionization in the high-field region near the drain junction (i.e., in the region $0 < y < l_d$). Using the local impact ionization approximation (i.e., the impact generated free carriers are at equilibrium withe the local states), the substrate current can be expressed by

$$I_{sub} = I_D\,(M - 1) = \int_0^{l_d} \alpha \exp\left(-\frac{\xi_C}{\xi}\right) dy \tag{1.120}$$

where M is the avalanche multiplication factor, α is the ionization coefficient, ξ_C is the critical field beyond which the impact ionization can occur, and ξ is the electric field in the high-field region. Several different values for α and E_C have been reported in the literature and are summarized in [26].

The question is what expression of ξ should be used in (1.120). There are two approaches of modeling ξ. The first suggests that ξ can be approximated by the maximum field ξ_{max}:

$$\xi_{max} = \frac{V_D - V_{Dsat}}{l_d} \tag{1.121}$$

The other believes that ξ can be represented by an effective field ξ_{eff}:

$$\xi_{eff} = \frac{V_D - V_{Dhot}}{l_d} \tag{1.122}$$

where V_{Dhot} is the onset drain voltage for the hot-carrier effect. Note that ξ is a function of the stress time because electrons captured in the oxide during the stress test can alter the field in the high field region as well as l_d.

An empirical I_{sub} model has been developed based on the approach of $\xi = \xi_{eff}$ and the impact ionization model in (1.120) [59]:

$$I_{sub} = \frac{\alpha}{\xi_C} I_D \left(V_D - V_{Dhot} \right) \exp\left(-\frac{\xi_C}{\xi_{eff}} \right) \tag{1.123}$$

To extract ξ_{eff}, (1.123) is rearranged to obtain another ξ_{eff} expression:

$$\xi_{eff} = \frac{\xi_C}{\ln\left(\dfrac{I_D \,\alpha\, (V_D - V_{Dhot})}{I_{sub}\, \xi_C} \right)} \tag{1.124}$$

For a given gate bias V_G, two important parameters, V_{Dhot} and l_d, can be extracted by the following steps [59]. The saturation drain voltage V_{Dsat} is first calculated from the I_D vs V_D measured data. This value will be used as the initial guess for determining V_{Dhot}. Next, iterating (1.121) and (1.124), and using point algorithm and measured I_{sub} vs voltage characteristics, V_{Dhot} and l_d can be found. As (1.121) implies, the relationship between ξ_{eff} and V_D is a straight line, and the slope and the intercept of the line to V_D axis give l_d and V_{Dhot} values, respectively.

Figure 1.39 shows the values of V_{Dhot} and l_d extracted from an MOS device at no stress and subject to two different stress conditions. It can be seen that both to V_{Dhot} and l_d increase with increasing gate voltage. Also, V_{Dhot} is less sensitive to the stress than l_d is. Based on these values, ξ_{eff} can be calculated, which is shown in Fig. 1.40. The locations of V_{Dsat} and V_{Dhot} on the I-V curves are clearly illustrated in Fig. 1.41. The results also show that the difference of the two voltages is increased as the gate voltage is increased.

Figure 1.42 compares the substrate current calculated from the model, using the extracted values of V_{Dhot}, l_d, and ξ_{eff}, and obtained from measurements under different stress conditions and for different gate and drain bias conditions. The agreement between the model and data is excellent.

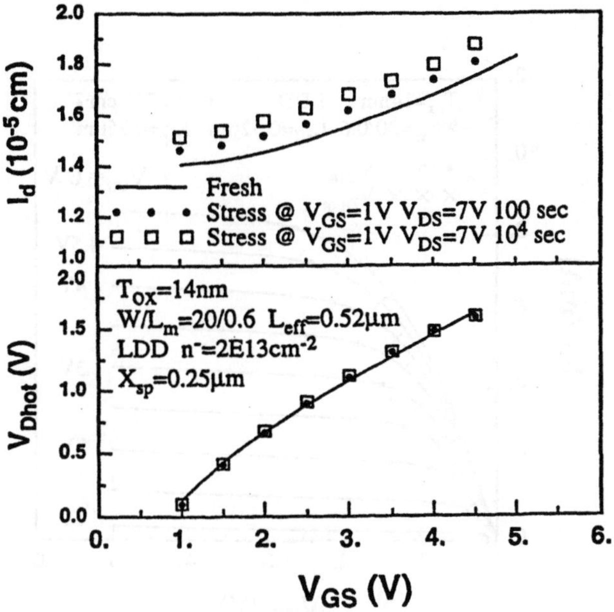

Figure 1.39 : Extracted values for V_{dhot} and l_d for an MOS device subject to three different stress conditions (Source: Yang et al. [59], reprinted with permission).

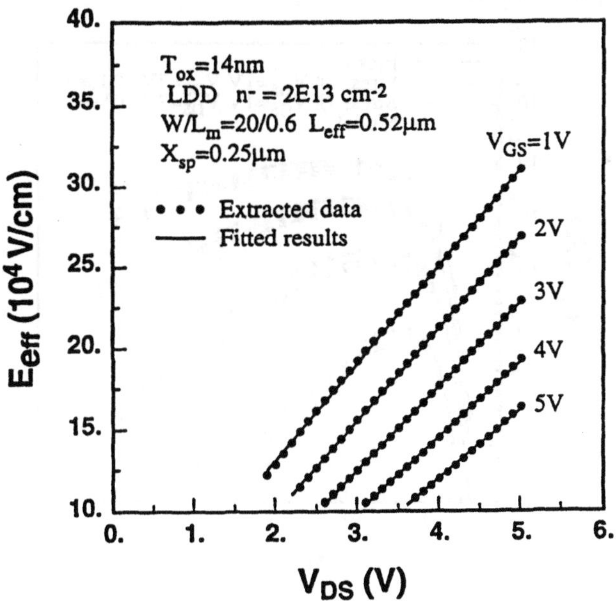

Figure 1.40 : Extracted E_{eff} as a function of the drain and gate bias conditions (Source Yang et al. [59]. Reprinted with permission).

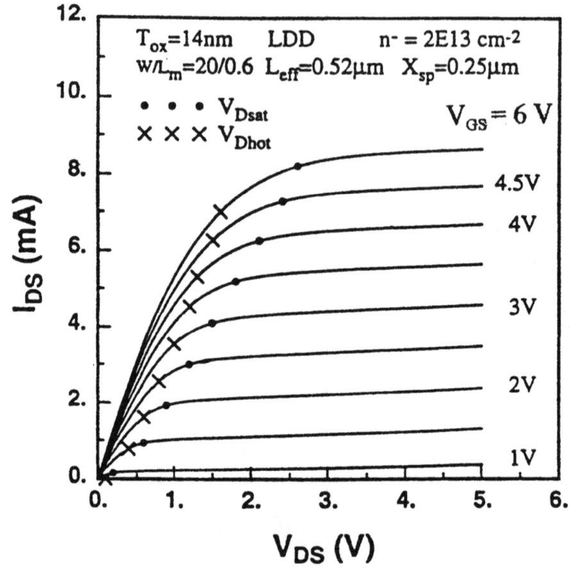

Figure 1.41 : Comparison of V_{Dsat} and V_{Dhot} on the I-V curves (Source: Yang et al. [59]. Reprinted with permission).

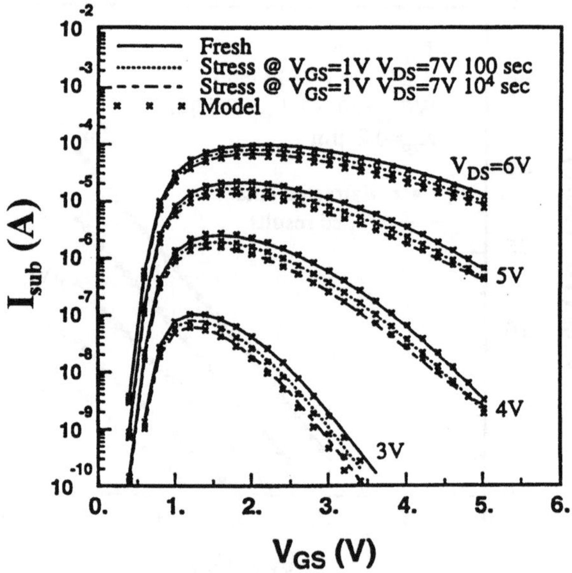

Figure 1.42 : Substrate current calculated from the model and obtained from measurement for three different stress conditions and different gate and drain bias conditions (Source: Yang et al. [59]. Reprinted with permission).

1.7.4 Hot-carrier effect on threshold voltage

The injected hot electrons can sometimes be trapped in the oxide. The trapped charge in the oxide can then shift the flatband voltage and thereby the threshold voltage V_T. Assuming that the trapped charge is concentrated in a sheet at an average distance of d' above the Si-SiO_2 interface, then

$$V_T = V_{T,C} - \Delta V_T = V_{T,C} - \frac{1}{C_{ox}} Q_h \frac{(d-d')}{d} \tag{1.125}$$

where $V_{T,C}$ is the threshold voltage without the trapped hot-carrier effect, ΔV_T is the threshold voltage shift due to trapped hot-carrier effect, C_{ox} is the oxide capacitance, d is the oxide thickness, and Q_h is the trapped hot-carrier charge in the oxide.

1.7.5 Modeling the MOSFET lifetime

Hot-carrier induced MOSFET lifetime degradation is one of the most important constraints for MOS device scaling, particularly for deep-submicron MOS devices because the supply voltage scales at a less aggressive rate than the device channel length [60]. Various models have been developed to describe the MOSFET degradation process [55,61], and it has been found that the substrate current I_{sub} in particular is a very useful parameter for monitoring the lifetime of MOSFET subject to hot-carrier stress [62]. This section presents an empirical model capable of predicting the MOSFET lifetime based on the measured I_{sub} characteristics.

A simple relationship between the transconductance and the stress time was established by Tekada and Suzuki [63]:

$$\frac{G_{m0} - G_m}{G_{m0}} = A\, t^n \tag{1.126}$$

where G_m is the transconductance after stressing, G_{m0} is the transconductance before stressing, t is the stress time, n is the slope of (G_{m0} - G_m) vs t plot, and A is proportional to $\exp(-\alpha/V_D)$, where α is an empirical parameter which can be determined from the experimental data. The stress condition is set at the worst possible voltage that can be applied to the drain (i.e., $V_D + 10\%V_D$), while the gate voltage is set such that the maximum substrate current is generated.

Usually, a 10% G_m degradation is defined as the MOSFET lifetime criterion, and the time to reach a 10% G_m degradation is called the device lifetime τ:

$$\tau = A_1 \ (I_{sub} \ )^{-m} \tag{1.127}$$

where $A_1 = B(L_{eff})^{1/n}$, B is another empirical parameter and L_{eff} is the MOSFET effective channel length, and m can be determined from the slope of τ vs I_{sub} plot. This equation gives a mean to determine the 10% G_m degradation lifetime for a specified device geometry, provided its substrate current is known.

The substrate current has the same dependency as (1.126) and can be expressed as

$$I_{sub} = C \ \exp\left(- \frac{\beta}{V_D} \right) \tag{1.128}$$

where C and β are empirical parameters. The substrate current can also be expressed in terms of the drain voltage and channel length:

$$I_{sub} = K_0 \ L_{eff}^{\ K} \ \exp\left(- \frac{\beta}{V_D} \right) \tag{1.129}$$

where the parameters K_0 and K can be easily obtained by plotting C's vs L_{eff} from (1.128). Putting (1.129) into (1.127) yields

$$\tau = D \ L_{eff}^{\ (\frac{1}{n} - \frac{K\alpha}{n\beta})} \ \exp\left(- \frac{\beta}{V_D} \right) \tag{1.130}$$

where $D = BK_0^{\alpha/n\beta}$. Equation (1.130) provides an explicit expression for the device lifetime in terms of L_{eff} and V_D.

The MOS devices used for the study have 0.6, 0.7, 1.0, and 1.5 μm channel lengths and 100 μm channel width. The devices were stressed at various drain voltages, and with the gate biased such that the maximum substrate current is obtained. From a plot of G_m degradation vs the stress time, the slope n in (1.126) was extracted to be close to 0.5. This value was then put into (1.127) to extract the value of m and A_1. m was found to be about 2.5, and a log-log plot of A_1 vs L_{eff} (line) fitted to measured data (symbols) is shown in Fig. 1.43, which can be used to extract the value for B in (1.127). The substrate current is a function of the drain voltage, as illustrated in Fig. 1.44 where substrate currents are plotted against $1/V_D$ (lines) fitted to the measurements (symbols). The slope of the curve in Fig. 1.44 is β, which was found to be -18.5. The intercept C in (1.128) is a function of L_{eff} and $C = K_0 L_{eff}^K$. From Fig. 1.44, the values of K and K_0 extracted were -0.8 and 0.3052, respectively. Also, α in (1.126) can be extracted to be about 68.5.

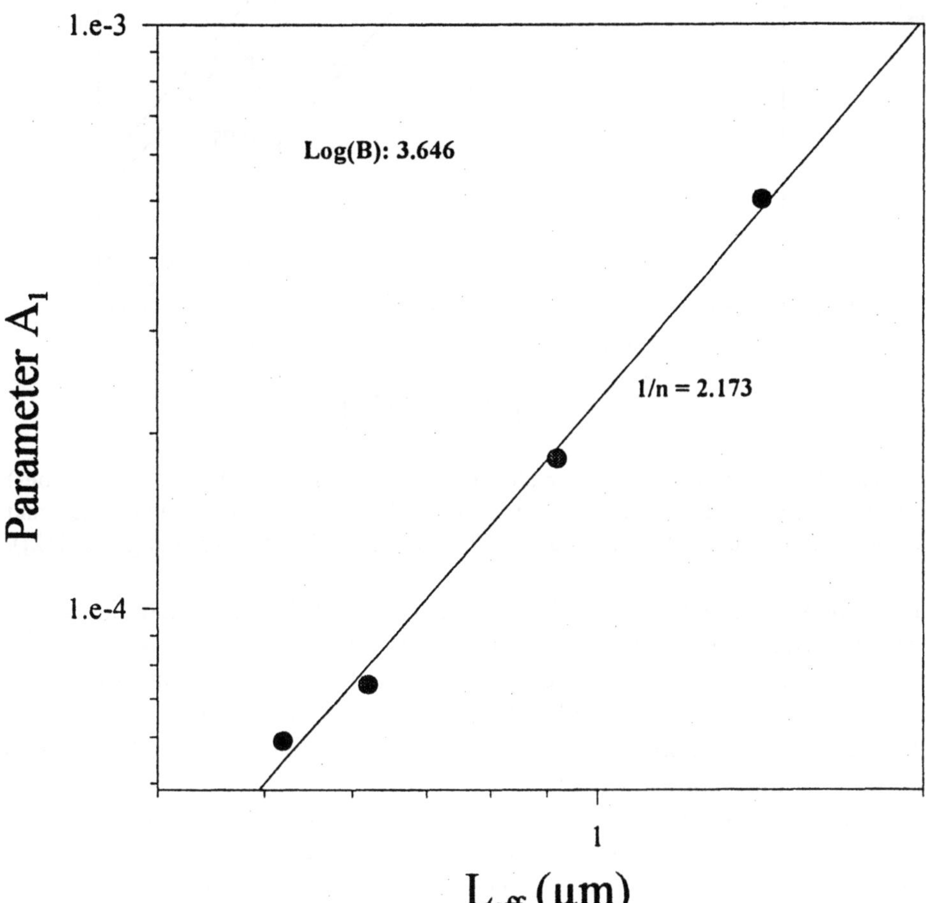

Figure 1.43 : Log-log plot of A_1 vs L_{eff} (line) fitted to the measurements (symbols). This can be used to extract parameter B (after Wong et al. [62]).

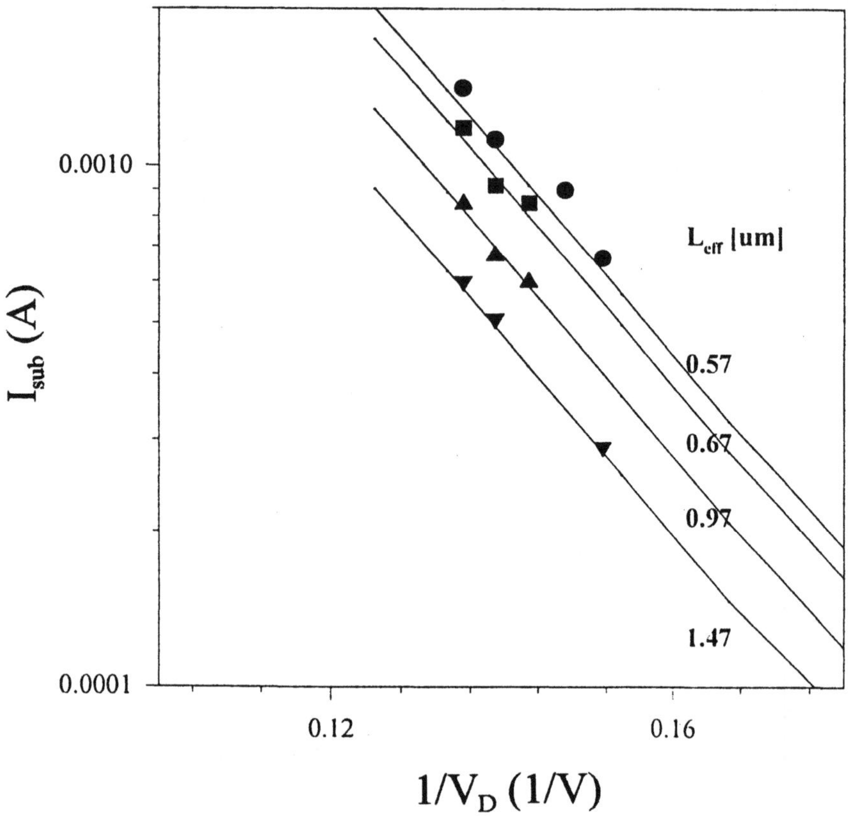

Figure 1.44 : Substrate current dependence on drain voltage and channel length (lines) fitted to measurements (symbols). N-channel MOSFETs with 100 μm channel width and channel length of 0.57, 0.67, 0.97, and 1.47 μm are considered (after Wong et al. [62]).

Given the values of n, α, β, B, K and K_0, and using 10% G_m degradation as the criterion, lifetimes of MOSFETs at various applied drain voltages can be predicted using (1.130). This is clearly illustrated in Figs. 1.45 and 1.46, where model predictions (lines) and data measured from the accelerated stress (symbols) are compared. Figure 1.45 shows the calculated and measured lifetimes of the MOS transistors as a function of the channel length using the drain voltage as a parameter, whereas Fig. 1.46 shows the same characteristics but as a function of the drain voltage using the channel length as a parameter. The results show that the lifetime decreases with increasing drain voltage and decreasing channel length. The agreement between the model and measurements is excellent, and it is indicated that the present model can accurately predict the lifetime of MOS devices beyond those being measured.

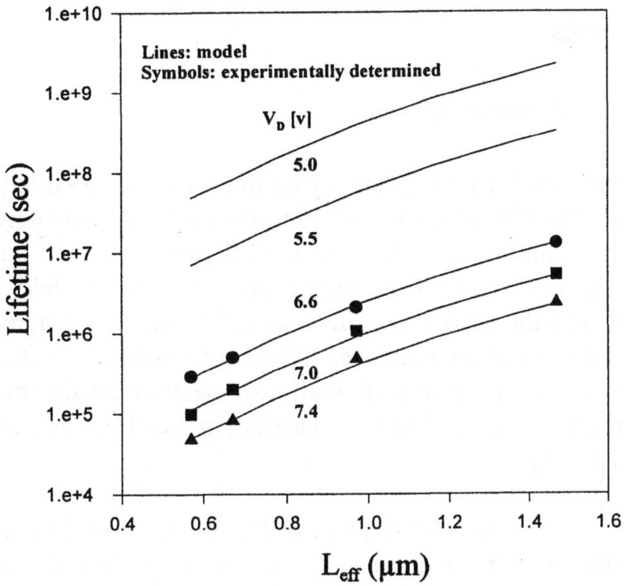

Figure 1.45 : Comparison of MOS lifetime vs the channel length calculated from the model (lines) and obtained from measurements (symbols) (after Wong et al. [62]).

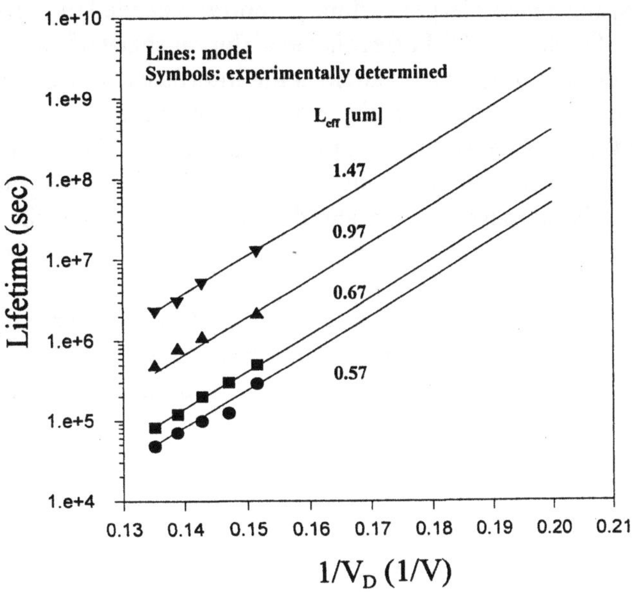

Figure 1.46 : Comparison of MOS lifetime vs the applied voltage calculated from the model (lines) and obtained from measurements (symbols) (after Wong et al. [62]).

1.8 Quantum mechanical effects in deep-submicron MOS devices

1.8.1 General concept

The scaling of MOSFETs relies heavily on the use of successively thinner gate dielectrics and higher level of channel doping density. As gate lengths approach deep submicron dimensions (less than 0.25 μm) based on such a scaling methodology, there can be a very large transverse electric field at the Si/SiO_2 interface, even near the threshold of inversion. This leads to a significantly large energy bands bending at the interface. It has been known that with a sufficiently large band bending, the potential well can become sufficiently narrow to quantize the motion of inversion layer carriers in the direction perpendicular to the interface [64-65].

The quantum mechanical (QM) picture differs in several aspects from the classical one [66-67]. First, the energy spectrum consists of a set of discrete energy levels (i.e., splitting of the energy levels into subbands), and the lowest of the allowed energy levels for electrons in the well does not coincide with the bottom of the conduction band. This effectively widens the bandgap for all temperatures, and hence a larger surface potential is needed for a given inversion layer charge. The second QM effect is the different shape of the wave function. The electron density n(x) has to vanish at the interface and the average distance <x> of the inversion carriers to the interface increases by an amount of Δx compared to the classical solution. This displacement of inversion layer charge away from the interface in effect increases the effective oxide layer thickness and consequently increases the threshold voltage and reduces the drain current level.

For a homogeneous channel doping, the potential can be approximated as $V(x) = \xi x$, where ξ is the electric field due to the charge in the depletion layer. Accounting for the QM effects, the extra band bending needed to create the state of inversion is

$$\Delta\psi_S = \frac{\Delta\epsilon - \Delta E_g}{q} + \xi\,\Delta x \tag{1.131}$$

where $\Delta\epsilon$ is the energy difference between the conduction band edge and the lowest allowed subband, and ΔE_g is the bandgap narrowing due to the heavy doping effect. Analytical expressions for $\Delta\epsilon$ and Δx can be found in [68]:

$$\Delta\epsilon = \left(\frac{h^2}{8\,\pi^2\,m^*}\right)^{1/3}\left(\frac{9}{8}\,\pi\,q\,\xi\right)^{2/3} \tag{1.132}$$

$$\Delta x = \frac{2\,\Delta\epsilon}{3\,q\,\xi} - \frac{k\,T}{q\,\xi} \tag{1.133}$$

where h is the Planck constant and m* is the electron effective mass. Here, the assumption that only the first allowed energy level is occupied has been used.

1.8.2 Approximated solution to Schrödinger equation

In order to obtain physically reasonable approximations for Schrödinger wave equation in the inversion layer, a number of assumptions are used. First, the effective mass approximation is applied, and the 3-D Schrödinger equation is decoupled into a 1-D form. Such an equation describes the envelop function $\zeta(x)$ perpendicular to the interface, which constrains Bloch waves traveling parallel to the interface (i.e., y-z plane). Another assumption is that the potential barrier to electrons in the potential well ($\approx$ 3.1 eV) is infinitely high at the interface. Based on these approximations, and using a parabolic band structure, the envelop function is found to satisfy Schrödinger equation [66]

$$\left(-\frac{h^2}{8\,\pi^2\,m^*}\frac{d^2}{dx^2} + q\,V(x)\right)\zeta_{ij}(x) = E_{ij}\,\zeta_{ij}(x) \tag{1.134}$$

where V is the electrostatic potential, and E_{ij} and ζ_{ij} are the eigenvalue and eigenfunction of the electron energy, respectively, of the jth subband in the ith valley. Once the eigenenergies of the 2-D system have been determined, the inversion layer electron density at a distance x below the Si/SiO_2 interface can be found by summing over all of the subbands:

$$n(x) = \frac{4\,\pi\,k\,T}{h}\sum g_i\,m^* \cdot \sum \frac{\ln\left(1 + \exp\left(\dfrac{E_{fn} - E_{ij}}{k\,T}\right)\right)}{1 + \exp\left(\dfrac{E_{fn} - E_{cl}}{k\,T}\right)}\,\zeta{ij}^2 \tag{1.135}$$

where g_i is the degeneracy of the ith valley, $m^* = 0.916\ m_0$, and E_{cl} is the energy level corresponding to the classical regime, above which the density of states is modeled classically.

If three subbands are considered (i.e., three-subband model [66]), the eigenfunction and eigenenergies are determined by assuming that the electrostatic potential V(x) has a triangular shape. This is a good approximation when the device is in depletion or weak inversion, but may be questionable for strong inversion condition. The results are the Airy function [69]

$$\zeta_{ij} = A_i \left(\frac{8\ \pi^2\ m^*\ q\ \xi_s}{h^2} \right)^{1/3} \left(x - \frac{E_{ij}}{q\ \xi_s} \right) \tag{1.136}$$

$$E_{ij} = \left(\frac{h^2}{8\ \pi^2\ m^*} \right)^{1/3} \left(1.5\ \pi\ q\ \xi_s (j + 3/4) \right)^{2/3} \tag{1.137}$$

Here ξ_s is the electric field at the interface, and E_{ij} is measured with respect to the conduction band edge at the Si/SiO_2 interface. When the device is in moderate or strong inversion, the Airy function does not describe the ground state (lowest state) eigenfunction accurately. To improve this, a modified model has been shown to provide a good estimate of the wavefunction of the lowest subband [70]:

$$\zeta_{10}(x) = b^{1.5} \frac{x}{\sqrt{2}} \exp\left(- \frac{b\ x}{2} \right) \tag{1.138}$$

where the parameter b is determined by minimizing the energy of the system using the wavefunction given in (1.138). Such an approach also yields the following approximated expression for the lowest subband energy:

$$E_{10} = \frac{2 \left(\frac{q^2\ h}{2\ \pi\ \epsilon_{si}\ \sqrt{m^*}} \right)^{2/3} \left(N_{dep} + \frac{55}{96} N_{inv} \right)}{\left(N_{dep} + \frac{11}{32} N_{inv} \right)^{1/3}} \tag{1.139}$$

Here N_{dep} and N_{inv} are the depletion and inversion carrier densities, respectively.

In the three-subband model proposed by Hareland et al. [66], the modified model given in (1.138)-(1.139) is selected to model the wavefunction of the first subband. For the next two higher subbands, the Airy function given in (1.136)-(1.137) is used. A classical regime is then used to model all of the subbands above the third subband.

1.8.3 Results and discussions

Figure 1.47 shows the threshold voltage shift ΔV_T due to the QM effects (i.e., V_T calculated from the QM model subtracted from that calculated from the classical model). Two different QM models, the three subband model [66] and the simplified model [67], are included, and they compare favorably with measurement. The device considered was a long-channel MOSFET with channel doping density ranging from $5x10^{17}$ to $5x10^{18}$ cm^{-3}. The long-channel device is used so that the shift is due to the QM effect only, rather than the short-channel effect. The physical origin of ΔV_T will be discussed later in this section, and the experimental extraction of ΔV_T will be discussed in Chapter 3.

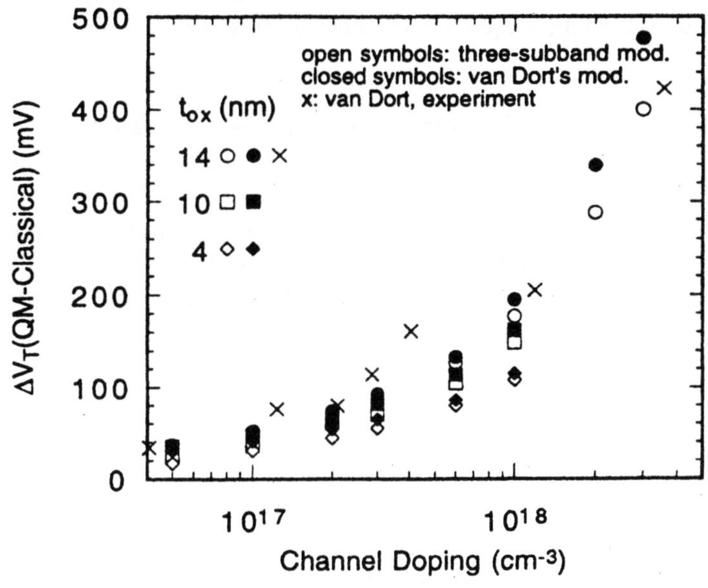

Figure 1.47 : Threshold voltage shift between the quantum mechanical and classical predictions for a range of oxide thickness and substrate doping density. Two quantum-mechanical models are included: three-subband model [66] and model by van Dort et al. [67] (Source: Hareland et al. [66]. Reprinted with permission).

Figure 1.48 shows the inversion layer density in the channel as a function of the depth calculated from the three-subband model and the self-consistent approach (i.e., direct solution of Schrödinger equation) for a MOSFET with an oxide thickness of 4 nm and a channel doping density of 7×10^{17} cm^{-3}.

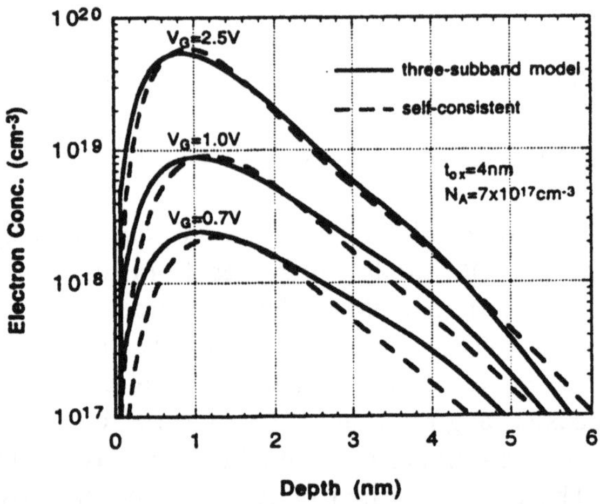

Figure 1.48 : Inversion layer concentration versus the depth into the bulk calculated from the three- subband model and self-consistent approach (i.e., direct numerical calculation of Schrödinger equation) (Source: Hareland et al. [66]. Reprinted with permission).

The QM effects are very important in such a device because of the very thin oxide and heavily doped substrate. The subband energies versus the gate voltage of this device are plotted in Fig. 1.49, indicating that the separation of the conduction band edge (reference energy) and first subband is much larger than that of the first and second subbands. The same trend applies to the higher subbands as well. The comparison of the classical and QM solutions of the inversion layer density is given in Fig. 1.50. The classical solution predicts a charge peak at the interface, whereas the quantum solution displays the peak away from the interface. Furthermore, the total inversion layer charge predicted by the classical solution is greater that by the quantum counterpart. Note that, when the QM effects are taking into account, the much smaller inversion carrier density near the interface is a direct consequence of the quantization of the energy in conduction band, which in effect increases the energy bandgap (i.e., by an amount between the conduction band edge and the subbands, see Fig. 1.49) and therefore decreases the free-carrier density in the region. As the distance from the interface is increased, the QM effects are less prominent, and the inversion carrier density is increased. At an even larger distance, the carrier density is decreased toward to bulk region.

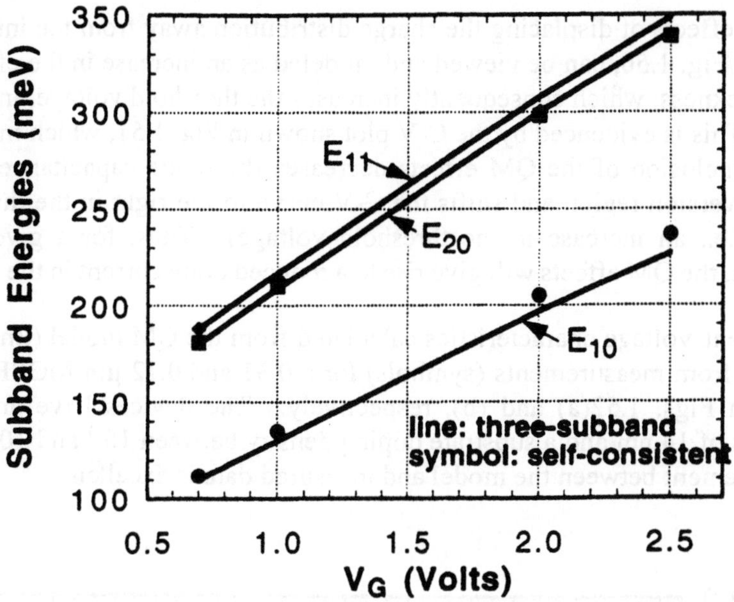

Figure 1.49 : Subband energies calculated from the three-subband model and self-consistent approach for a MOSFET with 4 nm oxide thickness and $7x10^{17}$ cm^{-3} channel doping (Source: Hareland et al. [66]. Reprinted with permission).

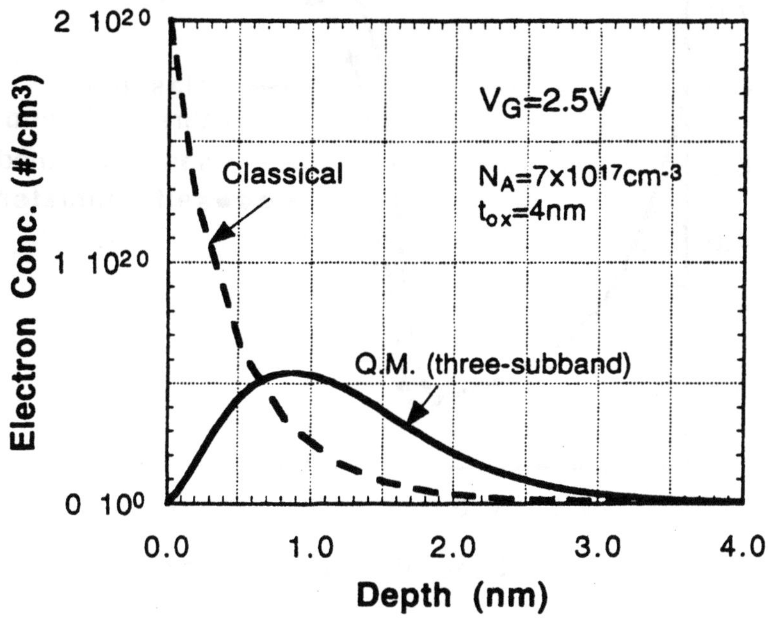

Figure 1.50 : Comparison of inversion layer density calculated from the QM model and classical modei (Source: Hareland et al. [66]. Reprinted with permission).

The QM effects of displacing the charge distribution away from the interface, shown in Fig. 1.50, can be viewed and modeled as an increase in the effective oxide thickness, which consequently increases the threshold voltage, in such a device. This is evidenced by the C-V plot shown in Fig. 1.51, which indicates that the inclusion of the QM effects decreases the oxide capacitance in the strong inversion region and shifts the C-V curve to the right in the threshold region (i.e., an increase in the threshold voltage). Thus, for a given bias condition, the QM effects will give rise to a reduced drain current in the device.

The current-voltage characteristics calculated from the QM model (lines) and obtained from measurements (symbols) for a 0.31 and 0.12 μm MOSFET are shown in Figs. 1.52(a) and (b), respectively. The devices have an oxide thickness of 14 nm and a substrate doping density between 10^{17} and 10^{18} cm^{-3}. The agreement between the model and measured data is excellent.

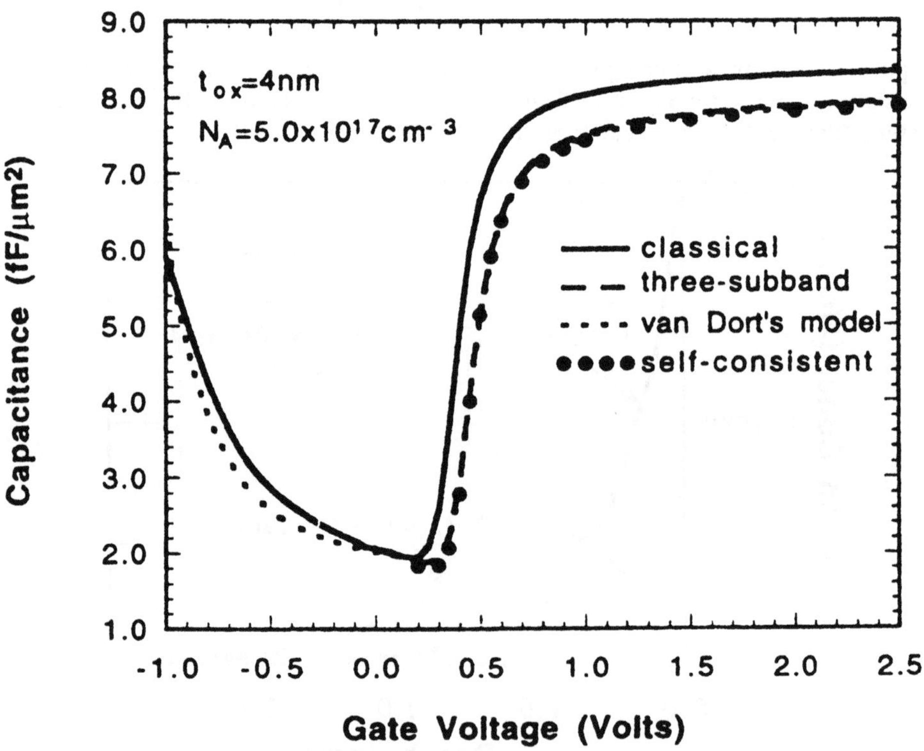

Figure 1.51 : Calculated C-V curves using the classical model, the three-subband model [66], model by van Dort et al. [67], and self-consistent approach (Source: Hareland et al. [66]. Reprinted with permission).

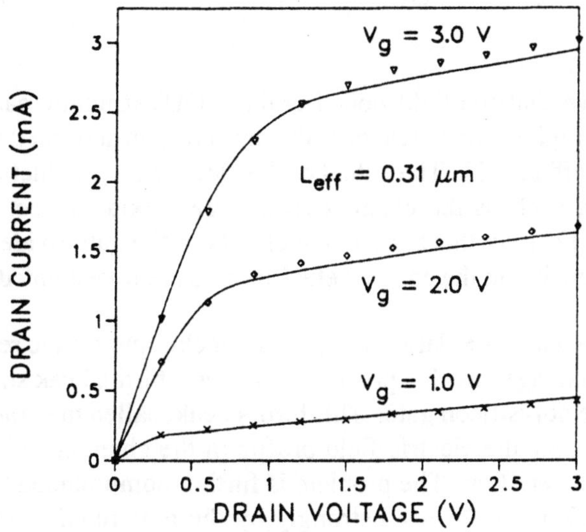

(a)

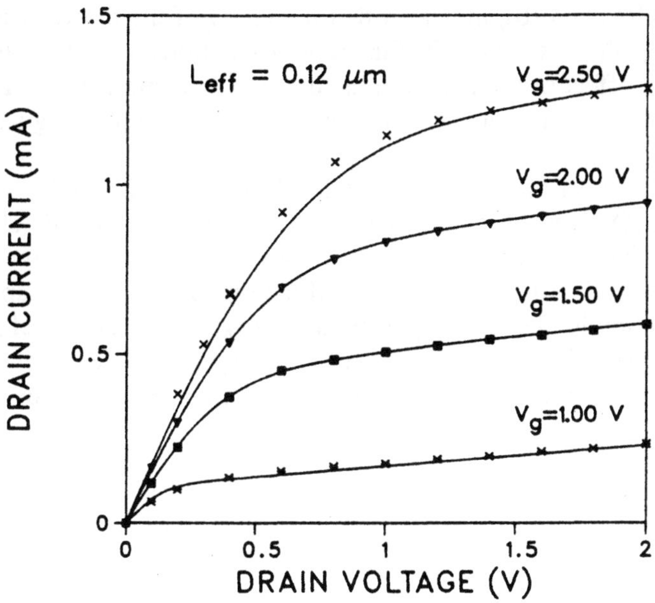

(b)

Figure 1.52 : Calculated (lines) and measured (symbols) current-voltage characteristics of (a) 0.31- μm and (b) 0.12-μm MOSFETs. The calculated results are obtained from a QM model [67]. The device has a n oxide thickness of 14 nm and substrate doping around 10^{17} cm^{-3} (Source: van Dort et al. [67]. Reprinted with permission).

1.9 Modeling the lightly-doped-drain (LDD) MOSFET

It is well known that the light-doped drain (LDD) structure can alleviate the problem of the high electric field near the drain region and can therefore reduce the hot-carrier effect [71]. The extent of the electric field reduction depends on several factors, such as the channel length, gate-oxide shape, and the LDD doping density [72]. In this section, the effects of the gate-oxide shape and the LDD doping density on the channel electric field are investigated and modeled.

The polysilicon gate reoxidation is a popular technique for the reduction of the LDD implant damage. Such a process produces a bird's beak shape of the gate oxide under the polysilicon gate. This bird's beak, called the graded gate oxide (GGO), can change the electric field profile in the channel and thus the LDD MOSFET characteristics. The problem is further compounded by the fact that the shape of GGO is influenced strongly by the gate reoxidation process; the GGO is thicker and longer when the gate reoxidation is done in wet rather than dry [73]. Results simulated from a two-dimensional process simulator, shown in Figs. 1.53(a) and (b), indicated that the GGO of the dry reoxidation has a linear shape, whereas the GGO of the wet reoxidation has a parabolic shape. In addition, the effect of the LDD doping density, in combination with the GGO effect, has to be taking into account for accurate LDD modeling and analysis.

Figure 1.54 shows the schematic of the LDD structure under study. Two different GGO shapes are considered: the linear and parabolic shapes. The model development makes use of the quasi-two-dimensional approach proposed by Elmansy and Boothroyd [74], in which the Gauss's law is applied to the rectangular box in the velocity saturation region of the channel (denoted by the dashed box in Fig. 1.54) and in the LDD region. A differential equation for the electric field at the surface of the channel can be expressed by

$$\frac{X_j}{\eta}\frac{d\xi_y(0,y)}{dy} = \frac{q\,X_j}{\epsilon_s}(\,N_A - N_D\,) + \frac{Q_m}{\epsilon_s} - \frac{\epsilon_{ox}}{\epsilon_s}\,\xi_x(0,y) \tag{1.140}$$

where $N_A - N_D$ is the net impurity density, Q_m is the mobile charge density, X_j is the depth of LDD junction, $\xi_x(0,y)$ is the vertical electric field at the surface, and η is a fitting parameter for the nonuniformity effect of the lateral field and the uncertainty of the finite channel depth and is assumed to be 1.2. At the beginning of the velocity saturation region (i.e., $y = -L_{sat}$, see Fig. 1.54), the vertical electric field at the surface is approximated by

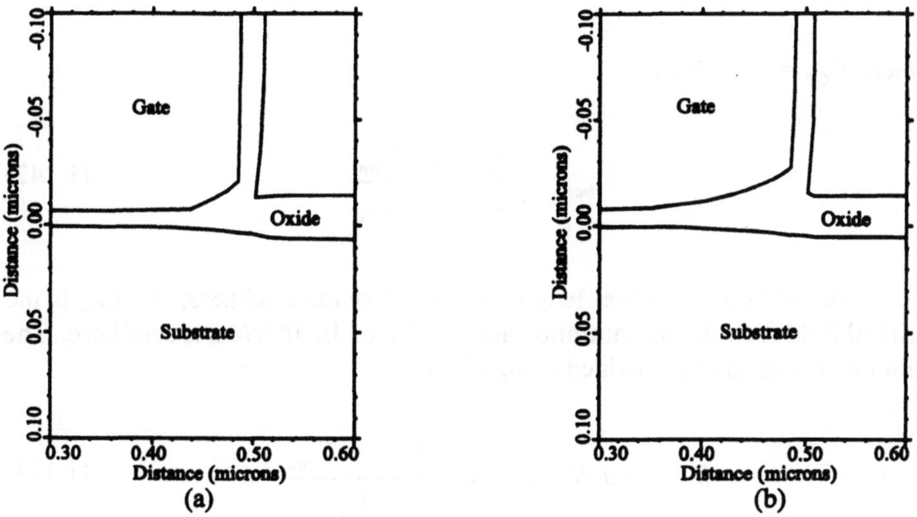

Figure 1.53 : Two-dimensional GGO shape obtained from process simulation for (a) dry reoxidation at 900 °C and 60 mins, and (b) wet reoxidation at 850 °C and 15 mins (Source: Kim et al. [72]. Reprinted with permission).

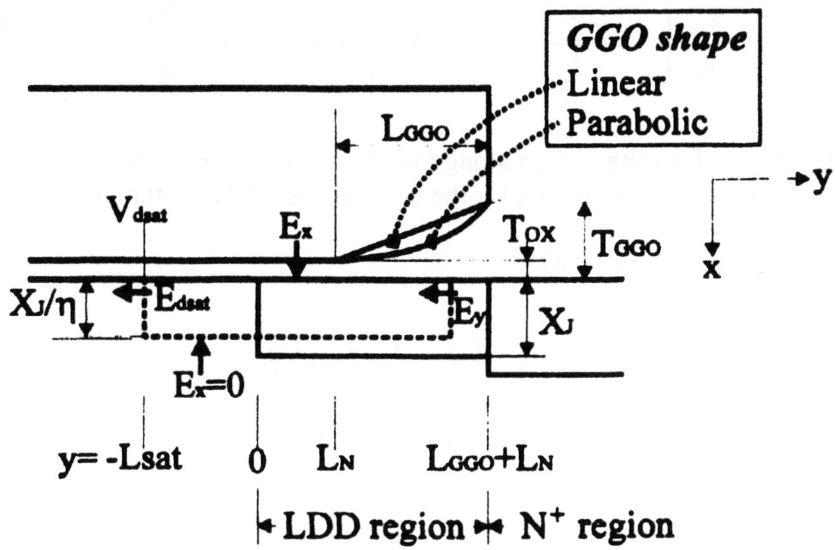

Figure 1.54 : Schematic of the LDD structure with the two types of GGO shape (Source: Kim et al. [72]. Reprinted with permission).

$$\xi_x(0,-L_{sat}) = \frac{V_{GT} - V_{dsat}}{T_{ox}} \qquad (1.141)$$

where $V_{GT} = V_{GS} - V_T$ and

$$V_{dsat} = \frac{V_{GT}\, L_{eff}\, \xi_{dsat}}{V_{GT} + L_{eff}\, \xi_{dsat}} \qquad (1.142)$$

L_{eff} is the effective channel length, T_{ox} is the oxide thickness, and ξ_{dsat} is the critical field for velocity saturation and a value of $4\text{x}10^4$ V/cm is used here. The mobile charge can be obtained using (1.141):

$$Q_m = -q\, N_A\, X_j + \epsilon_{ox} \frac{V_{GT} - V_{dsat}}{T_{ox}} \qquad (1.143)$$

Putting (1.143) into (1.140) yields [72]

$$\frac{X_j}{\eta} \frac{d\xi_y}{dy} = \frac{\epsilon_{ox}}{\epsilon_s} \left(\frac{V_{GT} - V_{dsat}}{T_{ox}} - \frac{V_{GT} - V(y)}{T(y)} \right) - \frac{q\, X_j\, N_D}{\epsilon_s\, \eta} \qquad (1.144)$$

where V(y) is the voltage drop along the channel (i.e., y direction) and T(y) is the oxide thickness accounting for the GGO shape, which is function of y.

For the case of a linear GGO,

$$T(y) = T_{ox} \left(\zeta \left(y - L_N \right) + 1 \right) \qquad (1.145)$$

where L_N is the starting point of the GGO (see Fig. 1.54) and

$$\zeta = \frac{\dfrac{T_{GGO}}{T_{ox}} - 1}{L_{GGO}} \qquad (1.146)$$

where T_{GGO} and L_{GGO} are the geometry parameters describing the shape of GGO, as shown in Fig. 1.54. For the case of a parabolic GGO,

$$T(y) = T_{ox} \left(\zeta \left(y - L_N \right) + 1 \right)^2 \tag{1.147}$$

where

$$\zeta = \frac{\left(\frac{T_{GGO}}{T_{ox}} \right)^{1/2} - 1}{L_{GGO}} \tag{1.148}$$

To obtain an analytic solution of (1.144), one can divide the velocity saturation region into three subregions; the velocity saturation region of the channel (region I), the region of the LDD region without GGO (region II), and the region of the LDD region with GGO (region III). In region I (i.e., $-L_{sat} < y < 0$), $N_D = 0$ and $T(y) = T_{ox}$, and (1.144) is reduced to

$$\frac{d\xi_y}{dy} = V(y) - \frac{V_{dsat}}{\iota^2} \tag{1.149}$$

$$\iota^2 = \frac{\epsilon_s T_{ox} X_j}{\eta \epsilon_{ox}} \tag{1.150}$$

Using the boundary conditions of $V(0, -L_{sat}) = V_{dsat}$ and $\xi_y(0, -L_{sat}) = \xi_{dsat}$, we have

$$\xi_y(y) = \xi_{sat} \cosh \left(\frac{y + L_{sat}}{\iota} \right) \tag{1.151}$$

In region II (i.e., $0 < y < L_N$), the solution of (1.144) is

$$\xi_y(y) = \xi_{sat} \cosh\left(\frac{y + L_{sat}}{\iota}\right) - m\,\iota^2 \sinh\left(\frac{y}{\iota}\right) \tag{1.152}$$

$$m = \frac{q\,\eta\,N_D\,\iota}{\epsilon_s} \tag{1.153}$$

The solutions for the linear and parabolic shape GGO are the same in the fist two regions. However, in region III (i.e., $L_N < y < L_{GGO}$), the lateral field is altered by the GGO shape and is given below [72]:

$$\frac{d\xi_y}{dy} = \frac{V_{GT} - V_{dsat}}{\iota^2} - \frac{V_{GT} - V(y)}{\iota^2\left(\zeta(y - L_N) + 1\right)} - \frac{q\,\eta\,N_D}{\epsilon_s} \tag{1.154}$$

for linear GGO and

$$\frac{d\xi_y}{dy} = \frac{V_{GT} - V_{dsat}}{\iota^2} - \frac{V_{GT} - V(y)}{\iota^2\left(\zeta(y - L_N) + 1\right)^2} - \frac{q\,\eta\,N_D}{\epsilon_s} \tag{1.155}$$

for parabolic GGO. The parameter ζ in (1.154) and (1.155) are defined in (1.146) and (1.148), respectively. Analytic solutions for $\xi_y(y)$ for the linear and parabolic GGO can be obtained from these two equations [72].

Figure 1.55 gives a comparison of the model calculations and MEDICI device simulation of the channel electric field versus the distance from the LDD edge for an LDD MOSFET with a mask channel length of 0.25 μm, oxide thickness of 7 nm, LDD junction depth of 0.1 μm, LDD doping density of 10^{18} cm^{-3}, and a linear GGO shape. The bias conditions are $V_{DS} = 4$ V and $V_{GS} = 3$ V. The agreement between the two results is excellent.

The effects of the GGO dimension on the channel electric field profile are illustrated in Figs. 1.56(a) and (b). Here a linear GGO with two different L_{GGO} (i.e., $L_{GGO} = 0.03$ μm in (a) and 0.08 μm in (b)), $T_{ox} = 7$ nm, $V_{GS} = 3$ V, and $V_{DS} = 4$ V are considered. Clearly, the maximum electric field, which occurs at the edge of the LDD region, decreases slightly with increasing T_{GGO} and/or increasing L_{GGO}. Thus, a shorter and thicker GGO is preferred for reducing the maximum channel field in the LDD MOSFET.

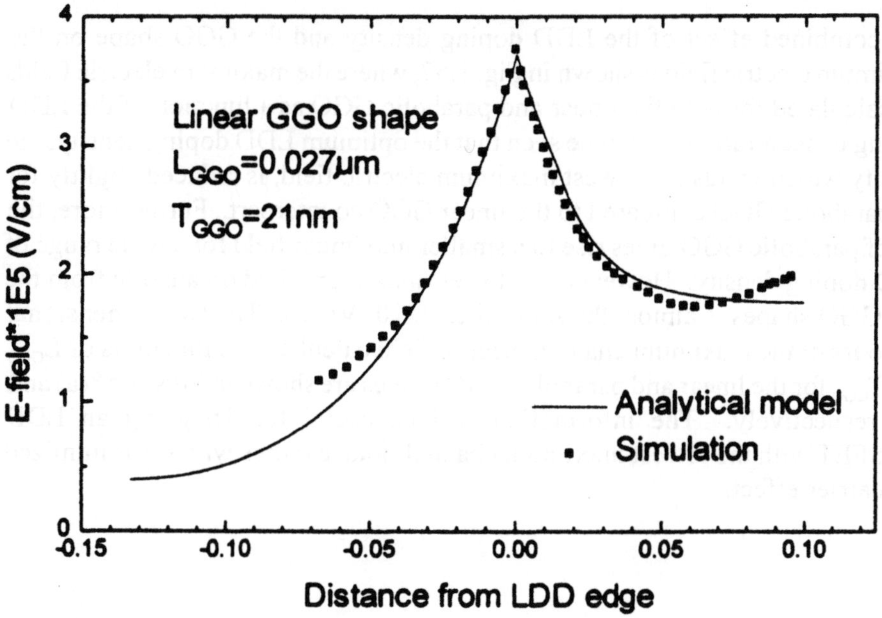

Figure 1.55 : Comparison of the channel electric field versus the distance from he LDD edge calculated from the model and simulated from two-dimensional device simulator MEDICI (Source: Kim et al. [72]. Reprinted with permission).

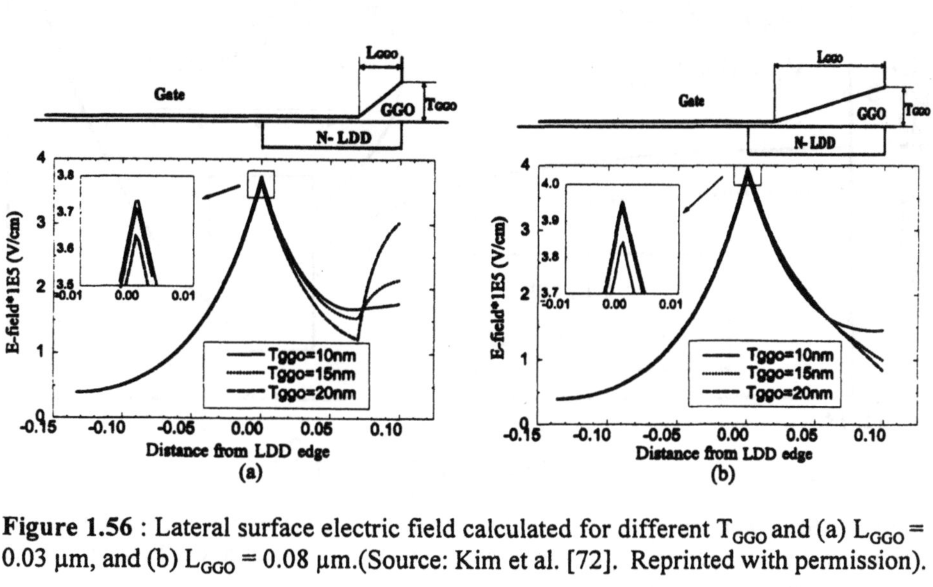

Figure 1.56 : Lateral surface electric field calculated for different T_{GGO} and (a) L_{GGO} = 0.03 μm, and (b) L_{GGO} = 0.08 μm.(Source: Kim et al. [72]. Reprinted with permission).

The combined effect of the LDD doping density and the GGO shape on the maximum electric field is shown in Fig. 1.57, where the maximum electric fields are calculated for both the linear and parabolic GGO as a function of the LDD doping concentration. It can be seen that the optimum LDD doping density, the density which yields the lowest maximum electric field, is reduced slightly for the parabolic GGO compared to the linear GGO counterpart. Furthermore, the use of parabolic GGO gives rise to a smaller maximum field for a wide range of LDD doping density. However, the lowest maximum field obtainable from the two GGO shapes is almost the same (i.e., 3×10^5 V/cm). The two-dimensional contours of the maximum channel electric field calculated as functions of L_{GGO} and T_{GGO} for the linear and parabolic GGO shapes are shown in Figs. 1.58(a) and (b), respectively. The information will be useful for designing an LDD MOSFET with the lowest maximum channel field and thus with the minimized hot-carrier effect.

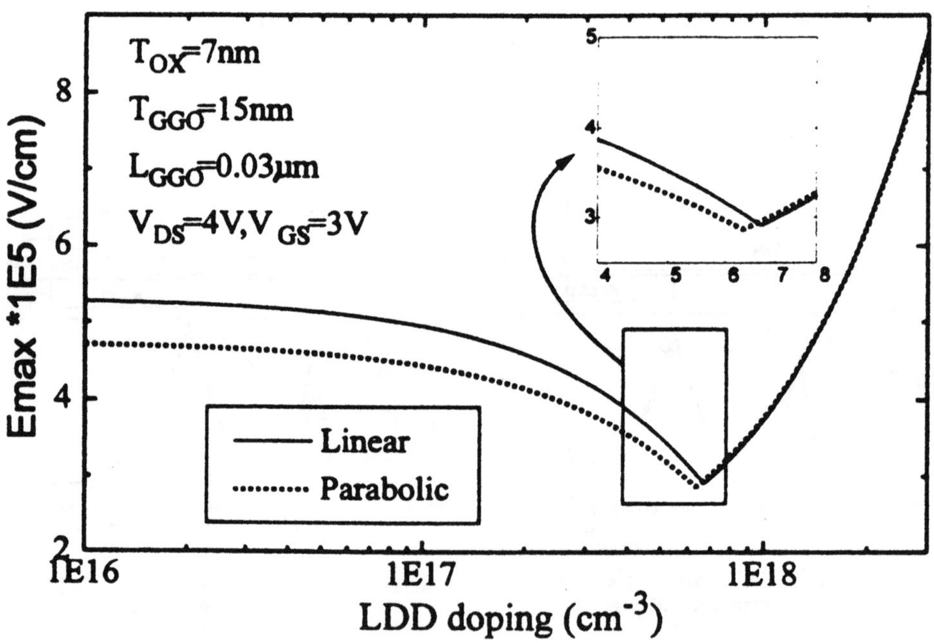

Figure 1.57 : Maximum electric field calculated from the model for the linear and parabolic GGO shapes as a function of the LDD doping concentration (Source: Kim et al. [72]. Reprinted with permission).

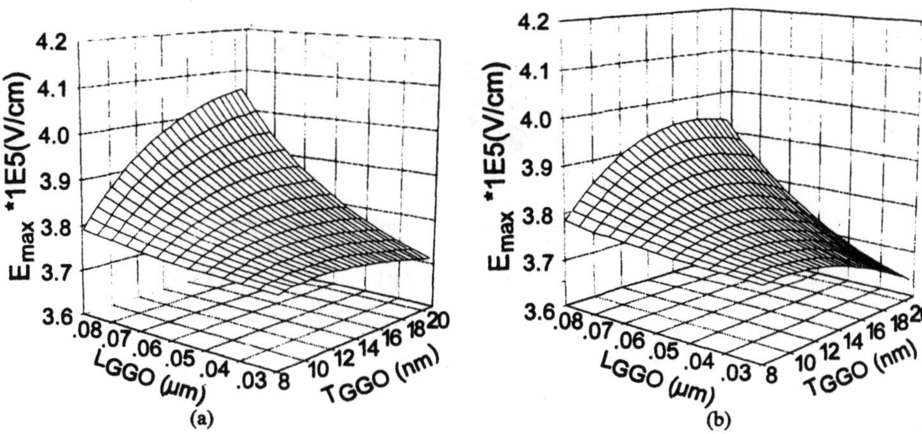

Figure 1.58 : Two-dimensional contour of the maximum electric field calculated for different L_{GGO} and T_{GGO} and for (a) linear GGO shape, and (b) parabolic GGO shape (Source: Kim et al. [72]. Reprinted with permission).

While the LDD can reduce the maximum field and hot-carrier effect near the drain junction, such a structure often introduces additional damage, which occurs inside the LDD spacer where trapped electrons can increase the parasitic drain resistance and subsequently reduce the drain current [75]. This is because the quality of the LDD spacer oxide is poorer than the gate oxide above the channel region, and this oxide is susceptible to hot-carrier stress.

The hot-carrier-induced trapped electrons in the spacer oxide cause depletion of free carriers at the surface of the underlying LDD region and reduction of free-carrier mobility, thus leading to an increase in the parasitic drain resistance. Figs. 1.59(a) and (b) show the measured and calculated current versus voltage characteristics of LDD MOSFETs with channel lengths of 0.9 and 0.6 μm, respectively, for the cases of without electron trapping (i.e., fresh) and with electron trapping (i.e., degraded) in the LDD spacer. It is evident that the drain current is reduced in the degraded device and that the drain current degradation is increased with increasing gate voltage.

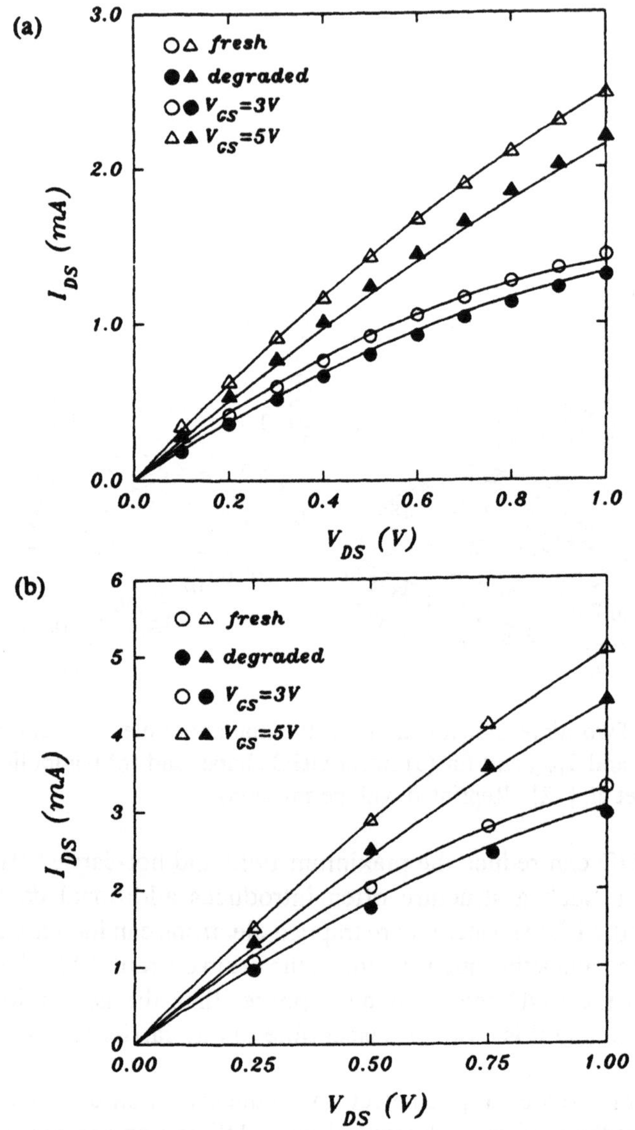

Figure 1.59 : Measured (symbols) and calculated (lines) current-voltage characteristics of LDD MOSFETs with channel length of (a) 0.9 μm and (b) 0.6 μm with (degraded) and without (fresh) the effects of electron trapping in the LDD spacer (after Liu et al. [75]).

The extraction of device parameters associated with LDD MOSFETs, such as the threshold voltage and effective channel length, will be addressed in Chapter 6.

1.10 Modeling the silicon-on-insulator (SOI) MOSFET

The fundamental benefits of the silicon-on-insulator (SOI) structure over the traditional bulk MOSFET have motivated abundant recent research work [76-80]. The main benefits include suppression of latch-up, higher circuit speed, lower power consumption, greater immunity to radiation, increase of the density, 3-D integration, and reduction of short-channel effects. Good review articles were presented recently by Jurczak [78] describing and comparing the various SOI's models, and by Alles [77] scrutinizing the motivations of using SOI in integrated circuits.

Figures 1.60(a) and (b) compare the CMOS inverters fabricated with bulk and SOI technologies. It can be seen that the area of the SOI inverter is smaller, and therefore the areal density is higher, compared to its bulk counterpart. Since SOI devices are thinner, they possess higher radiation immunity. Also, 3-D integration can be more easily achieved with SOI using more semiconductors layers; i.e., transistors can be fabricated on top of transistors. Furthermore, the SOI device has fewer parasitic capacitances, thus leading to a higher circuit speed.

The dashed lines in Fig. 1.60(a) indicate the path of a parasitic device (i.e., p^+-n-p-n^+ device) from V_{DD} to V_{SS} in bulk CMOS. If such a device, called the semiconductor controlled rectifier (SCR), starts to conduct, then the CMOS will suffer a permanent damage because there will be a short-circuit between the two power supplies. This failure, called the latch-up, can occur in bulk CMOS because the bases of the two bipolar in the SCR, p^+-n-p and n-p-n^+, have relatively low doping densities. On the contrary, the SOI CMOS shown in Fig. 1.60(b) does not have any SCR between the two power supplies and the parasitic bipolar transistors have a very low current gain due to the high doping densities in their base.

Probably the most important motivation today for using the SOI device is the lower power consumption, especially in the portable electronics arena. We have seen in Section 1.1 that the supply voltage is reduced in order to decrease the power consumption in the chip. For example, in a Pentium processor, a power of 16 W is one of the strongest limitations to its speed. If the supply voltage is reduced, the threshold voltage must also be reduced. However, the degree of the reduction of the supply and threshold voltages is limited by the subthreshold

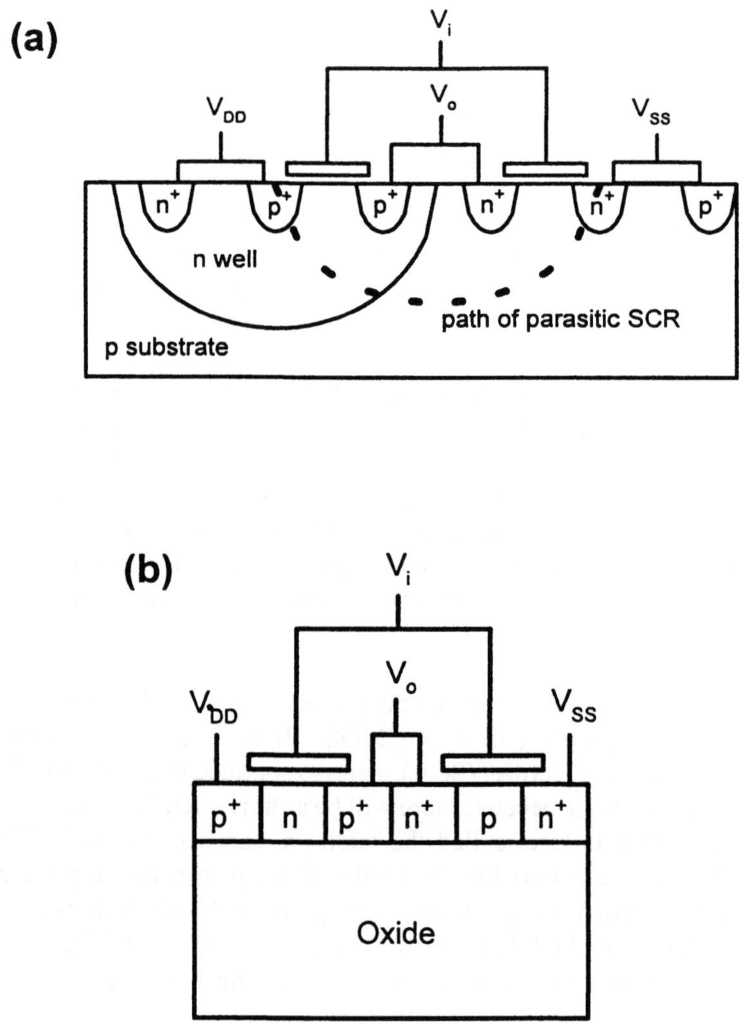

Figure 1.60 : CMOS inverters fabricated with (a) bulk, and (b) SOI technologies.

slope, which is defined as the gate voltage required to increase the drain current by one order of magnitude in the weak inversion. Figure 1.61 shows the log(I_D) versus V_{GS} characteristics for the SOI and bulk MOSFETs having two different threshold voltages, demonstrating that the SOI device has a larger subtheshold slope and thus a lower leakage current than its bulk counterpart. This allows the use of an SOI MOSFET with a small threshold voltage, thus the use of a smaller supply voltage, without having to be concerned with a significant leakage current. On the other hand, for the bulk MOSFET, a large threshold voltage, and thus a large supply voltage, is needed to ensure a small leakage current in the device.

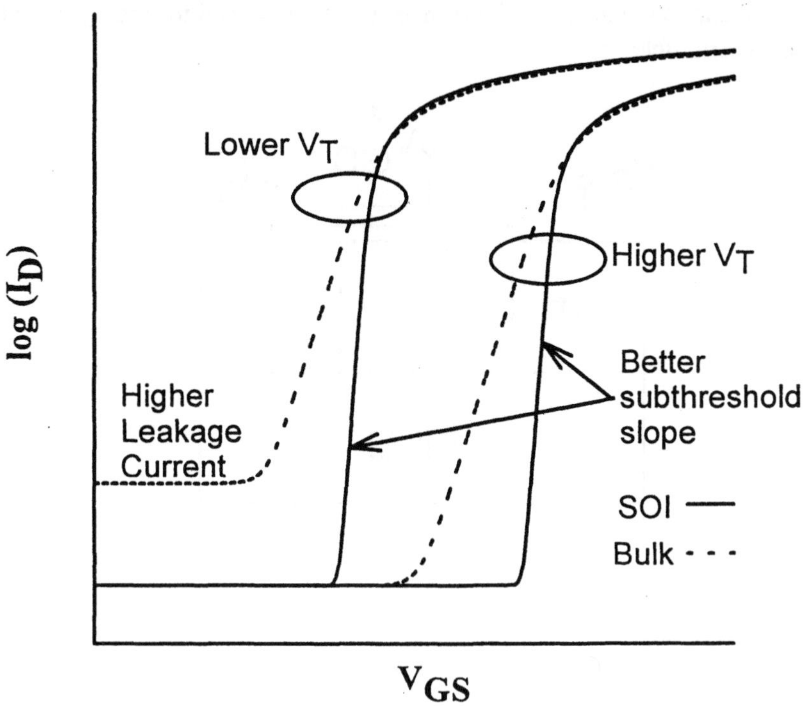

Figure 1.61 : Drain current versus gate voltage characteristics of the SOI and bulk MOSFETs having two different threshold voltages. The advantages of the SOI over bulk device, larger subthreshold slope and lower leakage current, are demonstrated.

Figure 1.62 gives the schematic of an SOI MOSFET. It can be seen that the main feature differentiating the SOI MOSFET from its bulk counterpart is the fact that the SOI MOSFET has both front and back oxide interfaces and therefore is subjected to charge coupling effects between the two gates. The bulk MOSFET can therefore be considered as a special case of an SOI MOSFET with a very large semiconductor film thickness. The mixed boundary condition at the front oxide-silicon interface, analogously to (1.52), is

$$V_{GS}^{f} - V_{FB}^{f} = \psi_{Sf} + \frac{\epsilon_s \xi_{Sf}}{C_{of}} \tag{1.156}$$

where V^f_{GS} is the front-gate voltage, V^f_{FB} is the front-flatband voltage, C_{of} is the front-oxide capacitance, ψ_{Sf} is the front-surface band bending and ξ_{Sf} is the front-surface electric field.

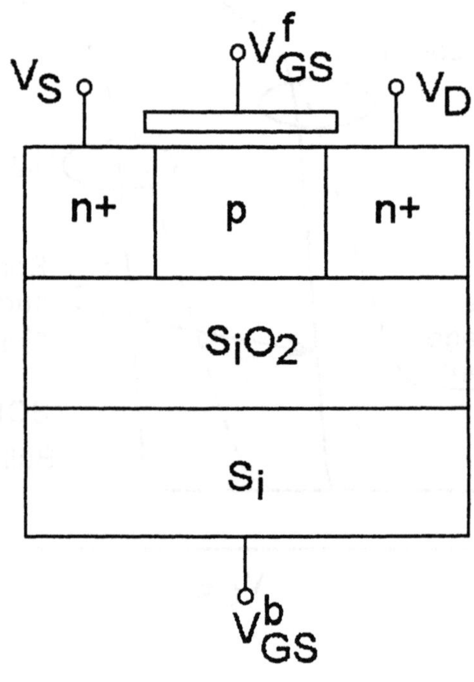

Figure 1.62 : A two-dimensional SOI MOSFET structure showing the top and bottom Si-SiO_2 interfaces.

On the other hand, at the back oxide-silicon interface, the boundary condition is

$$V_{GS}^{b} - V_{FB}^{b} = \psi_{Sb} - \frac{\epsilon_s \xi_{Sb}}{C_{ob}} \tag{1.157}$$

where V^b_{GS} is the back-gate voltage, V^b_{FB} is the back-flatband voltage, C_{ob} is the back-oxide capacitance, ψ_{Sb} is the back-surface band bending and ξ_{Sb} is the back-surface electric field.

Evaluating (1.49) and at the front-interface ($x = 0$, $\psi = \psi_{Sf}$ and $\xi = \xi_{Sf}$) and at the back-interface ($x = t_b$, $\psi = \psi_{Sb}$ and $\xi = \xi_{Sb}$), we obtain:

$$\xi_{Sf}^2 - F^2(\psi_{Sf}, V) = \xi_{Sb}^2 - F^2(\psi_{Sb}, V) \equiv \alpha \tag{1.158}$$

where, unlike the bulk MOSFET, α is not equal to zero and is a parameter that quantifies the charge coupling between the front- and back-gates. Finally, the semiconductor film thickness t_b can be calculated using the following relationship:

$$t_b = \int_{\psi_{Sb}}^{\psi_{Sf}} \frac{d\psi}{\xi} \tag{1.159}$$

The values of ψ_{Sf}, ψ_{Sb}, ξ_{Sf} and ξ_{Sb} can be calculated numerically from (1.156)-(1.159).

The drain current for the SOI MOSFET can be expressed by the following single-integral equation [79]:

$$I_D = \mu_n \frac{W}{L_{eff}} \Big[C_{of} \left((V_{GS}^f - V_{FB}^f)(\psi_{SfL} - \psi_{Sfo}) - \frac{(\psi_{SfL}^2 - \psi_{Sfo}^2)}{2} \right)$$

$$\frac{Q_b \, n_o^2}{\beta} \left(\beta V_{DS} + e^{-\beta V_{DS}} - 1 \right)$$

$$+ \epsilon_s \int_{\psi_{Sbo}}^{\psi_{Sfo}} \xi(\psi, V{=}0) \, d\psi - \epsilon_s \int_{\psi_{SbL}}^{\psi_{SfL}} \xi(\psi, V{=}V_{DS}) \, d\psi \qquad (1.160)$$

$$+ \frac{\epsilon_s \, t_b \, (\alpha_L - \alpha_o)}{2}$$

$$C_{ob} \left((V_{GS}^b - V_{FB}^f)(\psi_{SbL} - \psi_{Sbo}) - \frac{(\psi_{SbL}^2 - \psi_{Sbo}^2)}{2} \right) \Big]$$

where $\psi_{Sf}(y = y_s) = \psi_{Sfo}$, $\psi_{Sf}(y = y_d) = \psi_{SfL}$, $\psi_{sb}(y = y_s) = \psi_{Sb}$, $\psi_{Sb}(y = y_d) = \psi_{SbL}$, $\alpha(y = y_s) = \alpha_o$, $\alpha(y = y_d) = \alpha_L$, and $L_{eff} = (y_d - y_s)$ is the effective channel length.

For a very large t_b, as would be the case for a bulk MOSFET, the charge coupling between the front- and back gate diminishes, and α_o and α_L approach zero. Also, for this case, there will be a point x_o inside the semiconductor at which $\psi(x = x_o) = \xi(x = x_o) = 0$. Taking the point x_o to be the back interface, we get $\psi_{Sbo} = \psi_{SbL} = 0$, and (1.160) reduces to (1.68). Therefore, Pierret-Shield's model for the bulk MOSFET, given in (1.68), can be considered as a special case for the general model given in (1.160).

REFERENCES

[1] J. E. Lilienfeld, "Methods and apparatus for controlling electric current," US Patent 1 745 175. Application filed Oct. 8, 1926, granted Jan. 18, 1930.

[2] J. E. Lilienfeld, "Device for controlling electrical current," US Patent 1 900 018. Application filed Mar. 28, 1928, granted Mar. 7, 1933.

[3] O. Heil, "Improvements in or relating to electrical amplifiers and other control arrangements and devices," British Patent 439 457. Application filed Mar. 4, 1935, granted Dec. 6, 1935.

[4] D. Kahng and M. M. Atalla,"Silicon-silicon dioxide field induced surface devices," IRE-AIEEE Solid-State Device Research Conference, Carnegie Institute of Technology, Pittsburgh, PA, 1960.

[5] A. H. Wilson, Proc. Royal Soc., vol. 133A, p. 458, 1931.

[6] R. S. Ohl, "Light-sensitive electrical device," US Patent 2 402 662. Application filed May 27, 1941, granted June 25, 1946.

[7] M. Riordan and L. Hoddeson, "The origins of the pn junction," IEEE Spectrum, vol. 34, pp. 46-51, June 1997.

[8] W. H. Brattain, "Discovery of the transistor effect," Adventures in Experimental Phys., vol. 5, pp. 1-31, 1976.

[9] W. B. Shockley, "The path to the conception of the junction transistor," IEEE Trans. Electron Dev., vol. ED-23, pp. 597-620, July 1976. Reprinted, vol. ED-31, pp. 1523-1546, Nov. 1984.

[10] J. Bardeen and W. H. Brattain, "Three-electrode circuit element utilizing semiconductive materials," US Patent 2 524 035. Application filed June 17, 1948, granted Oct. 3, 1950.

[11] J. Bardeen and W. H. Brattain, "The transistor, a semi-conductor triode," Phys. Rev., vol. 74, pp. 230-231, July 1948.

[12] W. B. Shockley, *Electron and Holes in Semiconductors*, Van Nostrand, New York, 1950.

[13] W. B. Shockley, "Circuit element utilizing semiconductive material," US Patent 2 569 347. Application filed June 26, 1948, granted Sept 25, 1951.

[14] C. T. Sah, "Evolution of the MOS transistors- from conception to VLSI," Proc. IEEE, vol. 76, pp. 1280-1326, Oct. 1988.

[15] T. P. Brody, "The thin film transistor-A late flowering bloom,". IEEE Trans. Electron Device, vol. ED-31, pp. 1614-1628, Nov. 1984.

[16] F. M. Wanlass and C. T. Sah, "Nanowatt logic using field-effect metal-oxide semiconductor triodes", IEEE Int. Solid State Cir. Conf., pp. 32-33, Feb. 1963.

[17] F. M. Wanlass, "Low stand-by power complementary field-effect circuitry", US Patent 3 356 858. Application filed June 18, 1963, granted Dec. 5, 1957.

[18] J. S. Kilby, "Invention of the integrated circuit," IEEE Trans. Electron Device, vol. ED-23, pp. 648-654, Jul. 1976.

[19] A. S. Grove, *Physics and Technology of Semiconductor Devices*, Wiley, New York, 1967.

[20] R. N. Noyce, "Semiconductor device-and-lead structure", US Patent 2 981 877. Application filed July 30, 1959, granted Apr. 25, 1961.

[21] G. E. Moore, "Cramming more components onto integrated circuits," Electronics, vol. 38, pp. 114-117, April 1965. See also a reprinted version in Proceedings of the IEEE, vol. 86, pp. 82-85, Jan. 1998.

[22] G. J. Myers, A. Y. C. Yu and D. L. House, "Microprocessor technology trends," Proc. IEEE, vol. 74, pp. 1605-1621, Dec. 1986.

[23] R. R. Schaller, "Moore's law: past, present and future," IEEE Spectrum, vol. 34, pp. 52-59, June 1997.

[24] A. Yu, "The future of microprocessor," IEEE Micro, vol. 16, pp. 46-53, Dec. 1996.

[25] D. W. Marquardt, "An algorithm for least-squares estimation of non-linear parameters," J. Soc. Ind. Appl. Math., vol. 11, pp. 431-441, 1963.

[26] J. J. Liou, *Advanced Semiconductor Device Physics and Modeling,* Artech House, Inc., Boston, 1994.

[27] R. H. Bube, *Electronic Properties of Crystalline Solids*, New York: Academic

Press, 1974.

[28] R. F. Pierret, *Semiconductor Device Fundamentals*, Reading: Addison Wesley, 1996.

[29] E. H. Nicollian and J. R. Brews, *MOS (Metal Oxide Semiconductor) Physics and Technology*, New York: Wiley, 1982.

[30] J. G. Fossum and A. Ortiz-Conde, "Effects of grain boundaries on the channel conductance of SOI MOSFETs," IEEE Trans. Electron Device, vol. ED-30, pp. 933-940, Aug. 1983.

[31] D. R. Frankl, *Electrical Properties of Semiconductor Surfaces*, Pergamon Press, Oxford, 1967.

[32] MICROTEC Manual, Siborg Systems, Inc., Waterloo, Ontario, 1996.

[33] D. R. Frankl, "Conditions of quasi-equilibrium in a semiconductor surface space-charge layer," Surface Sci, vol. 3, pp. 101-108, Apr., 1965.

[34] R. S. C. Cobbold, Theory and Applications of Field Effects Transistors, Wiley-Interscience, New York, 1970.

[35] P. Richman, *MOS Field-Effect Transistors and Integrated Circuits*, Wiley-Interscience, New York, 1973.

[36] C. T. Hsing, D. P. Kennedy, A. D. Sutherland, and K. M. van Vliet, "Quantum mechanical determination of the potential and carrier distributions in the inversion layer of metal-oxide-semiconductor devices," Phys. Status. Solidi (a), vol. 56, pp. 129-141, Nov. 1979.

[37] R. H. Kingston and S. F. Neustadter, "Calculation of the space-charge, electric field and free carrier concentration at the surface of a semiconductor," J. Appl. Phys., vol. 26, pp. 718, 1955.

[38] M. Shur, *Physics of Semiconductor Devices*, Prentice-Hall, Englewood Cliffs, NJ, 1990.

[39] S. M. Sze, *Physics of Semiconductor Devices*, 2nd ed., Wiley, New York, 1981.

[40] Y. P. Tsividis, *Operation and Modeling of the MOS Transistor*, McGraw-Hill, New York, 1987.

[41] E. H. Nicollian and J. R. Brews, *MOS Physics and Technology*, Wiley, New York, 1982.

[42] H. C. Pao and C. T. Sah, "Effects of diffusion current on characteristics of metal-oxide (insulator)-semiconductor transistors," Solid-St. Electron., vol. 9, pp. 927-937, Oct. 1966.

[43] R. F. Pierret and J. A. Shields, "Simplified long-channel MOSFET theory", Solid-St. Electron., vol. 26, pp. 143-147, Feb. 1983

[44] G. Baccarani, M. Rudan and G. Spadini, "Analytical IGFET model including drift and diffusion currents", IEE J. Solid-St. Electron. Device, vol. 2, pp. 62-68, March 1978.

[45] J. R. Brews, "A charge-sheet model of the MOSFET," Solid-St. Electron., vol. 21, pp. 345-355, Feb. 1978.

[46] J. R. Brews, "Physics of the MOS transistor," in D. Kahng, Ed., *Applied Solid State Science,* Suppl. 2A., New York: Academic Press, 1981.

[47] P. Antognetti and G. Massobrio (eds.), *Semiconductor Device Modeling with SPICE,* New York: McGraw-Hill, 1988.

[48] T. A. Fjeldly, B. J. Moon, and M. Shur, "Analytical solution of generalized diode

equation", IEEE Trans. Electron Dev., vol. ED-38, pp. 1976-1977, Aug. 1991.

[49] K. Lee, M. Shur, T. A. Fjeldly, T. Ytterdal, *Semiconductor Device Modeling for VLSI*, Prentice-Hall, Englewood Cliffs, NJ, 1993.

[50] M. Shur, T. A. Fjeldly, T. Ytterdal, and K. Lee, "Unified MOSFET model", Solid -State Electron., vol. 35, pp. 1795-1802, Dec. 1992.

[51] Y. P. Tsividis and K. Suyama, "MOSFET modeling for analog circuit CAD: problems and prospects," IEEE J. Solid State Circ., vol. 29, pp. 210-216, March 1994.

[52] R. Narayanan, Z. Latif, A. Ortiz-Conde, J. J. Liou, L. Golovanova, W. Wong and F. J. García Sánchez, "A model for reverse short-channel effects in MOSFETs", Int.Caracas Conf. on Cir. Dev. and Sys., Caracas, Venezuela, pp. 294-297, Dec. 1995.

[53] R. Narayanan, Z. Latif, A. Ortiz-Conde, J. J. Liou, L. Golovanova, W. Wong and F. J. García Sánchez, "On the reverse short-channel effects of submicrom MOSFETs", Proc. of Southcon, Orlando, Florida, pp. 345-349, June 1996.

[54] T. H. Ning, P. W. Cook, R. H. Dennard, C. M. Osburn, S. E. Shuster, and H. N. Yu, "1-μm MOSFET VLSI technology: part IV--hot-electron design constraints," IEEE Trans. Electron Devices, vol. ED-26, pp. 520-533, 1979.

[55] P. E. Cottrell, R. R. Troutman, and T. H. Ning, "Hot electron emission in n-channel IGFETs," IEEE Trans. Electron Devices, vol. ED-26, p. 520, 1979.

[56] C. Hu, S. C. Tam, F. C. Hsu, P. K. Ko, T. Y. Chan, and K. W. Terrill, "Hot-electron- Induced MOSFET degradation-model, monitor, and improvement," IEEE Trans. Electron Devices, vol. ED-32, pp. 375, 1985.

[57] C. T. Huang, C., T. Wang, C. N. Chen, M. C. Chang, and J. Fu, "Modeling hot-electron gate current in Si MOSFET's using a coupled drift-diffusion and Monte Carlo method," IEEE Trans.Electron Devices, vol. 39, pp. 2562, 1992.

[58] M. El-Banna and M. El-Nokali, "A simple analytical model for hot-carrier MOSFET's," IEEE Trans. Electron Devices, vol. 36, pp. 979, 1989.

[59] J. -J. Yang, S. Chung, P. -C. Chou, C. -H. Chen, and M. -S. Lin, "A new approach to modeling the substrate current of pre-stressed and post-stressed MOSFETs," IEEE Trans. Electron Devices, vol. 42, p. 1113, 1995.

[60] C. Hu, "MOSFET scaling in the next decade and beyond," Semiconductor International, vol. 17, p. 105, 1994.

[61] J. Chung, K. N. Quader, C. G. Sodini, P. K. Ko, and C. Hu, "The effect of hot electron degradation on analog MOSFET performance," IEDM Dig., p. 553, 1990.

[62] W. Wong, A. Ice, and J. J. Liou, "An empirical model for the characterization of hot- carrier induced MOS device degradation," Solid-St. Electron., to appear.

[63] E. Takeda and N. Suzuki, "An empirical model for device degradation due to hot-carrier injection," IEEE Electron Device Lett., vol. EDL-4, p. 111, 1983.

[64] J. R. Schrieffer, in Semiconductor Surface Physics, R. H. Kingston, Ed., University of Penn. Press, p. 55, 1957.

[65] F. Stern and W. E. Howard, "Properties of semiconductor surface inversion layers in the electric quantum limit," Phy. Rev., vol. 163, p. 816, 1967.

[66] S. A. Hareland, S. Krishnamurthy, S. Jallepalli, C. F. Yeap, K. Hasnat, A. F. Tasch, Jr., and C. M. Maziar, "A computationally efficient model for inversion

layer quantization effects in deep submicron N-channel MOSFETs," IEEE Trans. Electron Devices, vol. 43, p. 90, 1996.

[67] M. J. Van Dort, P. H. Woerlee, A. J. Walker, C. A. H. Juffermans, and H. Lifka, "Influence of high substrate doping levels on the threshold voltage and the mobility of deep-submicrometer MOSFETs," IEEE Trans. Electron Devices, vol. 39, p. 932, 1992.

[68] J. A. Pals, "A general solution of the quantization in a semiconductor surface inversion layer in the electric quantum limit," Phys. Lett. A, vol. 39, p. 101, 1972.

[69] M. Abramowitz and I. A. Stegun, Eds., Handbook of Mathematical Functions, Washington DC: U.S. GPO, 1964.

[70] F. F. Fang and W. E. Howard, "Negative field-effect mobility on (100) Si surface," Phys. Rev. Lett., vol. 16, p. 797, 1966.

[71] S. Ogura, P. J. Jsang, and W. W. Walker, "Design and characterization of the lightly doped drain-source (LDD) insulated gate field-effect transistor," IEEE Electron Devices, vol. ED-27, p. 1359, 1980.

[72] J. -S. Kim, K. -S. Seo, and H. -J. Yoo, "An analytical model for the effect of graded gate oxide on the channel electric field in MOSFETs with lightly doped drain structure," Solid-St. Electron., vol. 41, p. 650, 1997.

[73] T. Mizuno, Y. Sawada, S. Shinozaki, and O. Ozawa, "A new degradation mechanism of current drivability of asymmetrical LDD MOSFET's," IEDM Tech. Dig., p. 250, 1985.

[74] Y. A. Elmansy and A. R. Boothroyd, "A simple two-dimensional model for IGFET operation in the saturation," IEEE Trans. Electron Devices, vol. ED-24, p. 254, 1977.

[75] S. Liu, M. Hu, and S. Jang, "An analytical, physics-based linear current-voltage model for ot-carrier damaged LDD MOSFETs," Solid-St. Electron., vol. 41, pp. 793-797, 1997.

[76] J. P. Colinge, *Silicon-On-Insulator Technology: Materials to VLSI*, Kluwer Academic, Boston, 1991.

[77] M. L. Alles, "Thin-film SOI emerges," IEEE Spectrum, vol. 34, pp. 37-45, June 1997.

[78] M. Jurczak, A. Jakunbowski and L. Lukasiak, "A review of SOI transistor models," Microelectronics Journal, vol. 28, pp. 173- 182, Feb. 1997.

[79] A. Ortiz-Conde, F. J. García Sánchez, P. E. Schmidt, and A. Sa-Neto, "The non-equilibrium inversion layer charge of the thin-film SOI MOSFET," IEEE Trans. Electron Device, vol. ED-36, pp. 1651-1656, Sept. 1989.

[80] A. Ortiz-Conde, R. Herrera, P. E. Schmidt, F. J. García Sánchez, and J. Andrian, "Long-channel silicon-on-insulator MOSFET theory", Solid-St. Electron., vol. 35, pp. 1291-1298, Sept. 1992.

Chapter 2

MOSFET simulation using device simulators

Semiconductor device simulation is a useful tool for predicting the behavior of semiconductor devices prior to their actual fabrication and thus can be used to reduce greatly the cost and time of device development cycle. Two approaches to device simulation have been used frequently. The first is the analytical approach which involves the use of simplified assumptions on both the device geometry and the mathematical models [1-2]. In this approach, a device is usually treated as a one-dimensional structure having a piecewise constant impurity doping concentration. The structure is further separated into quasi-neutral and space-charge regions which have different physical characteristics. Local solutions of the two adjacent regions are then linked using the boundary conditions, leading to closed-form expressions applicable for simulating the device terminal characteristics. The other approach, called the numerical approach, solves directly and numerically the fundamental semiconductor device equations such as the Poisson equation, electron and hole continuity equations, and electron and hole current equations [3-5]. Realistic device structures and all relevant device physics can be incorporated into such an approach, but with the expense of more extensive computational time and development of numerical algorithm. The analytical approach was discussed in details in Chapter 1, and the numerical approach will be the focus of this chapter.

2.1 Introduction to device simulators

Several semiconductor device simulators developed based on the numerical approach are commercially available. They include, but not limited to, SEDAN [6], BIPOLE [7], BAMBI [8], MINIMOS [9], MEDICI [10], DAVINCI [11], ATLAS [12], and MICROTEC [13]. SEDAN is a one-dimensional device simulator (i.e., device specified in the simulator is in one dimension only). Given the device make-up such as doping profiles and layer thicknesses, SEDAN can be used to simulate silicon bipolar or field-effect devices consisting of up of five metallurgical junctions.

BIPOLE is a quasi-two-dimensional device simulator developed for simulation of bipolar transistors. It was first developed by University of Waterloo and later being supported by the Technology Modeling Associates, Inc. The numerical calculation is based on the variable boundary regional approach using one-dimensional charge transport equations. Two-dimensional and quasi-cylindrical edge effects in the bipolar transistors are handled by combining the vertical one-dimensional analysis with a coupled one-dimensional horizontal analysis of the charge transport equations in the base region.

BAMBI is a two-dimensional device simulator. It solves the two-dimensional Poisson and continuity equations simultaneously. BAMBI does not include models for physical parameters such as the electron and hole mobilities. Instead, these parameters are defined by external functions specified by the user.

MINIMOS is a two- and three-dimensional device simulator for field-effect transistors. The numerical approach is the same as the other device simulators, but MINIMOS builds the consistent solution of the semiconductor equations in a hierarchical manner, starting with a relatively simple model which is subsequently refined by taking into account more comprehensive but complicated physical mechanisms.

MEDICI, an enhanced version of PISCES developed at Stanford University, is a powerful two-dimensional device simulator. It can be used to simulate Si bipolar and field-effect devices, as well as unconventional devices consisting of heterojunctions (i.e., devices fabricated with different semiconducting materials). All models and material parameters can be modified by the user on a region to region basis. In addition, with its non-uniform triangular simulation grid structure, MEDICI can model arbitrary device geometries with both planar and non-planar surface topographies.

DAVINCI is basically evolved from MEDICI, but with the three-dimensional features included in the simulator. When using DAVINCI the user will need to specify the three-dimensional device structure and make-up, and three-dimensional results, such as free-carrier density and current contours in the device, can be generated.

ATLAS supported by Silvaco International is another versatile tool for device simulation. The modular architecture of ATLAS allows the simulations of all types of semiconductor devices (Si, GaAs, and heterojunction devices) in two- and three-dimensional environments.

All the device simulators mentioned above were designed for the UNIX environment, and MICROTEC is the only device simulator which can be run on a PC platform with a 386 or higher microprocessor and 8 Mbyte memory. It allows two-dimensional simulation, but only limited to Si bipolar and MOS devices. In addition, unlike the other device simulators which can be used under steady-state, small-signal, and transient conditions, MICROTEC is only capable for steady-state analysis. Despite apparent simplicity, MICROTEC is attractive for educational purposes due to its ease of use, efficient and flexible graphics tools.

The above mentioned device simulators are summarized in Table 2.1. Here, we will focus on MEDICI device simulator, which has been used widely by students and researchers at universities and device engineers in semiconductor industry.

Name	Capability	Developer
SEDAN	One-dimensional, all Si devices, dc, ac, and transient analysis	Tech. Modeling Asso.
BIPOLE	Quasi-two-dimensional, bipolar devices only, dc, ac, and transient analysis	University of Waterloo
BAMBI	Two-dimensional, all Si devices, dc, ac, and transient analysis	Tech. University Vienna
MINIMOS	Two- and three-dimensional, field-effect devices only, dc, ac, and transient analysis	Tech. University Vienna
PISCES	Two-dimensional, all Si devices, dc, ac, and transient analysis	Stanford University
MEDICI	Two-dimensional, all Si and GaAs devices, dc, ac, and transient analysis	Tech. Modeling Asso.
DAVINCI	Three-dimensional, all Si and GaAs devices, dc, ac, and transient analysis	Tech. Modeling Asso.
ATLAS	Two- and three-dimensional, all Si and GaAs devices, dc, ac, and transient analysis	Silvaco International
MICROTEC	Two-dimensional, all Si devices, dc analysis only	Siborg Systems

Table 2.1 : Existing Semiconductor Device Simulators

2.2 Description of MEDICI device simulator

2.2.1 Basic semiconductor equations

Basically, MEDICI is a software package which solves numerically the following five basic semiconductor device equations:

$$\epsilon \nabla^2 \psi = -q\left(p - n + N_D^+ - N_A^-\right) - \rho_s \tag{2.1}$$

is the Poisson equation,

$$\frac{\partial n}{\partial t} = \frac{1}{q} \nabla \bullet \mathbf{J_n} - U_n \tag{2.2}$$

$$\frac{\partial p}{\partial t} = -\frac{1}{q} \nabla \bullet \mathbf{J_p} - U_p \tag{2.3}$$

are the electron and hole continuity equations, and

$$\mathbf{J}_n = -q\mu_n n \nabla \phi_n \tag{2.4}$$

$$\mathbf{J}_p = -q\mu_p p \nabla \phi_p \tag{2.5}$$

are the electron and hole current equations. In the above equations, boldface indicates the parameter is a vector, ϵ is the dielectric permittivity, ψ is the electrostatic potential, n and p are the electron and hole concentrations, N^-_D and N^+_A are the ionized donor and acceptor impurity concentrations, ρ_S is the surface charge density which may be present due to the fixed charge in insulating materials or charged interface states, J_n and J_p are the electron and hole current densities, U_n and U_p are the electron and hole net recombination rates, μ_n and μ_p are the electron and hole mobilities, and ϕ_n and ϕ_p are the electron and hole quasi-Fermi potentials. The primary function of MEDICI is to solve the partial differential equations in (2.1)-(2.3) self-consistently for ψ, n, and p. These solutions can then be put into (2.4)-(2.5) to find the electron and hole currents of the device. It should be mentioned that MEDICI, like other conventional device simulators, does not account for the quantum mechanical effects, which become important in deep-submicron MOS devices. However, as will be discussed in the next section, MEDICI includes a feature which solves, in addition to the above classical equations, a set of energy balance equations.

Such a feature yields the so-called semi-classical solution with an accuracy lies between the classical and quantum solutions.

2.2.2 Physical mechanisms

The five basic device equations can be solved only if the physical mechanisms imbedded in the equations are appropriately described and modeled.

(a) Free-carrier mobilities

The electron and hole mobilities account for scattering mechanisms (i.e., collision process among electrons, holes, and impurity atoms) associated with the free-carrier charge transport in the device. There are four main scattering processes: (1) ionized impurity scattering; (2) neutral impurity scattering; (3) acoustic phonon scattering; and (4) nonpolar optical scattering. The details discussions of these scattering mechanisms can be found in [1].

Since the free-carrier mobilities depend strongly on the magnitude of electric field, mobility model in MEDICI consists of low-field and high-field mobility components. For the low-field mobility model, six choices are available. The simplest one is to choose low-field electron and hole mobilities (μ_{0n} and μ_{0p}) as constant throughout the device structure. This is the default if no other low-field mobility model is selected. The next level is the doping concentration-dependent mobility model, which assumes the values of electron and hole mobilities depend only on the doping concentration and can be determined using the table look-up (see Table 2.2). Other more sophisticated models include the temperature as well as concentration dependencies. For example, the Arora mobility model [14] gives

$$\mu_{0n} = \frac{88\left(\frac{T}{300}\right)^{-0.57} + 1252\left(\frac{T}{300}\right)^{-2.33}}{1+\left(\frac{N}{1.26x10^{17}}\left(\frac{T}{300}\right)^{2.4}\right)^{\alpha}} \tag{2.6}$$

$$\mu_{0p} = \frac{54.3\left(\frac{T}{300}\right)^{-0.57} + 407\left(\frac{T}{300}\right)^{-2.33}}{1+\left(\frac{N}{2.35x10^{17}}\left(\frac{T}{300}\right)^{2.4}\right)^{\alpha}} \tag{2.7}$$

Table 2.2

Mobility versus Impurity Concentration for Silicon and Gallium Arsenide (T=300 K)

Concentration (cm^{-3})	Mobility in Silicon (cm^2/V-s)		Mobility in GaAs (cm^2/V-s)	
	Electrons	Holes	Electrons	Holes
1.0E14	1350.0	495.0	8000.0	390.0
2.0E14	1345.0	495.0	7718.0	380.0
4.0E14	1335.0	495.0	7445.0	375.0
6.0E14	1320.0	495.0	7290.0	360.0
8.0E14	1310.0	495.0	7182.0	350.0
1.0E15	1300.0	491.1	7300.0	340.0
2.0E15	1248.0	487.3	6847.0	335.0
4.0E15	1200.0	480.1	6422.0	320.0
6.0E15	1156.0	473.3	6185.0	315.0
8.0E15	1115.0	466.9	6023.0	305.0
1.0E16	1076.0	460.9	5900.0	302.0
2.0E16	960.0	434.8	5474.0	300.0
4.0E16	845.0	396.5	5079.0	285.0
6.0E16	760.0	369.2	4861.0	270.0
8.0E16	720.0	348.3	4712.0	245.0
1.0E17	675.0	331.5	4600.0	240.0
2.0E17	524.0	279.0	3874.0	210.0
4.0E17	385.0	229.8	3263.0	205.0
6.0E17	321.0	203.8	2950.0	200.0
8.0E17	279.0	186.9	2747.0	186.9
1.0E18	252.0	178.0	2600.0	170.0
2.0E18	182.5	130.0	2060.0	130.0
4.0E18	140.6	90.0	1632.0	90.0
6.0E18	113.6	74.5	1424.0	74.5
8.0E18	99.5	66.6	1293.0	66.6
1.0E19	90.5	61.0	1200.0	61.0
2.0E19	86.9	55.0		
4.0E19	83.4	53.7		
6.0E19	78.8	52.9		
8.0E19	71.6	52.4		
1.0E20	67.8	52.0		
2.0E20	52.0	50.8		
4.0E20	35.5	49.6		
6.0E20	23.6	48.9		
8.0E20	19.0	48.4		
1.0E21	17.8	48.0		

Table 2.2 : Mobility versus impurity concentration for Silicon and Galium Arsenide (T=300 K)

where T is the absolute temperature of semiconductor (i.e., T = 300 K at room temperature), N is the total doping concentration, and

$$\alpha = 0.88\left(\frac{T}{300}\right)^{-0.146} \tag{2.8}$$

Three high-field mobility models are available in MEDICI. For example, the Caughey-Thomas model [15] takes into account dependence on electric field parallel to the direction of current flow:

$$\mu_n = \frac{\mu_{0n}}{\left(1 + \frac{\mu_{0n} E_{p,n}}{v_{sat}}\right)^{\frac{1}{\beta_n}}} \tag{2.9}$$

$$\mu_p = \frac{\mu_{0p}}{\left(1 + \frac{\mu_{0p} E_{p,p}}{v_{sat}}\right)^{\frac{1}{\beta_p}}} \tag{2.10}$$

where μ_n and μ_p are the high-field electron and hole mobilities, $E_{p,n}$ and $E_{p,p}$ are the electron and hole electric fields parallel to the current direction, and v_{sat} is the saturation drift velocity:

$$v_{sat} = \frac{2.4x10^7}{\left(1 + 0.8\, exp\left(\frac{T}{600}\right)\right)} \tag{2.11}$$

A more comprehensive model called the Hewlett-Packard model [10] is also available, which accounts for dependence of mobilities on electric fields both parallel and perpendicular to the direction of current flow.

In addition to the carrier mobility models for the bulk of semiconductor mentioned above, several mobility models for the insulator-semiconductor interfaces are also available in MEDICI. Since the carrier mobilities can be substantially lower along the interfaces than in the bulk, mobility degradation factors **GSURFN** and **GSURFP** (the capital and boldface letters indicate the parameters are user accessible and can be changed from their default values by the user) can be multiplied to the low-field mobility. That is,

$$\mu_{S,n} = \textbf{\textit{GSURFN}}\ \mu_{0n} \tag{2.12}$$

$$\mu_{S,p} = \textbf{\textit{GSURFP}}\ \mu_{0p} \tag{2.13}$$

where $\mu_{S,n}$ and $\mu_{S,p}$ denote electron and hole mobilities at the semiconductor surface. MEDICI also allows the selection of other more sophisticated surface mobility models, including a model which takes into account phonon scattering, surface roughness scattering, and impurity scattering.

Apparently, using different mobility models can give rise to different simulation results, as evidenced by the current-voltage characteristics of a MOSFET simulated using different mobility models shown in Fig. 2.1.

(b) Free-carrier recombination

Recombination and generation are processes whereby electrons and holes are destroyed and created, respectively. They are restoring mechanisms in semiconductors by which the excess carriers are stabilized when a perturbation is applied. At thermal equilibrium, the recombination rate R equals the generation rate G. At nonequilibrium, however, R ≠ G, which results in a net recombination rate U. MEDICI supports both Shockley-Read-Hall (SRH) and Auger recombination statistics [1-2]. Their model expressions are

$$U_{SRH} = \frac{pn - n_i^2}{\tau_p\left(n + n_i\ exp\left(\frac{ETRAP}{kT}\right)\right) + \tau_n\left(p + n_i\ exp\left(-\frac{ETRAP}{kT}\right)\right)} \tag{2.14}$$

$$U_{Auger} = \textbf{\textit{AUGN}}\left(pn^2 - nn_i^2\right) + \textbf{\textit{AUGP}}\left(np^2 - pn_i^2\right) \tag{2.15}$$

where n_i is the intrinsic free-carrier concentration, τ_n and τ_p are the electron and hole lifetimes, and **ETRAP**, **AUGN**, and **AUGP** are user accessible parameters, which can be changed from their default values by the user. Electron and hole lifetimes may be impurity concentration dependent and can be expressed as

$$\frac{\textbf{\textit{TAUN0}}}{\tau_n} = \textbf{\textit{AN}} + \textbf{\textit{BN}}\frac{N}{NSRHN} + \textbf{\textit{CN}}\frac{N}{NSRHN} \tag{2.16}$$

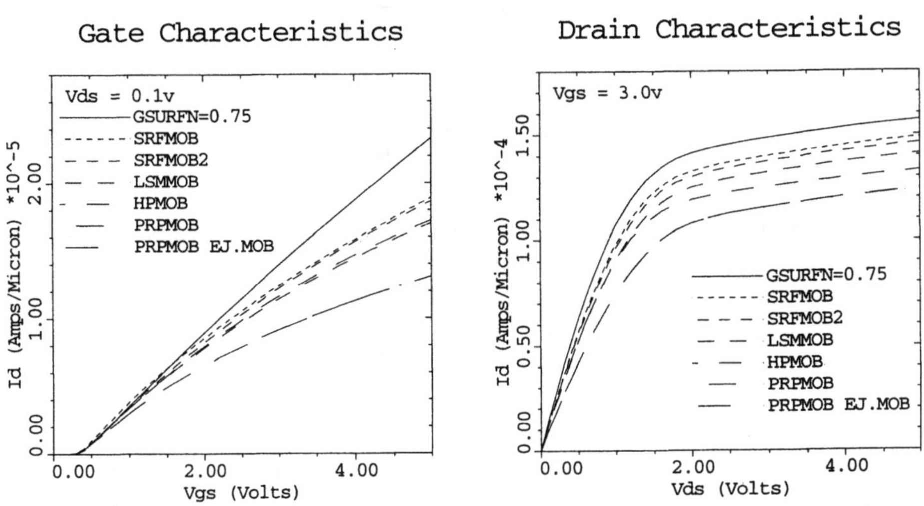

Figure 2.1 : MEDICI simulation results for an n-channel MOSFET using several different mobility models. The legends GSURFN denotes a mobility reduction at the interface by a specified factor (i.e., 0.75), SRFMOB denotes the surface mobility model, SRFMOB2 denotes the enhanced surface mobility model, LSMMOB denotes the Lombardi surface mobility model, HPMOB denotes the Hewlett-Packard mobility model, PRPMOB denotes the perpendicular electric field mobility model, and PRPMOB EJ.MOB denotes the mobility model using electric field components parallel and perpendicular to current flow (Source: [10]. Reprinted with permission).

$$\frac{TAUP0}{\tau_n} = AP + BP\frac{N}{NSRHP} + CP\frac{N}{NSRHP} \tag{2.17}$$

A more generalized SRH recombination model that includes tunneling at strong electric fields is also available.

(c) Energy bandgap and effective density of states

The energy bandgap (E_g) and effective density of states in the conduction and valence bands (N_C and N_V) are important parameters in determining the free-carrier statistics in semiconductors. These parameters are temperature dependent and can be expressed as

$$E_g(T) = E_g(300) + EGALPH\left(\frac{300^2}{300 + EGBETA} - \frac{T^2}{T + EGBETA}\right) \tag{2.18}$$

$$N_C(T) = N_C(300)\left(\frac{T}{300}\right)^{1.5} \tag{2.19}$$

$$N_V(T) = N_V(300)\left(\frac{T}{300}\right)^{1.5} \tag{2.20}$$

The intrinsic free-carrier concentration n_i is related to the above parameters as

$$n_i = \sqrt{N_C\,N_V}\;exp\left(\frac{E_g}{2kT}\right) \tag{2.21}$$

An E_g model accounting for the bandgap narrowing effect due to heavy doping in semiconductors is also available in MEDICI. When the doping level of semiconductor exceeds about 10^{17} cm^{-3}, a downward shift of conduction band edge with nearly parabolic density of states occurs due to the increasing electron-impurity ion interaction. The valence band also moves up by approximately the same amount. Carrier-carrier interaction also gives rise to a shift in the two band edges. The effective energy bandgap is thus reduced as the doping concentration is increased. To account for the heavy doping effect, $E_g(300)$ in (2.18) needs to be replaced by

$$E_g(300) = 0.009\left(ln\left(\frac{N}{10^{17}}\right) + \sqrt{ln^2\left(\frac{N}{10^{17}}\right) + 0.5} \right) \tag{2.22}$$

(d) Impact ionization

When the electric field in a semiconductor is sufficiently large, a free electron entering the region can gain enough kinetic force from the field and, when colliding with a bound electron, can create an electron-hole pair to contribute to the current. Such a process is called the impact ionization. In MEDICI, the generated free carriers due to impact ionization can be included self-consistently in the solution of the device equations. The impact-ionization generation rate G^{II} for electron-hole pairs can be expressed as

$$G^{II} = \frac{J_n}{q} N.IONIZA\ exp\left(\left(\frac{ECN.II}{En,p}\right)^{EXN.II}\right) + \frac{J_p}{q} P.IONIZA\ exp\left(-\left(\frac{ECP.II}{Ep,p}\right)^{EXN.II}\right) \tag{2.23}$$

(e) Lattice temperature

Conventionally, and by default in MEDICI simulation, the temperature of semiconductor lattice is assumed equal to the ambient temperature. However, lattice heating due to current flow can be important at high current levels, and the lattice temperature can become much higher than the ambient temperature [2]. The lattice temperature model is available in MEDICI to account for such an effect. To compute the spatially dependent lattice temperature T, MEDICI solves the heat flow equation:

$$\zeta \frac{\partial T}{\partial t} = H + \nabla(\lambda(T)\nabla T) \tag{2.24}$$

where ζ is the mass density of the material, c is the specific heat of the material, H is the heat generation, and λ is the thermal conductivity of the material. The heat generation is modeled using the expression

$$H = J_n \bullet E + J_p \bullet E + H_U \tag{2.25}$$

where the heat H_U due to electron-hole recombination and generation is given by

$$H_U = (U_{SRH} + U_{Auger} - G^{II})E_g \tag{2.26}$$

(f) Energy balance equations

The electron and hole current equations are by default described by the drift-diffusion model, which assumes that the temperatures of the free carriers is the same as the lattice temperature (i.e., lattice temperature is not the same as the ambient temperature if the option of lattice temperature model described in (e) is used). This is no longer valid in small-geometry devices in which the electric field is normally very high and the electron and hole temperatures can be much higher than the lattice temperature [16-17]. To account for such an effect, MEDICI has an option called the hydrodynamic model which incorporates the energy balance equations into drift-diffusion equations:

$$\boldsymbol{J}_n = q\mu_n n\boldsymbol{E} + q\mu_n \nabla u^{Tn} n + \frac{n u^{Tn} \nabla u^{Tn} \partial\mu_n(u^{Tn})}{\nabla u^{Tn}} \tag{2.27}$$

$$\boldsymbol{J}_p = q\mu_p p\boldsymbol{E} - q\mu_p \nabla u^{Tp} p - \frac{p u^{Tp} \nabla u^{Tp} \partial\mu_p(u^{Tp})}{\nabla u^{Tp}} \tag{2.28}$$

where u^{Tn} and u^{Tp} are the temperatures of electrons and holes, respectively, which can be much higher than lattice and ambient temperatures if the system is subject to a large perturbation. The last terms on the right hand side of (2.27) and (2.28) account for the nonequilibrium states of electrons and holes, respectively, in high-field regions.

(g) Transient simulation

The preceding sections have focused on the MEDICI descriptions for steady-state (dc) device simulation. To carry out transient device simulation, MEDICI uses the time discretization of the Poisson equation in (2.1) and electron and hole continuity equations in (2.2)-(2.3). Because of the extremely stiff nature of (2.1)-(2.3), strong stability requirements are placed on any transient integration scheme. In addition, it is most convenient to use a simple first-order backward difference formula (BDF1) [18] so that only the solution at the most recent time step is required. Based on the BDF1, (2.2) and (2.3) are discretized as

$$\frac{n_k - n_{k-1}}{\Delta t_k} = \frac{1}{q} \nabla \bullet J_n(\psi_k, n_k, p_k) - U_n(\psi_k, n_k, p_k) \equiv F_n(k) \tag{2.29}$$

$$\frac{p_k - p_{k-1}}{\Delta t_k} = -\frac{1}{q} \nabla \bullet J_p(\psi_k, n_k, p_k) - U_p(\psi_k, n_k, p_k) \equiv F_p(k) \tag{2.30}$$

where $\Delta t_k = t_k - t_{k-1}$ and subscripts k and k-1 denote at time t_k and t_{k-1}, respectively. As an alternative to BDF1, a second-order backward difference formula (BDF2) [18] could be used:

$$\frac{\dfrac{n_k(2-\gamma)}{(1-\gamma)} - \dfrac{n_{k-1}}{\gamma(1-\gamma)} + \dfrac{n_{k-2}(1-\gamma)}{\gamma}}{t_k - t_{k-2}} = F_n(k) \tag{2.31}$$

$$\frac{\dfrac{p_k(2-\gamma)}{(1-\gamma)} - \dfrac{p_{k-1}}{\gamma(1-\gamma)} + \dfrac{p_{k-2}(1-\gamma)}{\gamma}}{t_k - t_{k-2}} = F_p(k) \tag{2.32}$$

where

$$\gamma = \frac{t_{k-1} - t_{k-2}}{t_k - t_{k-2}} \tag{2.33}$$

MEDICI employs a variable order method. That is, at a given time point, the program checks the local truncation error (LTE) for both the BDF1 and BDF2 methods, and selects the method which yields the largest time step for the purpose of reducing the number of time steps.

(h) Small-signal analysis

In addition to dc and transient analysis, MEDICI also allows ac small-signal analysis to be performed. Specifically, starting from a dc bias condition, a sinusoidal input of given amplitude and frequency can be applied to a device from which small-signal terminal current-voltage characteristics can be calculated. An ac sinusoidal voltage is applied to an electrode I with the form of

$$V_i = V_{i0} + v_i e^{jw\tau} \tag{2.34}$$

where V_{i0} is the dc bias, v_i is the small-signal amplitude, ω is the angular frequency, and V_i is the total voltage. The ac solutions to the Poisson and electron and hole continuity equations are [19]:

$$\psi_i = \psi_{i0} + \psi_{iac}\, e^{jw\tau} \tag{2.35}$$

$$n_i = n_{i0} + n_{iac} e^{jw\tau} \tag{2.36}$$

$$p_i = p_{i0} + p_{iac} e^{jw\tau} \tag{2.37}$$

where ψ_{i0}, n_{i0}, and p_{i0} are the dc potential and carrier concentrations at node I, whereas ψ_{iac}, n_{iac}, and p_{iac} are the respective ac values. Putting these solutions into the Poisson and electron and hole continuity equations and expanding the resulting equation as a Taylor series of first-order only, one obtains nonlinear equations of the form [19]

$$F(\psi, n, p) = F(\psi_0, n_0, p_0) + \psi\, e^{jw\tau} \frac{\partial F}{\partial w} + n\, e^{jw\tau} \frac{\partial F}{\partial n} + p e^{jw\tau} \frac{\partial F}{\partial p} \tag{2.38}$$

for each of the three partial differential equations. Given the dc solution at a desired bias condition, these ac nonlinear equations can be splitted into real and imaginary parts and solved using a matrix method.

(i) Circuit analysis

A useful feature available in MEDICI is the ability of carrying out circuit simulation of a module including devices and a circuit. This option allows up to ten MEDICI device models to be imbedded in a circuit composed of SPICE lumped element circuit models. In MEDICI circuit analysis, the Kirchhoff equation describes the circuit, the semiconductor equations (i.e., Poisson, continuity, etc.) describe the devices, and they are solved as a coupled set.

Example shown in Fig. 2.2 illustrates how the circuit and device equations are constructed. In the figure, the circuit considered consists of two voltage controllers, G1 and G2, and one current controller, H1. The circuit is connected to a semiconductor device, in which the triangular mesh structures (to be discussed in the next section) are indicated by the solid lines (i.e., ab, bc, cd, etc.) and the perpendicular bisectors of the sides of the triangles are indicated by the dashed lines (i.e., h1, h2, etc.).

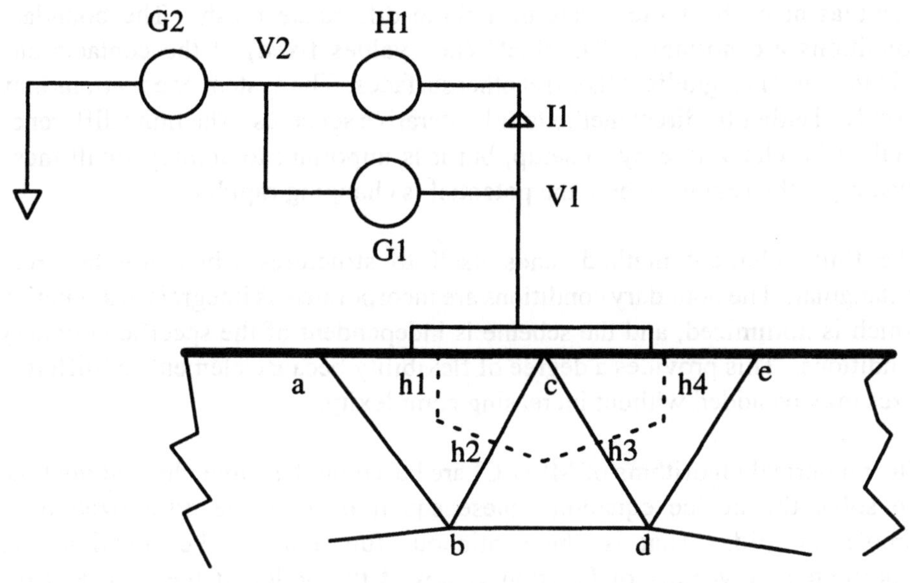

Figure 2.2 : Schematic showing the connection of device and circuit for circuit analysis using MEDICI (Source: [10]. Reprinted with permission).

The equation at node V2 is a simple Kirchhoff current equation:

$$0 = G2(V2) + G1(V2 - V1) + I1 \tag{2.39}$$

A loop equation is needed to accommodate the current controlled source:

$$0 = H1(I1) + V1 - V2 \tag{2.40}$$

Finally the equation at V1 containing circuits and device quantities is

$$0 = G1(V1 - V2) - I1 + h1 \bullet J_{ca} + h2 \bullet J_{\phi} + h3 \bullet J_{cd} + h4 \bullet J_{ce} \tag{2.41}$$

The terms J_{ca}, J_{cb}, J_{cd}, and J_{ce} are the current densities passing along the indicated mesh lines. These current densities are dependent upon the potential, electron, and hole concentrations at each of the nodes in the semiconductor device.

2.3 Numerical algorithms

Most device simulation programs use finite difference or finite element methods to solve the five basic semiconductor device equations. In the finite difference method, the derivatives of the five equations are replaced by finite difference formulas at each of the nodes in a finite difference mesh. The boundary conditions are normally "Dirichlet" (i.e., values fixed) at the contacts and "Neumann" (i.e., gradient fixed) at other surfaces. The system of equations may be solved either by direct methods or by iterative schemes. The finite difference method is relatively easy to setup, but it is important to employ small mesh spacing in the regions where the potential is changing rapidly.

The finite element method lends itself to structures which are far from rectangular. The boundary conditions are incorporated as integrals in a function which is minimized, and the scheme is independent of the specific boundary conditions. This provides a degree of flexibility because elements of different sizes may be added without increasing complexity.

The numerical algorithms of MEDICI are based on the finite element method. To solve the device equations, these equations must be discretized on a simulation grid. That is, the continuous functions of the equations are represented by vectors of function values at the nodes of the grid, and the differential operators are replaced by suitable difference operators. The key to discretizing the differential operators on a general grid is the box method [20]. Each equation is integrated over a small volume enclosing each node, yielding 3N nonlinear algebraic equations, where N is the number of grid points, for the unknown potentials and free-carrier concentrations. From the user's point of view, the discretization process is completely automatic.

2.3.1 Numerical methods

The discretization of the semiconductor device equations gives rise to a set of coupled nonlinear algebraic equations, which must be solved by a nonlinear iteration method. Two iteration approaches are available in MEDICI: Gummel's method and Newton's method. In Gummel's method, the equations are solved sequentially. First, the Poisson equation is solved assuming fixed quasi-Fermi potentials. Then the new potential is substituted into the continuity equations, which are linear and can be solved directly. The new free-carrier concentrations are substituted back into the charge term of Poisson equation and another iteration begins. At each state only one equation is being solved.

In Newton's method, all of the variables are allowed to change during each iteration. As a result, the Newton algorithm is very stable and the solution time is nearly independent of the bias condition. The disadvantage of Newton's method is that for large grids, the size of the matrixes need to be solved increases by two or three times, depending the selection of single- or two-carrier simulation. Thus, Newton's method does become rather expensive both in time and memory for complicated device structures. Furthermore, Gummel's method becomes increasingly slow as the power level of the device increases, and may cease to converge for some cases. It is important to point out that no single method is optimal in all cases.

2.3.2 Matrix solvers

An essential part of the Newton or Gummel iteration is the solution of linear matrixes. In MEDICI, the user has a choice of several matrix solvers. These being the direct method, the incomplete Cholesky conjugate gradient method (ICCG), and the incomplete LU conjugate gradient squared method (ILUCGS). In the direct method, a form of Gaussian elimination known as the LU decomposition is used (L and U represent the lower and upper triangular matrixes of the matrix, respectively). The direct method has the advantage of always finding the correct solution to a linear problem. The disadvantage of the method is that for the very large problems, the memory and CPU requirements become intractable.

The ICCG solver uses the method of conjugate gradients to iteratively minimize the residual. It has the advantage of requiring very little memory. Typically, the ICCG has good convergence properties, and systems with tens of thousands of equations can be solved in less than 50 iterations. The main drawback of ICCG is that it can only be used to solve the Poisson equation while Gummel's method is selected as the iteration method.

The ILUCGS solver uses the conjugate gradient squared method and incomplete LU decomposition, instead of the conjugate gradient method and Cholesky decomposition used in the ICCG. The ILUCGS, unlike ICCG, allows the solutions of all device equations while using Gummel's method. In addition, ILUCGS is robust in its convergence and requires manageable amount of memory. It has the disadvantage of being slower than the other methods as well as having a degraded convergence behavior for floating body problems, such as silicon-on-insulator MOSFETs, or devices with nearly isolated p/n junction.

2.3.3 Initial guesses

Several types of initial guesses are used in MEDICI. The first is the charge neutral assumption used to obtain the equilibrium bias point. This is the starting point of any device simulation. Any later solution with applied bias needs other types of initial guess. For example, with a previous initial guess, the solution currently loaded can be used as the initial guess for the next bias point. Another type of initial guess is obtained by projection, which uses an extrapolation of two previous solutions to the new bias. The last type of initial guess is used after performing a regrid. Such a guess is an interpolation of the solution on a coarse grid to the new grid, and can be used to start the solution of the same bias point on the new grid.

2.3.4 Convergence requirement

A solution is considered converged and iterations will be terminated when either the X norm (the difference between the update and previous values) or the RHS norm (the difference between the left and right hand sides of the equations) falls below a certain tolerance. For the X norm, the default tolerance is 10^{-5} kT/q for the potential and 10^{-5} relative change in the carrier concentration. For the RHS norm, the default tolerance is 10^{-26} C/μm for the Poisson equation and 5×10^{-18} A/μm for the continuity equations. By default, MEDICI uses a combination of the X norm and RHS norm to determine convergence, and therefore alleviates the difficulty of choosing one method or the other. Basically, the program assumes that a solution is converged when either the X norm or RHS norm tolerances are satisfied at each node in the device. This greatly reduces the number of iterations required to obtain a solution, as compared to when either X norm or RHS norm is used alone, without sacrificing the accuracy of the solution.

2.3.5 Summary

A large number of solution methods have been presented, but only a small subset of the possible combinations is needed and used in day-to-day MEDICI simulations. Generally speaking, Newton's iteration method with the direct method linear-matrix solver is by far the most stable method of solution. Unfortunately, it can be expensive in CPU time and memory, particularly for complicated device structures and/or two-carrier simulations. For these cases, the Gummel method offers an attractive alternative.

2.4 Grid in MEDICI

The correct allocation of grid is a crucial issue in device simulation. The number of nodes N_p in the grid has a direct influence on the time and accuracy of simulation; a small N_p will reduce the CPU time and memory but increase the likelihood of divergence, whereas a large N_p will give rise to an opposite effect. Because the different parts of a device have different electrical behavior, it is usually necessary to allocate fine grid in some regions and coarse grid in others.

MEDICI supports a general irregular grid structure, which permits the analysis of arbitrarily shape devices, as well as allows the refinement of particular regions with minimum impact on others. The MEDICI also gives the user a choice of the Cartesian or cylindrical coordinate system. Since MEDICI is a two-dimensional simulator, simulations are performed in the xy-plane when using the Cartesian, with the device behavior assumed spatially independent in the z-direction.

MEDICI provides a regridding mechanism which automatically refines an initial grid wherever key variables (i.e., doping density, electrostatic potential, etc.) vary rapidly. The initial grid is generated by user specifying a coarse rectangular mesh using input statements. The coarse mesh is then refined using the regrid capability until it is fine enough to accurately represent the structure. The mesh begins as a set of nonuniformly spaced x- and y-lines comprising a simple rectangle. The rectangle can then be distorted to track non-planar geometry or match the doping profile. Mesh lines may be terminated inside the device, and redundant nodes removed from the grid. Material and electrodes regions can then be specified as a union of possibly distorted rectangles.

Mesh refinement is a key factor in simulation accuracy and convergence. Several procedures are available in MEDICI for smoothing meshes with poorly shaped elements, and in particular with very obtuse elements. It is difficult to triangular a general region without obtuse elements. Obtuse elements have two undesirable side effects on device simulation. The first is that they exacerbate any inherent roughness in a solution, making contour plots more difficult to interpret. The second, which is more serious and less commonly known, is that they can cause any solution technique to fail whenever the sum of opposite angles in a pair of elements exceeds 180 degrees. Although every step of grid generation can introduce obtuse elements, the following two steps in particular can cause such a problem:

1. Distorting a rectangular mesh unavoidably introduces a large number of very obtuse elements.

2. When a grid containing high aspect ratio elements is refined, very obtuse elements can be created in the transition region between high and low grid density.

Two main techniques are available to treat these difficulties, namely node smoothing and element smoothing. With node smoothing, several iterative passes are carried out during which each node is moved to a position which improves the angles of the elements surrounding it. Node smoothing is suitable only for an initial irregular grid. For a refined grid, such a technique tends to redistribute fine grid away from that dictated by the physical phenomena in the semiconductor device. Nor should it be applied to a distorted rectangular grid. With element smoothing, on the other hand, each adjoining pair of elements is examined, and if necessary the diagonal of the quadrilateral is flipped. This has the effect of stabilizing the discretization. Element smoothing is desirable in almost all cases, and should be performed both on the initial grid and on subsequent regrids.

To illustrate the grid refinement, we consider a doping profile in a silicon given in Fig. 2.3(a) and show in Figs. 2.3(b)-(f) the different levels of mesh smoothing for such a doping profile. First, the initial grid of the structure is shown in Fig. 2.3(b), where the distance on the vertical scale represents the distance on the horizontal scale in Fig. 2.3(a). Fig. 2.3(c) shows the grid after a refinement on doping without mesh smoothing, whereas Fig. 2.3(d) is the same grid, but with smoothing, which results in a better grid shape in the bulk (i.e., lower region of the structure). It can be seen that in both Figs. 2.3(c) and (d), more points have been added in the upper region of the structure, and the mesh maintains good angles throughout the structure. However, some extra nodes in the upper region may be redundant and can be deleted. The grid without these extra nodes is given in Fig. 2.3(e). If mesh smoothing is turned off in the upper region, then the grid becomes the one shown in Fig. 2.3(f).

2.5 Example of MOSFET simulation

Since the focus of this book is the MOSFET, we will present in this section examples of MOSFET simulation using MEDICI. The use of MEDICI is illustrated by going through some of the analysis that might be performed on an n-channel MOSFET device.

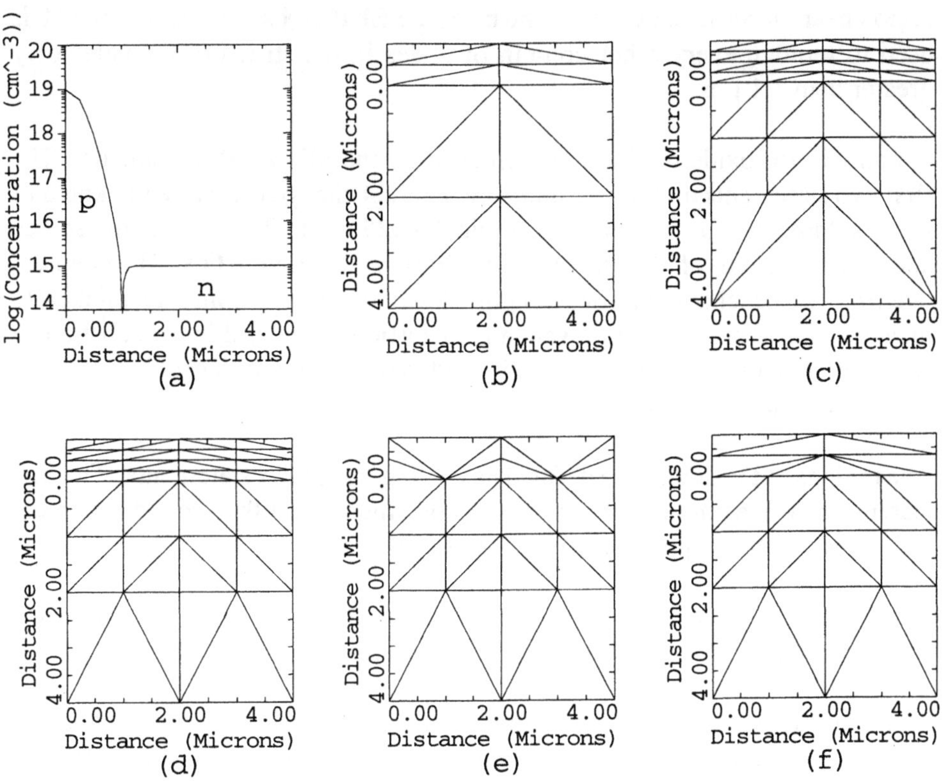

Figure 2.3 : (a) Vertical doping profile in the silicon; (b) Initial grid for the structure; (c) Grid after refinement on the doping, but without mesh smoothing; (d) Grid after refinement on the doping and with mesh smoothing; (e) Grid after refinement, with smoothing, and with extra nodes in the upper region deleted; and (f) Grid after refinement, no smoothing, and with extra nodes in the upper region depleted (Source: [10]. Reprinted with permission).

2.5.1 Generation of the MOSFET structure

The input file shows in Fig. 2.4 creates the simulation structure for the n-channel MOSFET with a channel length of 1.5 μm. The device structure is created by the use of mesh, and distortions of the mesh are also used to give the device its designed surface topography. The first step is to define an initial mesh, as is done in lines 3 through 8 of the input file. Such a mesh will later be refined. The **X.MESH** and **Y.MESH** statements specified the horizontal and vertical spacing of mesh lines, respectively. A spacing of 0.125 μm in the horizontal direction from x = 0 to x = 3 μm is specified. In the vertical direction, the first two statements define a surface oxide with a thickness of 0.025 μm. This is following by the definition of spacing of 0.125 and 0.25 μm for the region of 0 to 1 μm and 1 to 2 μm, respectively. Since only a fine mesh is needed in the region near the surface but not in the bulk, the **ELIMIN** statement on line 10 is used to remove every other column of nodes in the structure for vales of y greater than 1.1 μm.

A nonuniform oxide thickness is achieved by the **SPREAD** statements. The first two statements on lines 12 and 13 in Fig. 2.4 change the oxide thickness of the first three grid lines from their original 0.025 μm to 0.1 μm over the source and drain regions of the device. In addition, the **ENCROACH** statement determines the characteristics length of the transition from the thicker to unchanged grid region. To prevent any distortion of the grid in the substrate, a third **SPREAD** statement on line 15 is used to place the fourth grid line at 0.125 μm, its original location.

The regions of the device are defined next using **REGION** statements. The first statement defines the entire structure to be silicon, and the second statement redefines the three uppermost grid lines to be oxide.

The **ELECTR** statements specify the four electrodes of the device: gate contact, substrate contact, source contact, and drain contact.

The **PROFILE** statements specify the impurity profiles in the device, which are created using Gaussian functions, although they could also have been read in from the output of a process simulator such as SUPREM. The first statement specifies a uniform p-type substrate, and the second statement introduces a p-type threshold adjustment. The last two statements define the n^+ source and drain regions.

Example of MOSFET Simulation

Input file:

```
TITLE      TMA MEDICI Example 1 - 1.5 Micron N-Channel MOSFET

COMMENT    Specify a rectangular mesh
MESH       SMOOTH=1
X.MESH     WIDTH=3.0  H1=0.125

Y.MESH     N=1   L=-0.025
Y.MESH     N=3   L=0.
Y.MESH     DEPTH=1.0  H1=0.125
Y.MESH     DEPTH=1.0  H1=0.250

COMMENT    Eliminate some unnecessary substrate nodes
ELIMIN     COLUMNS  Y.MIN=1.1

COMMENT    Increase source/drain oxide thickness using SPREAD
SPREAD     LEFT   WIDTH=.625  UP=1  LO=3  THICK=.1  ENC=2
SPREAD     RIGHT  WIDTH=.625  UP=1  LO=3  THICK=.1  ENC=2

COMMENT    Use SPREAD again to prevent substrate grid distortion
SPREAD     LEFT   WIDTH=100   UP=3  LO=4  Y.LO=0.125

COMMENT    Specify oxide and silicon regions
REGION     NUM=1  SILICON
REGION     NUM=2  OXIDE     IY.MAX=3

COMMENT    Electrodes: #1=Gate, #2=Substrate, #3=Source, #4=Drain
ELECTR     NUM=1  X.MIN=0.625  X.MAX=2.375  TOP
ELECTR     NUM=2  BOTTOM
ELECTR     NUM=3  X.MAX=0.5  IY.MAX=3
ELECTR     NUM=4  X.MIN=2.5  IY.MAX=3

COMMENT    Specify impurity profiles and fixed charge
PROFILE    P-TYPE  N.PEAK=3E15  UNIFORM     OUT.FILE=MDEX1DS
PROFILE    P-TYPE  N.PEAK=2E16  Y.CHAR=.25
PROFILE    N-TYPE  N.PEAK=2E20  Y.JUNC=.34  X.MIN=0.0  WIDTH=.5
+          XY.RAT=.75
PROFILE    N-TYPE  N.PEAK=2E20  Y.JUNC=.34  X.MIN=2.5  WIDTH=.5
+          XY.RAT=.75
INTERFAC   QF=1E10

PLOT.2D    GRID  TITLE="Example 1 - Initial Grid"  FILL  SCALE

COMMENT    Regrid on doping
REGRID     DOPING  LOG  IGNORE=2  RATIO=2  SMOOTH=1  IN.FILE=MDEX1DS
PLOT.2D    GRID  TITLE="Example 1 - Doping Regrid"  FILL  SCALE

COMMENT    Specify contact parameters
CONTACT    NUM=1  N.POLY

COMMENT    Specify physical models to use
MODELS     CONMOB  FLDMOB  SRFMOB2

COMMENT    Symbolic factorization, solve, regrid on potential
SYMB       CARRIERS=0
METHOD     ICCG  DAMPED
SOLVE

REGRID     POTEN  IGNORE=2  RATIO=.2  MAX=1  SMOOTH=1
+          IN.FILE=MDEX1DS  OUT.FILE=MDEX1MS
```

Figure 2.4 : MEDICI input file for generating and plotting the MOSFET grid structures (Source: [10]. Reprinted with permission).

```
43... PLOT.2D     GRID  TITLE="Example 1 - Potential Regrid"  FILL  SCALE

44... COMMENT     Solve using the refined grid, save solution for later use
45... SYMB        CARRIERS=0
46... SOLVE       OUT.FILE=MDEX1S

47... COMMENT     Impurity profile plots
48... PLOT.1D     DOPING  X.START=.25  X.END=.25  Y.START=0  Y.END=2
   ... +          Y.LOG  POINTS  BOT=1E15  TOP=1E21  COLOR=2
   ... +          TITLE="Example 1 - Source Impurity Profile"
49... PLOT.1D     DOPING  X.START=1.5  X.END=1.5  Y.START=0  Y.END=2
   ... +          Y.LOG  POINTS  BOT=1E15  TOP=1E17  COLOR=2
   ... +          TITLE="Example 1 - Gate Impurity Profile"
50... PLOT.2D     BOUND  TITLE="Example 1 - Impurity Contours"  FILL  SCALE
51... CONTOUR     DOPING  LOG  MIN=16   MAX=20   DEL=.5  COLOR=2
52... CONTOUR     DOPING  LOG  MIN=-16  MAX=-15  DEL=.5  COLOR=1  LINE=2
```

Figure 2.4 : Continue

At this point the device structure has been defined, and all that remains is to refine the grid so that it is adequate for simulation. Figure 2.5 shows that device grid structure before the refinement.

The first grid refinement is requested with the **REGRID** statement on line 32 in Fig. 2.4. This statement causes an existing triangle to be divided into four congruent triangle whenever the impurity concentrations at the nodes of the triangle differ by more than two order of magnitude (specified by **RATIO** = 2 on line 32). Also, the **IGNORE** parameter is set equal to region 2 so that neither grid refinement nor smoothing will be performed in the oxide. The resulting grid structure is shown in Fig. 2.6 in which the junction regions are clearly discerning with the increased grid density.

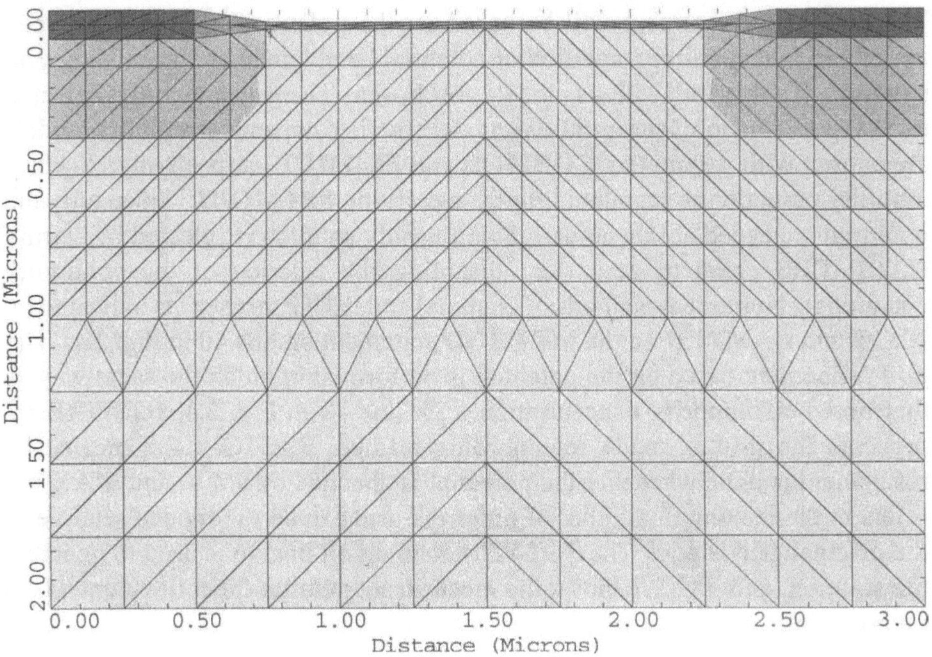

Figure 2.5 : Initial grid structure of the MOSFET generated by the PLOT.2D statement at line 30 in the simulation input file (Source: [10]. Reprinted with permission).

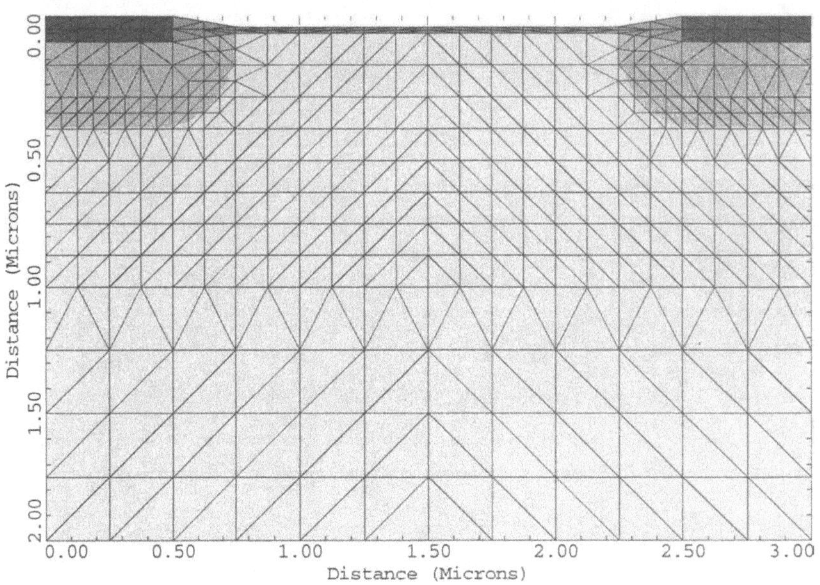

Figure 2.6 : Grid structure after the refinement based on doping concentration generated by the PLOT.2D statement at line 33 in the simulation input file (Source: [10]. Reprinted with permission).

The next grid refinement will be based on the potential difference between nodes and thus requires a solution of potential be obtained on the existing grid structure. To this end, various models are chosen. The gate material is selected to be n^+ polysilicon. Concentration and electric field dependent mobility models are chosen with parameters **CONMOB** and **FLDMOB**, respectively. Surface mobility reduction is accounted for by specifying **SRFMOB2**. Since only the potential is needed at this point, a Poisson-only solution is selected by setting **CARRIERS** equal to zero (i.e., disabling the solution of the continuity equations). Numerical methods of Gummel and ICCG are used, as indicated by **ICCG** and **DAMPED** on the **METHOD** statement on line 40 in Fig. 2.4. The grid refinement based on the potential is performed in much the same way as that based on impurity concentration. On line 42 in Fig. 2.4, **IGNORE** =2 prevents the grid in oxide region being refined, **RATIO** = 0.2 means the refinement is done whenever the potential is changed by 0.2 V, and MAX = 1 refers to the maximum number of times the grid can be subdivided relative to the original grid is one. The **SOLVE** statement on line 46 is used to generate the solution, and Fig. 2.7 shows the mesh structure after the refinement based on potential.

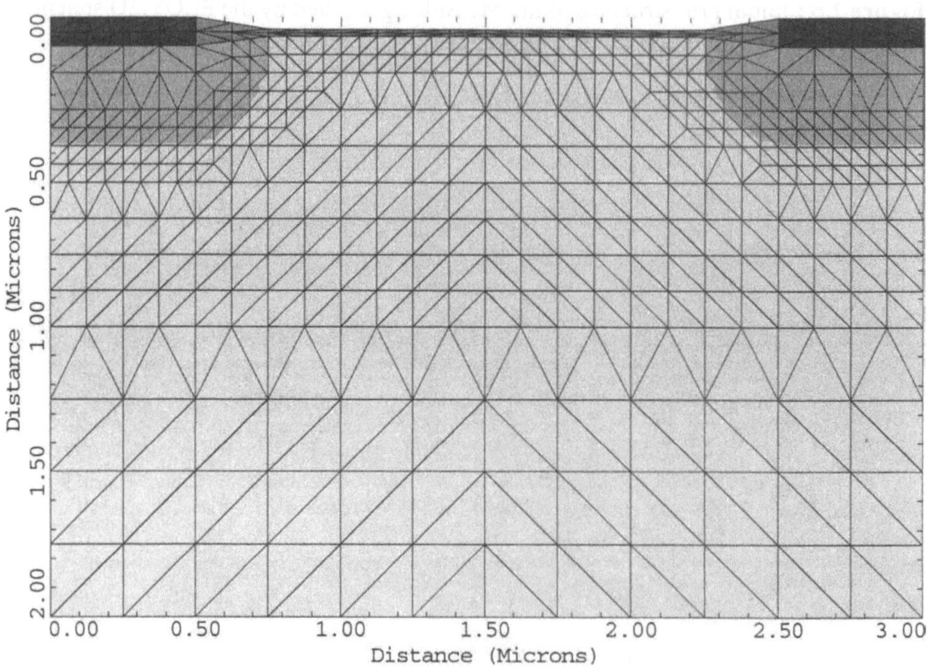

Figure 2.7 : Grid structure after the refinement based on doping concentration and potential generated by the PLOT.2D statement at line 43 in the simulation input file (Source: [10]. Reprinted with permission).

Figuress 2.8 and 2.9 plot the impurity concentration distributions of the device taken at two different locations, and Fig. 2.10 plots the two-dimensional impurity concentration contours. These plots are generated as a consequence of the plot statement at the end of the input file in Fig. 2.4.

2.5.2 Simulation of gate characteristics

An output file for simulating the drain current versus gate voltage characteristics is shown in Fig. 2.11. First, the mesh and solution files in Fig. 2.4 must be loaded, as is done by **MESH** and **LOAD** statements on lines 4 and 6, respectively, in Fig. 2.11. On the **SYMB** statement, Newton iteration method and single-carrier simulation have been selected. On the **SOLVE** statement, a bias of 0.1 V is applied to electrode 4 (drain), and electrode 1 (gate) is stepped to 2 V in 0.2 V increments. The resulting I-V curve is plotted in Fig. 2.12, the characteristic of which is useful in determining the threshold voltage of the MOSFET.

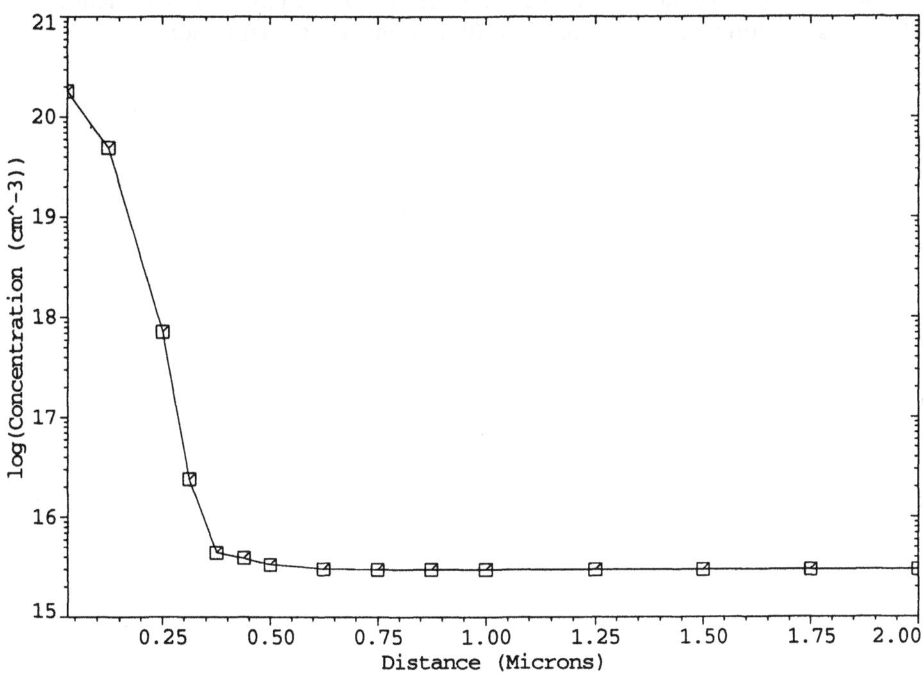

Figure 2.8 : Impurity doping profile taken at x = 0.25 μm (in the source region) and from the surface to the bulk (y = 0 to 2 μm) generated by the PLOT.1D statement at line 48 in the simulation input file (Source: [10]. Reprinted with permission).

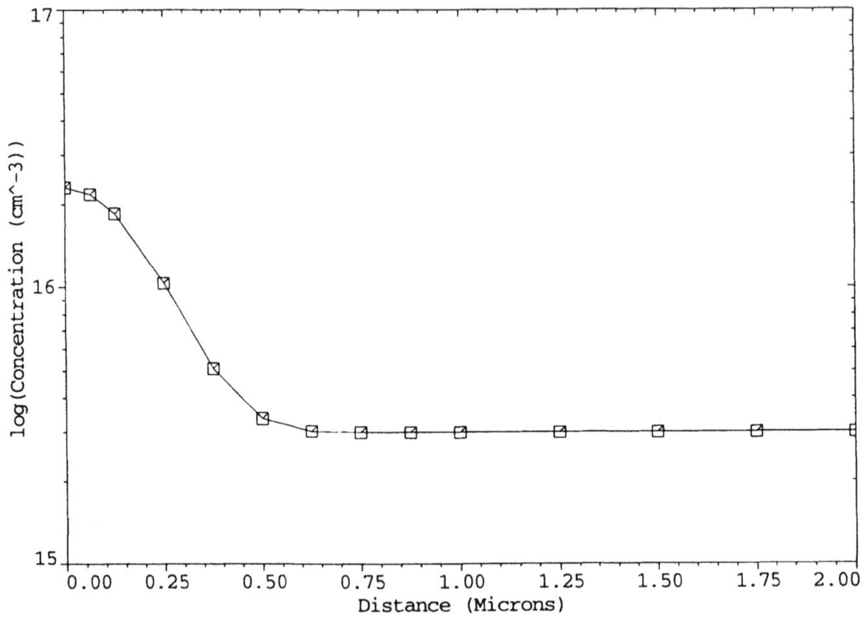

Figure 2.9 : Impurity doping profile taken at x = 1.5 μm (in middle of gate region) and from the surface to the bulk (y = 0 to 2 μm) generated by the PLOT.1D statement at line 49 in the simulation input file (Source: [10]. Reprinted with permission).

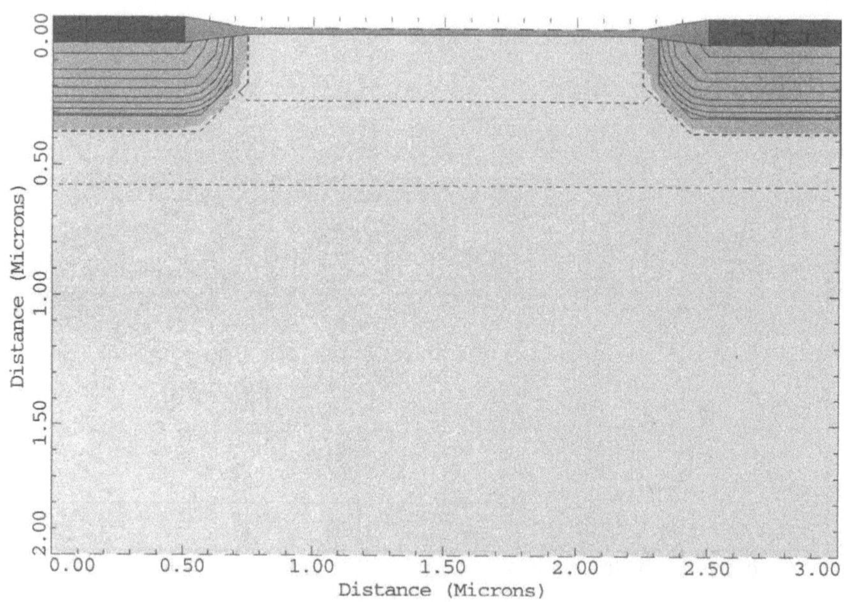

Figure 2.10 : Two-dimensional impurity concentration contours in the MOSFET generated by the PLOT.2D and CONTOUR statements at lines 50 through 52 in the simulation input file (Source: [10]. Reprinted with permission).

```
 1... TITLE      TMA MEDICI Example 1G - 1.5 Micron N-Channel MOSFET
 2... COMMENT    Calculate Gate Characteristics

 3... COMMENT    Read in simulation mesh
 4... MESH       IN.FILE=MDEX1MS

 5... COMMENT    Read in saved solution
 6... LOAD       IN.FILE=MDEX1S

 7... COMMENT    Use Newton's method for the solution
 8... SYMB       NEWTON  CARRIERS=1  ELECTRONS

 9... COMMENT    Setup log file for I-V data
10... LOG        IVFILE=MDEX1GI

11... COMMENT    Solve for Vds=0.1 and then ramp gate
12... SOLVE      V4=.1
13... SOLVE      V1=.2  ELEC=1  VSTEP=.2  NSTEP=9

14... COMMENT    Plot Ids vs. Vgs
15... PLOT.1D    Y.AXIS=I4  X.AXIS=V1  POINTS  COLOR=2
  ... +          TITLE="Example 1G - Gate Characteristics"
16... LABEL      LABEL="Vds = 0.1v"  X=1.6  Y=0.7E-6
```

Figure 2.11 : MEDICI output file for simulation of drain current versus gate voltage characteristics (Source: [10]. Reprinted with permission).

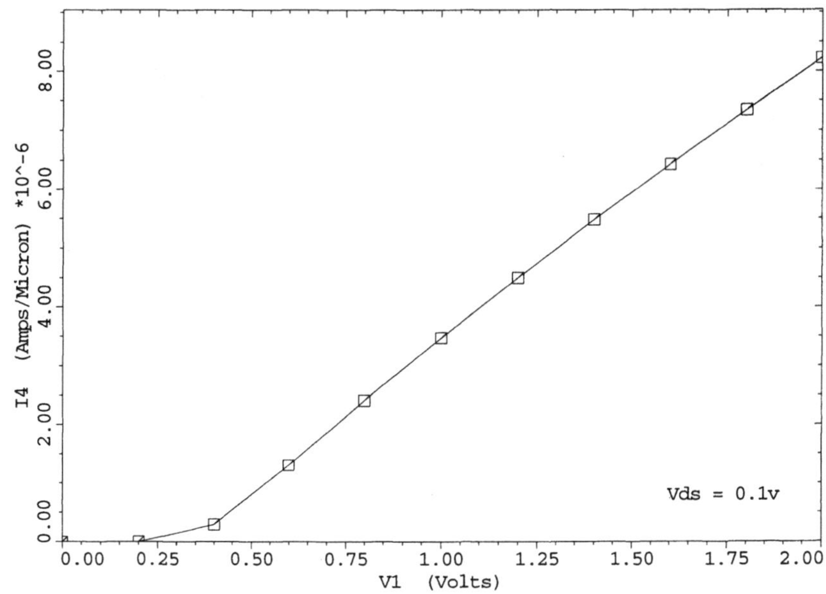

Figure 2.12 : The drain current versus gate voltage characteristics simulated for a drain voltage of 0.1 V generated by the PLOT.1D statement at lines 15 through 16 in the I_D - V_G simulation output file (Source: [10]. Reprinted with permission).

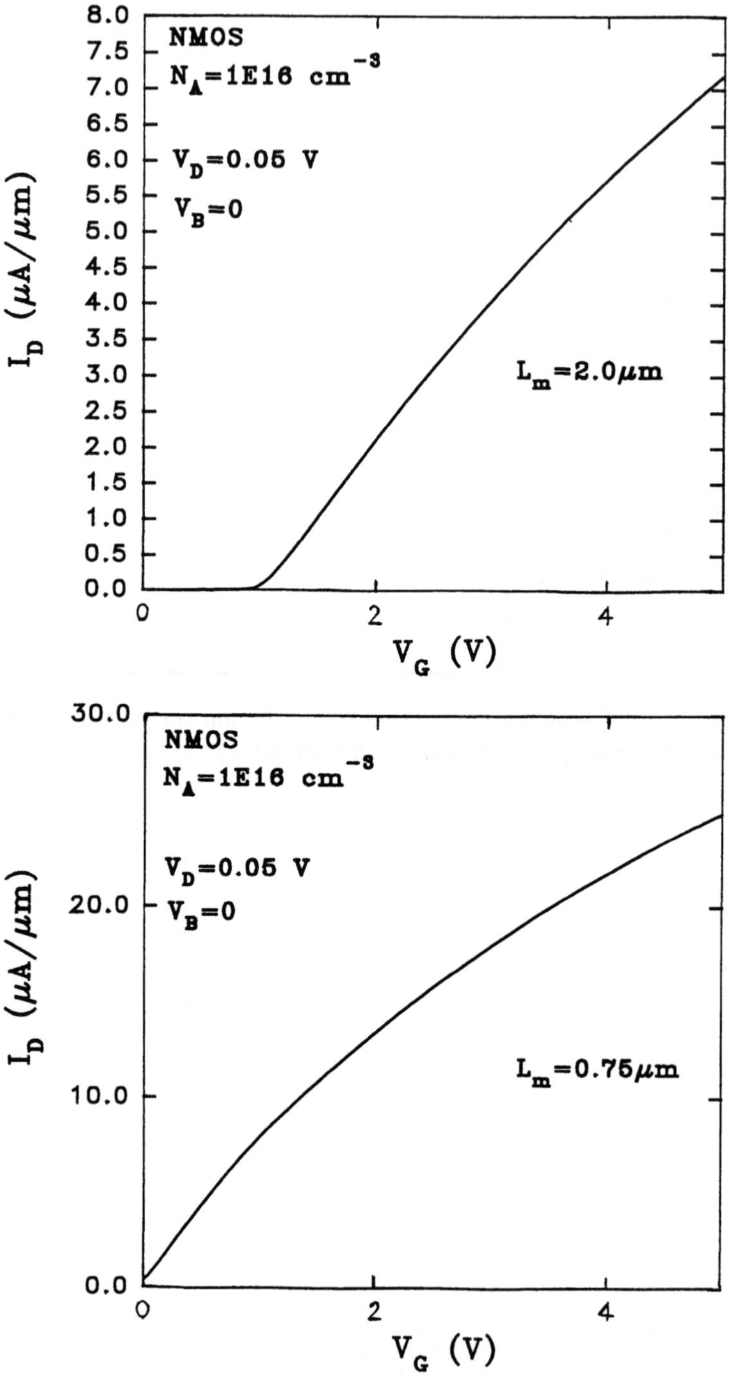

Figure 2.13 : The drain current versus gate voltage characteristics simulated for two MOSFETs having identical device make-up but different mask channel lengths of (a) 2 μm, and (b) 0.75 μm.

It is imperative to point out that the current, among other characteristics, simulated from MEDICI has a unit of A/μm, where μm is the dimension of the device width. This results because MEDICI is a two-dimensional simulator and considers only the x and y coordinates (i.e., cross section) of the device structure. The current (in Amp) of the device can be obtained by multiplying the simulation results (in A/μm) by the device width (in μm).

To compare the gate-current characteristics of a long- and short-channel MOSFET, we show in Figs. 2.13(a)-(b) the simulated I_G versus V_G curves for two MOSFETs have different mask channel lengths of 2 μm and 0.75 μm but otherwise identical device make-up. It can be seen that the linear relationship between I_D and V_G found in the long-channel device disappears in the short-channel device. Moreover, for the short-channel MOSFET, the intercept of the curve at the V_G axis has been shifted to the left, indicating the threshold voltage has been reduced for such a device.

2.5.3 Simulation of drain characteristics

An output file for simulating the drain current versus drain voltage characteristics is shown in Fig. 2.14. After loading the mesh and solution files generated previously, a zero-carrier (i.e., Poisson-only) solution with a gate bias of 3 V is first simulated (lines 8-10 in Fig. 2.14). This is followed by the use of Newton's method to solve for electrons specified in the **SYMB** statement. Finally, the drain is ramped from 0 to 3 V with 0.2 V steps, as defined in the **SOLVE** statement. The resulting I-V curve is shown in Fig. 2.15, and Fig. 2.16 plots the two-dimensional potential contours for the drain voltage of 3.0 V.

Figs. 2.17(a) and (b) show the I_D versus V_{DS} characteristics simulated for two MOSFETs having different channel lengths of 2 and 0.75 μm, but otherwise identical device make-up. There are two important differences in the two figures. First, the saturation region of the short-channel device has a relatively large slope, making it more difficult to distinguish the linear and saturation regions of such a device. This is due to the channel-length modulation effect. Second, at the same gate voltage, the short-channel has a higher drain current than its long-channel counterpart. This arises mainly from the fact that the threshold voltage of the short-channel device is reduced due to the short-channel effect (discussed in Chapter 1).

```
 1... TITLE     TMA MEDICI Example 1D - 1.5 Micron N-Channel MOSFET
 2... COMMENT   Calculate Drain Characteristics

 3... COMMENT   Read in simulation mesh
 4... MESH      IN.FILE=MDEX1MS

 5... COMMENT   Read in initial solution
 6... LOAD      IN.FILE=MDEX1S

 7... COMMENT   Do a Poisson solve only to bias the gate
 8... SYMB      CARRIERS=0
 9... METHOD    ICCG  DAMPED
10... SOLVE     V1=3.0

11... COMMENT   Use Newton's method and solve for electrons
12... SYMB      NEWTON  CARRIERS=1  ELECTRON

13... COMMENT   Setup log file for IV data
14... LOG       IVFILE=MDEX1DI

15... COMMENT   Ramp the drain
16... SOLVE     V4=0.0  ELEC=4  VSTEP=.2  NSTEP=15

17... COMMENT   Plot Ids vs. Vds
18... PLOT.1D   Y.AXIS=I4  X.AXIS=V4  POINTS  TOP=1.6E-4  COLOR=2
  ... +         TITLE="Example 1D - Drain Characteristics"
19... LABEL     LABEL="Vgs = 3.0v"  X=2.4  Y=0.1E-4

20... COMMENT   Potential contour plot using most recent solution
21... PLOT.2D   BOUND  JUNC  DEPL  FILL  SCALE
  ... +         TITLE="Example 1D - Potential Contours"
22... CONTOUR   POTENTIA  MIN=-1  MAX=4  DEL=.25  COLOR=6
23... LABEL     LABEL="Vgs = 3.0v"  X=0.2  Y=1.6
24... LABEL     LABEL="Vds = 3.0v"
```

Figure 2.14 : MEDICI output file for simulation of drain current versus drain voltage characteristics (Source: [10]. Reprinted with permission).

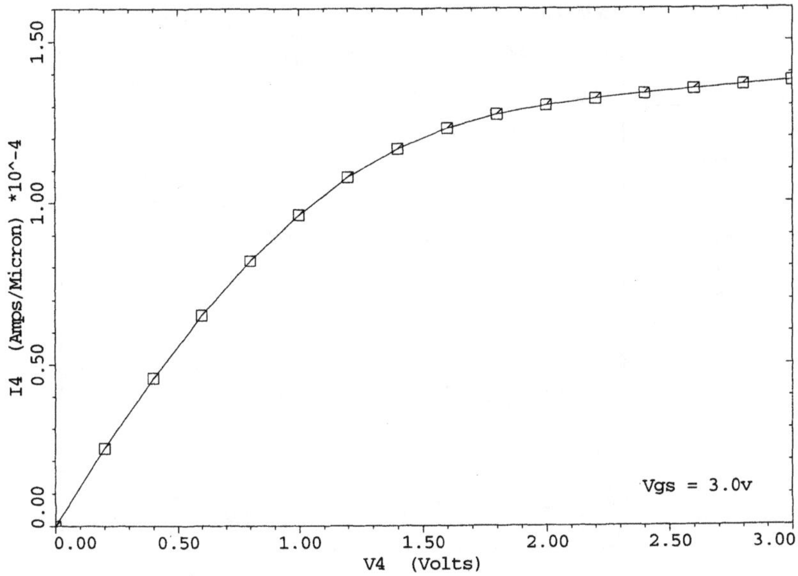

Figure 2.15 : The drain current versus drain voltage characteristics simulated for a gate voltage of 3.0 V generated by the PLOT.1D statement at lines 18 through 19 in the I_D-V_D simulation output file (Source: [10]. Reprinted with permission).

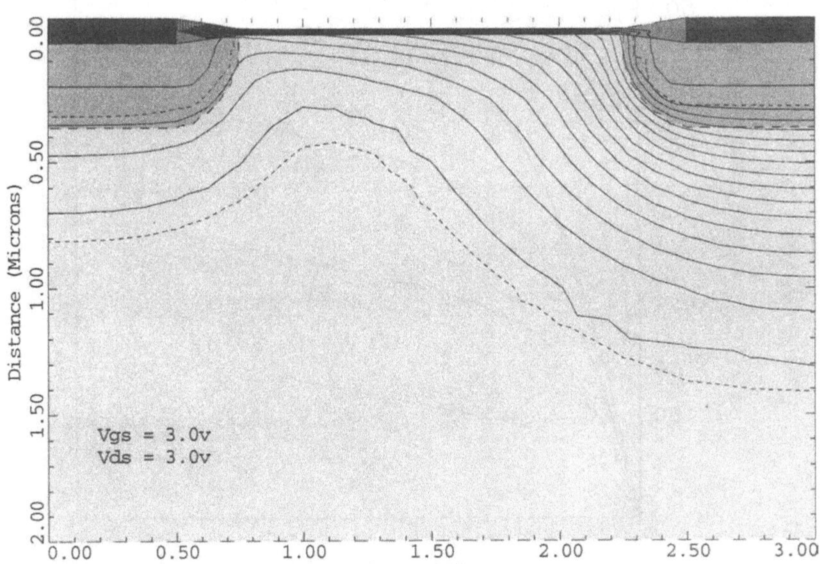

Figure 2.16 : Two-dimensional potential contours of the MOSFET simulated for a gate voltage of 3.0 V and a drain voltage of 3.0 V generated by the PLOT.2D and CONTOUR statements at lines 21 through 24 in the I_D-V_D simulation output file (Source: [10]. Reprinted with permission).

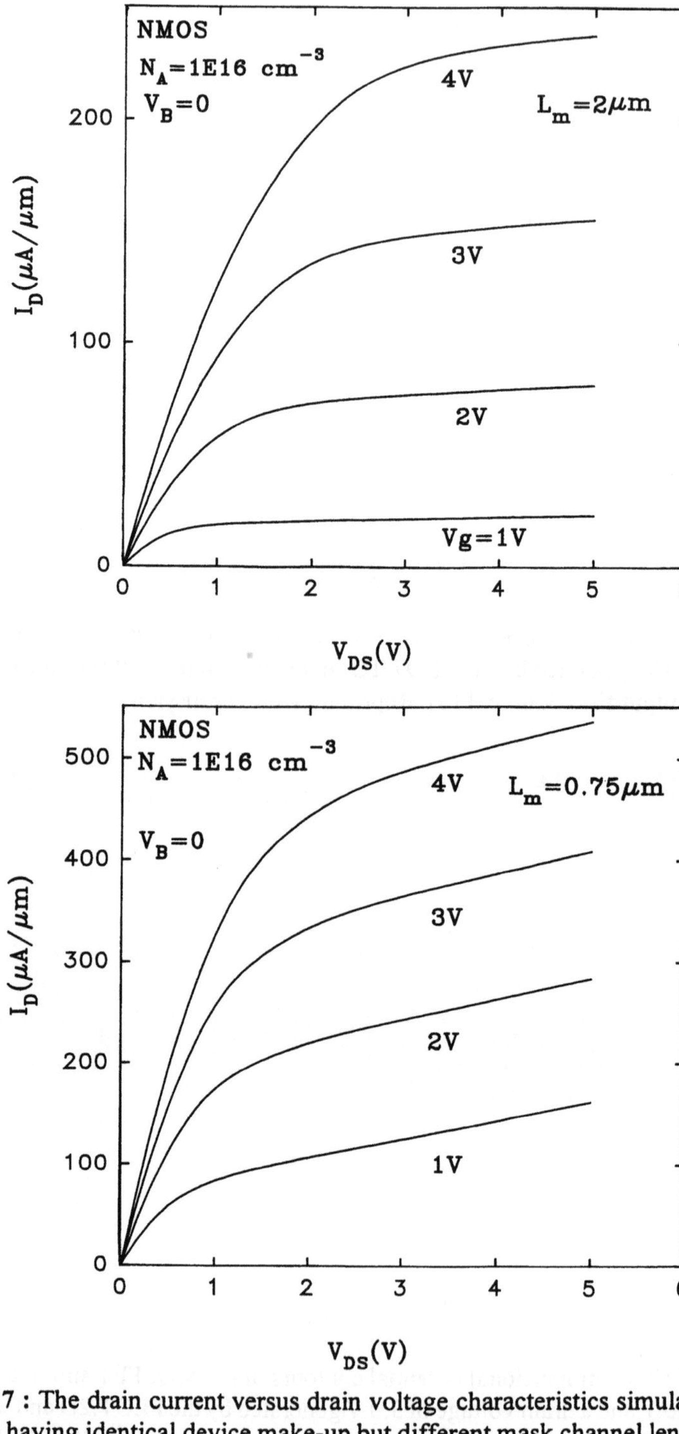

Figure 2.17 : The drain current versus drain voltage characteristics simulated for two MOSFETs having identical device make-up but different mask channel lengths of (a) 2 μm, and (b) 0.75 μm.

2.5.4 Simulation of other characteristics

As the channel length is decreased, the electric field near the drain junction is increased, and the effect of free-carrier transport (called hot-carrier effect) in this high-field region becomes important. MEDICI simulation can generate results for impact ionization rate in the MOSFET biased under a relatively large gate and drain voltage, as shown in Fig. 2.18. It is shown most of the impact ionization occurs at the silicon-oxide interface in the vicinity of the drain junction. Such a mechanism can then lead to a large substrate current and gate current under certain bias conditions, as illustrated in Figs. 2.19 and 2.20, respectively. These simulation results are highly useful in characterizing the hot-carrier effect in the MOSFET.

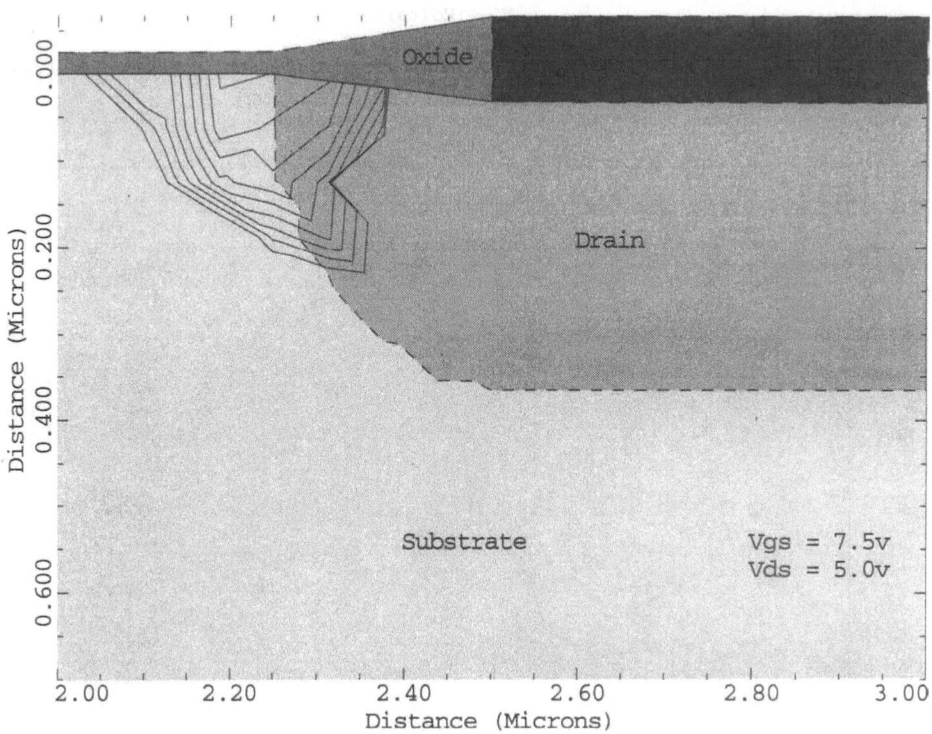

Figure 2.18 : Impact ionization contours simulated for a gate voltage of 7.5 V and a drain voltage of 5.0 V (Source: [10]. Reprinted with permission).

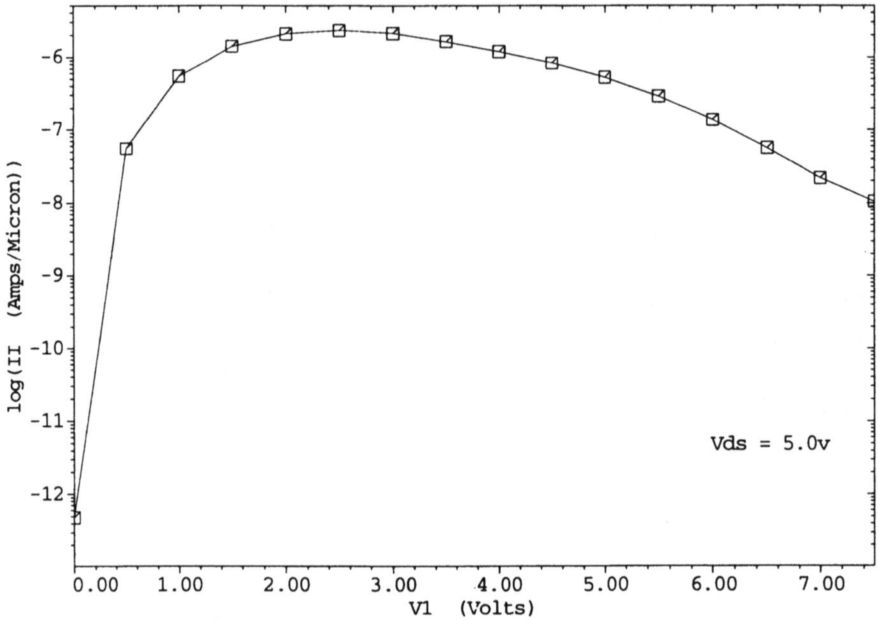

Figure 2.19 : The substrate current versus gate voltage characteristics simulated for a drain voltage of 5.0 V (Source: [10]. Reprinted with permission).

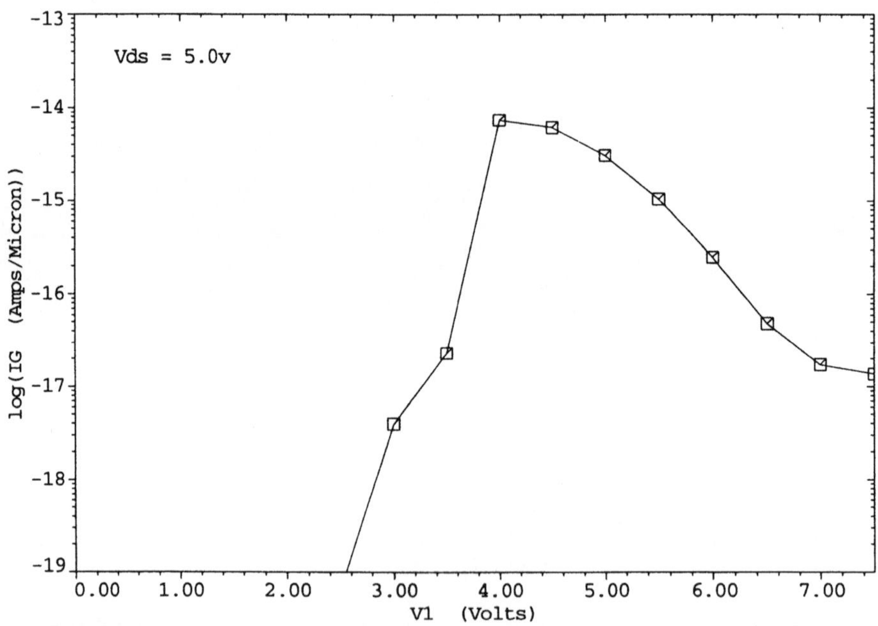

Figure 2.20 : The gate current versus gate voltage characteristics simulated for a drain voltage of 5.0 V (Source: [10]. Reprinted with permission).

Other physical insights can also be provided by MEDICI simulation. Figs. 2.21(a) and (b) show the two-dimensional electron current contours and the electron current vectors, respectively, in the 0.75-μm MOSFET. The information give both the magnitude, distribution, and direction of the electron current in the entire device structure. The three-dimensional electric field contours for the 0.75 μm MOSFET biased at $V_G = 0$ and $V_G = 5$ V are shown in Figs. 2.22(a) and (b), respectively. It can be seen that the main difference in the two cases is that in Fig. 2.23(b) a fairly uniform electric field caused by the nonzero gate voltage is present in the channel region, whereas in Fig. 2.22(a) the electric field near the drain is high due to the drain voltage (i.e., $V_D = 5$ V) but the electric field near the source is low due to the absence of gate voltage. The effect of the body voltage on the electric field is illustrated in Figs. 2.23(a) and (b) for the cases of $V_B = 0$ and $V_B = -3$ V, respectively. The body voltage changes slightly the electric field profile in the channel but increases considerably the electric field in the vicinity of the source junction.

The effects of fast interface states can be accounted for in MEDICI simulation as well. This is invoked using the **INTERFACE** statement and using the parameters **A.DONOR** and **N.ACCEPTOR**, which are the densities of fast electron-donor and electron-acceptor states, respectively. The donor states are positively charged above the electron quasi-neutral Fermi potential and are neutral below, whereas the acceptor states are neutral above the electron quasi-neutral potential and negatively charged below. Figs. 2.24(a), (b), and (c) show the MOSFET energy band diagrams plotted versus the vertical distance (i.e., from the surface to bulk) for the cases of zero interface states (zero charge), $5x10^{11}$ cm^{-2} donor states (positive states), and $5x10^{11}$ cm^{-2} acceptor states (negative states), respectively. It can be seen that the donor states increase the band bending, and the acceptor states decrease the band bending. The simulated drain current characteristics with and without the interface states are plotted in Fig. 2.25.

Transient and ac (i.e., small-signal) simulations can also be obtained. Figure 2.26 shows the drain current characteristics of two MOSFETs (1 and 0.75 μm) which are turned off with a gate bias changing from 3 to 0 V at t = 0. The snap shots of the electron concentrations in the 0.75-μm MOSFET taken at t = 0.1 psec and 2.0 psec are illustrated in Figs. 2.27(a) and (b), respectively. The MOSFET transconductance and output conductance, which are two figures of merit for ac analysis, are shown in Figs. 2.28(a) and (b), respectively.

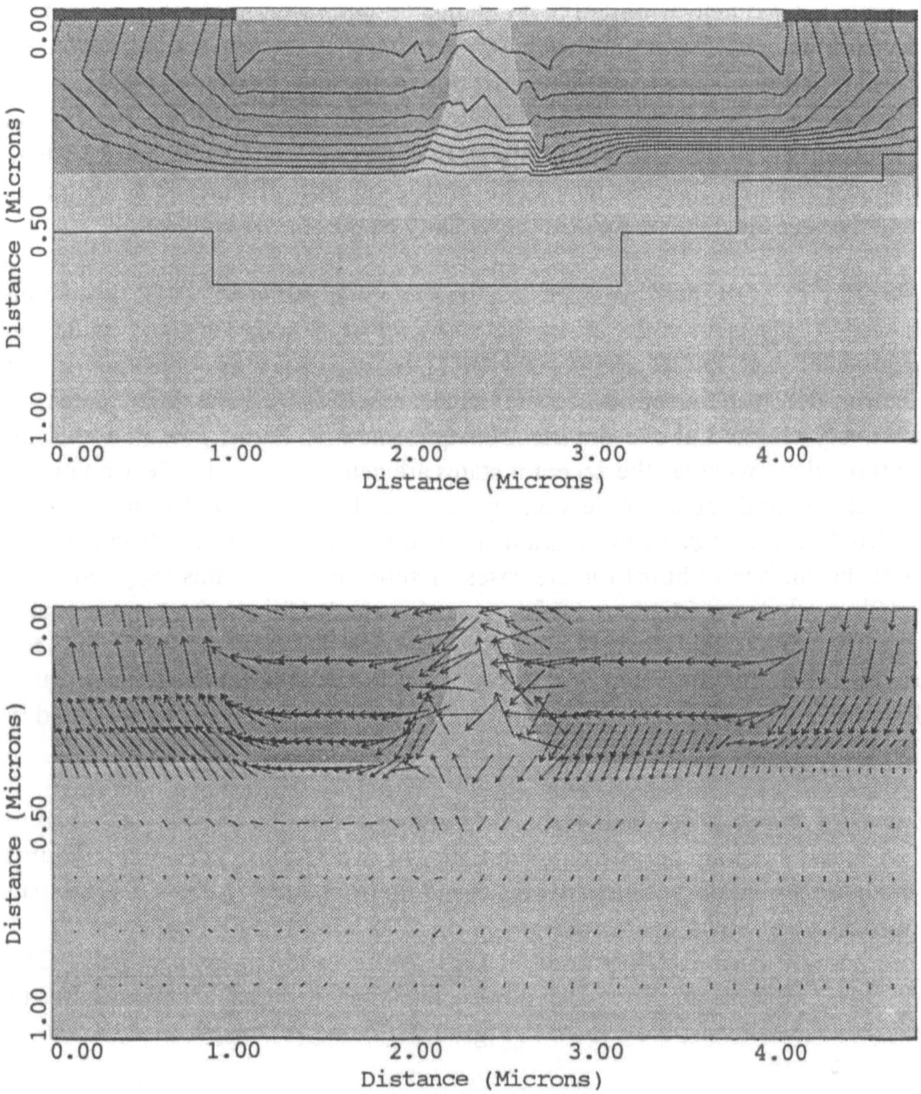

Figure 2.21 : Two-dimensional electron current (a) contours and (b) vectors simulated for a 0.75-μm MOSFET biased at $V_D = 5$ V and $V_G = 3$ V.

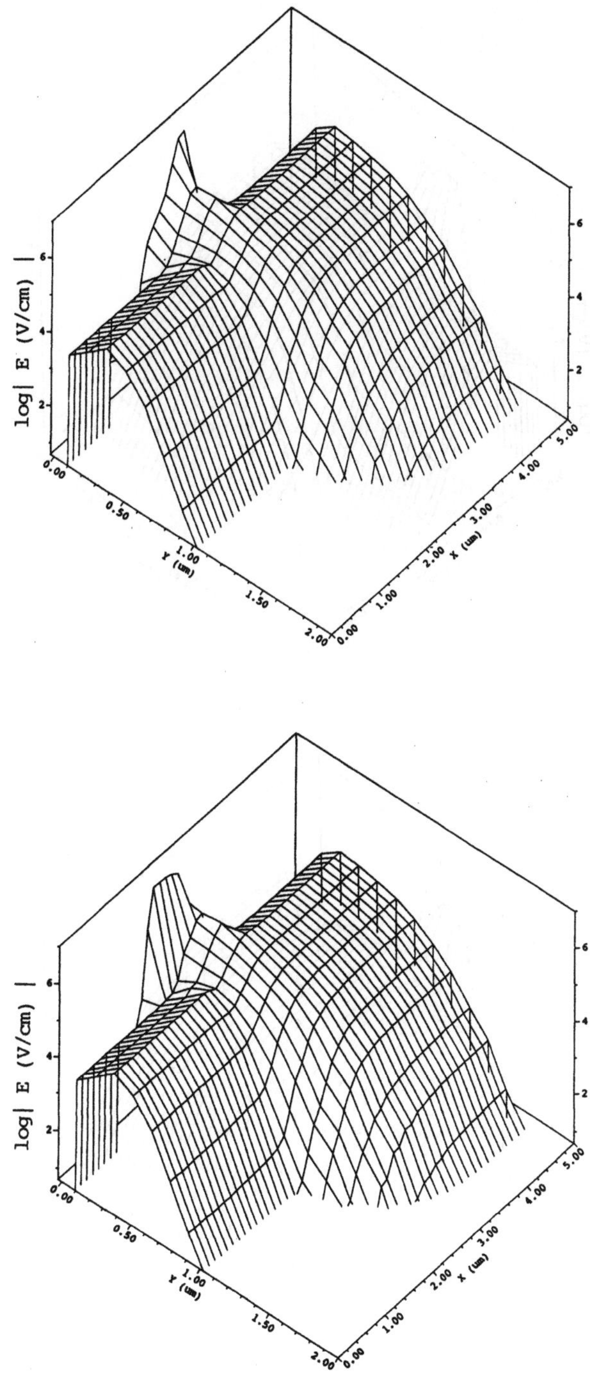

Figure 2.22 : Three-dimensional electric field contour in the 0.75 μm MOSFET biased at $V_D = 5$ V and (a) $V_G = 0$ and (b) $V_G = 5$ V.

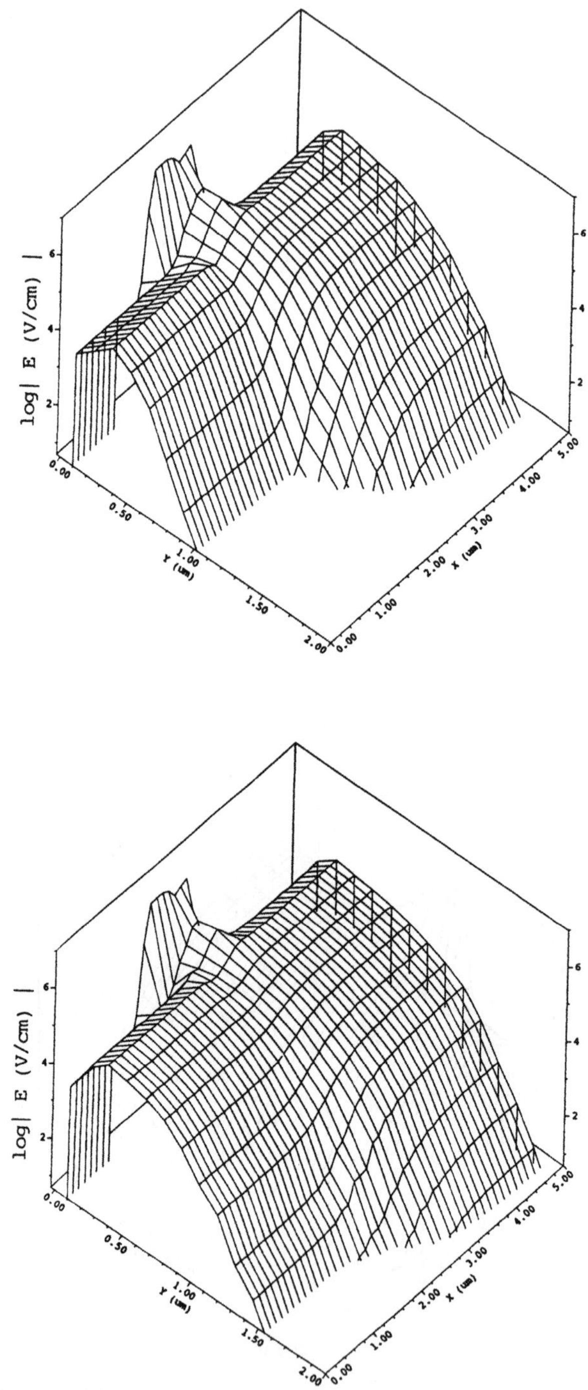

Figure 2.23 :Three-dimensional electric field contour in the 0.75 μm MOSFET biased at $V_D = 5$ V, $V_G = 3$ V, and (a) $V_B = 0$ and (b) $V_B = -3$ V.

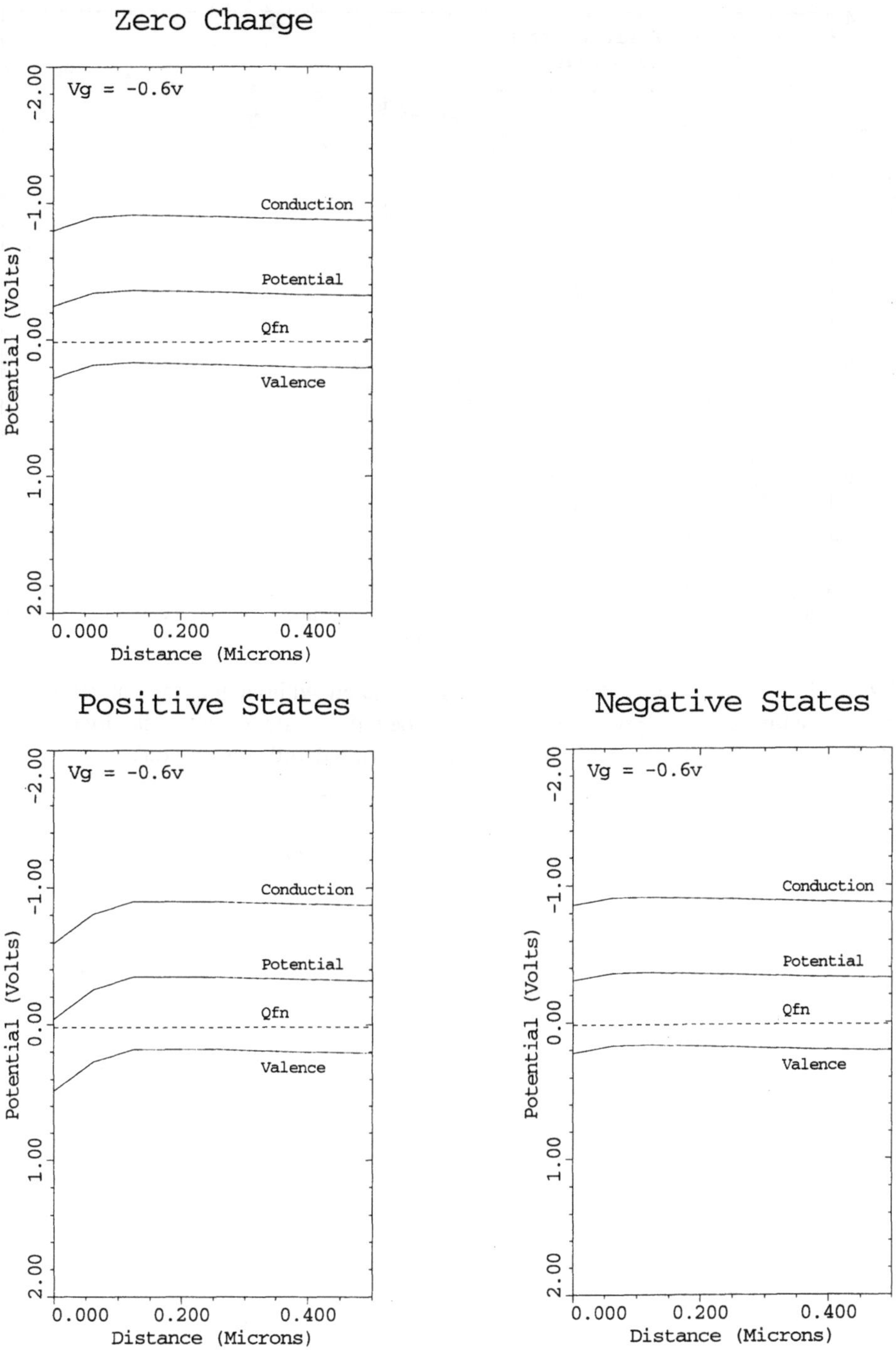

Figure 2.24 : Energy band diagrams versus the vertical direction simulated for (a) zero interface states; (b) donor-type interface states; and (c) acceptor-type interface states (Source: [10]. Reprinted with permission).

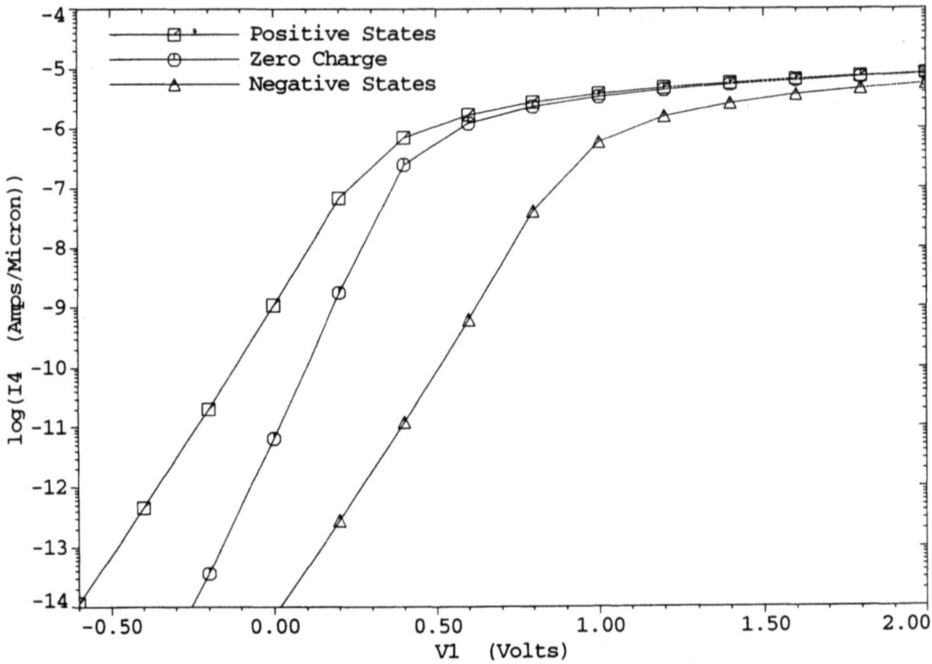

Figure 2.25 : Drain current characteristics simulated without interface states (zero charge), with acceptor-type interface states (negative states), and with donor-type interface states (positive states) (Source: [10]. Reprinted with permission).

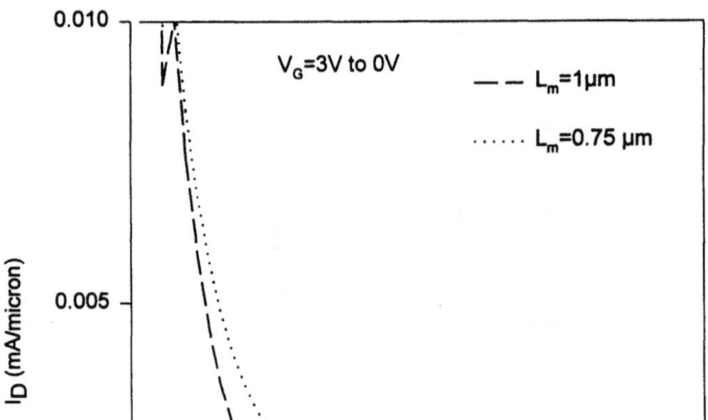

Figure 2.26 : The drain current turn-off transient characteristics simulated for MOSFETs with two different channel lengths of 1 and 0.75 μm.

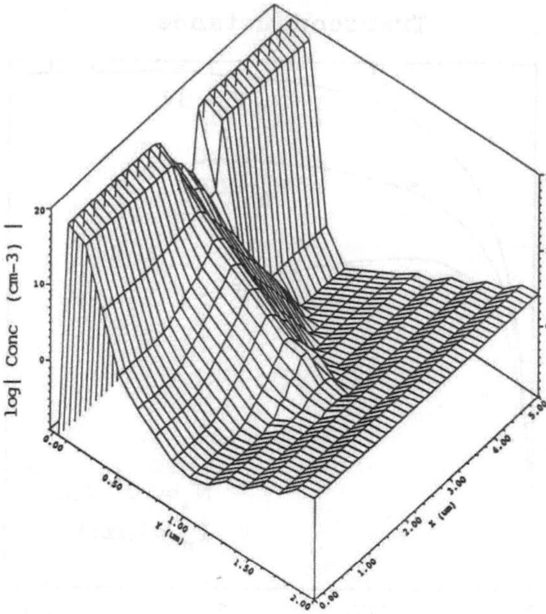

t = 0.1 psec

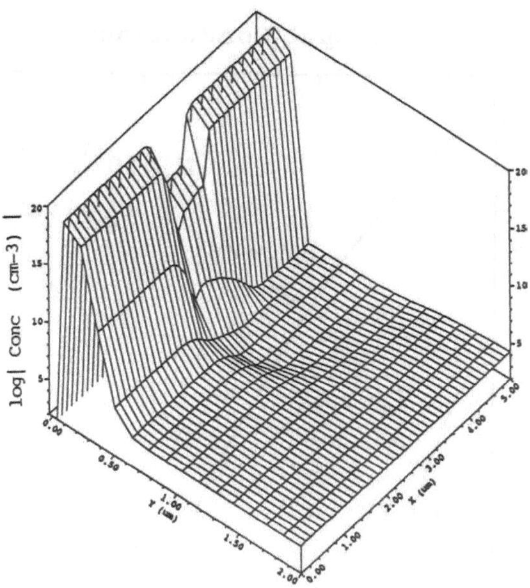

t = 2 psec

Figure 2.27 : Three-dimensional electron concentration contours in the MOSFET simulated at (a) t = 0.1 psec, and (b) t = 2 psec.

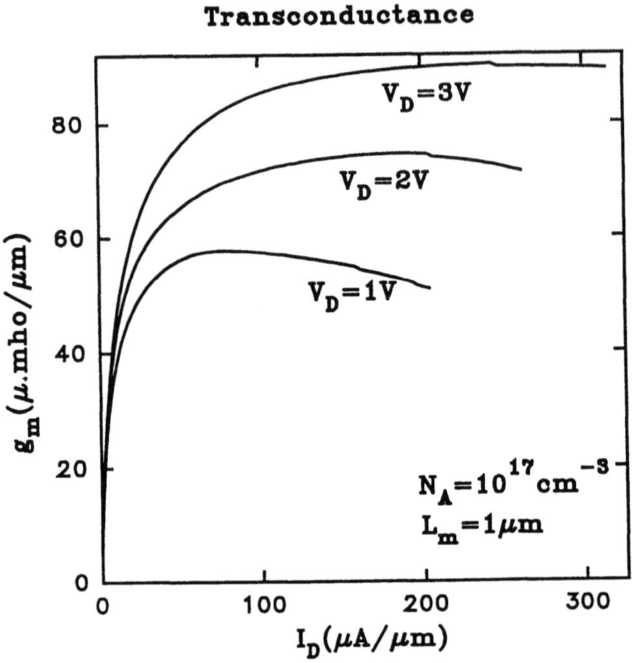

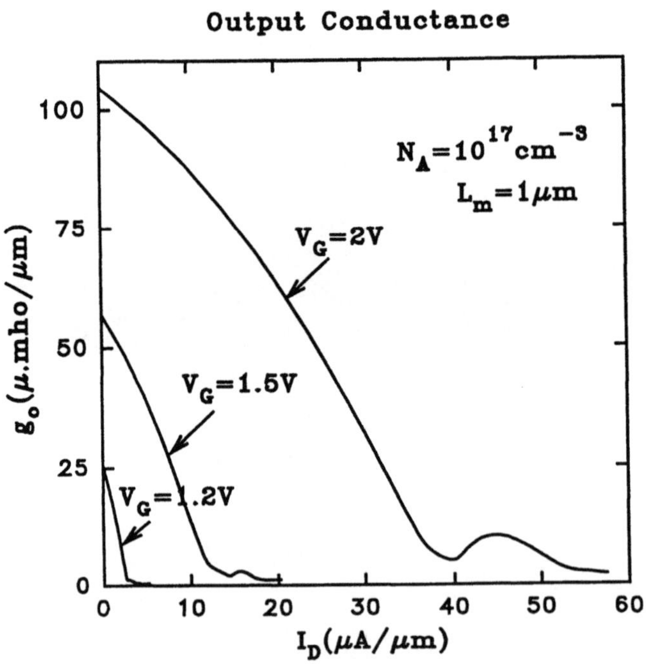

Figure 2.28 : (a) Transconductance and (b) output conductance versus drain current characteristics.

As mentioned in Sec. 2.1, MEDICI has the capability of including the lattice heating effect. Figure 2.29 shows the lattice temperature of a MOSFET biased at a gate voltage of 10 V and drain voltage of 6 V. Clearly, the highest lattice temperature occurs in the channel region where a large current is present, and the lattice temperature decreases toward the ambient temperature (i.e., 300 K) in the substrate. The effect of lattice heating on the I-V curve is shown in Fig. 2.30. It is evident that such an effect decreases the drain current, particularly in the operation region where the drain voltage is high.

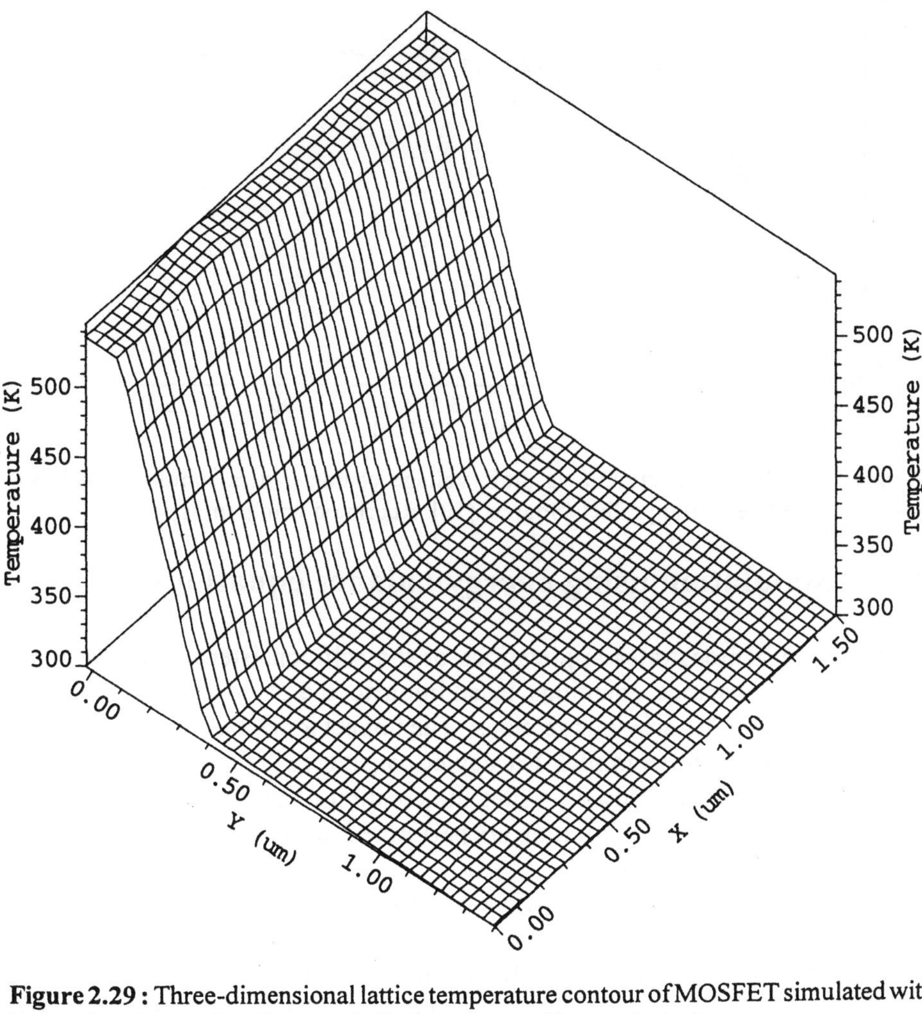

Figure 2.29 : Three-dimensional lattice temperature contour of MOSFET simulated with lattice- heating option (Source: [10]. Reprinted with permission).

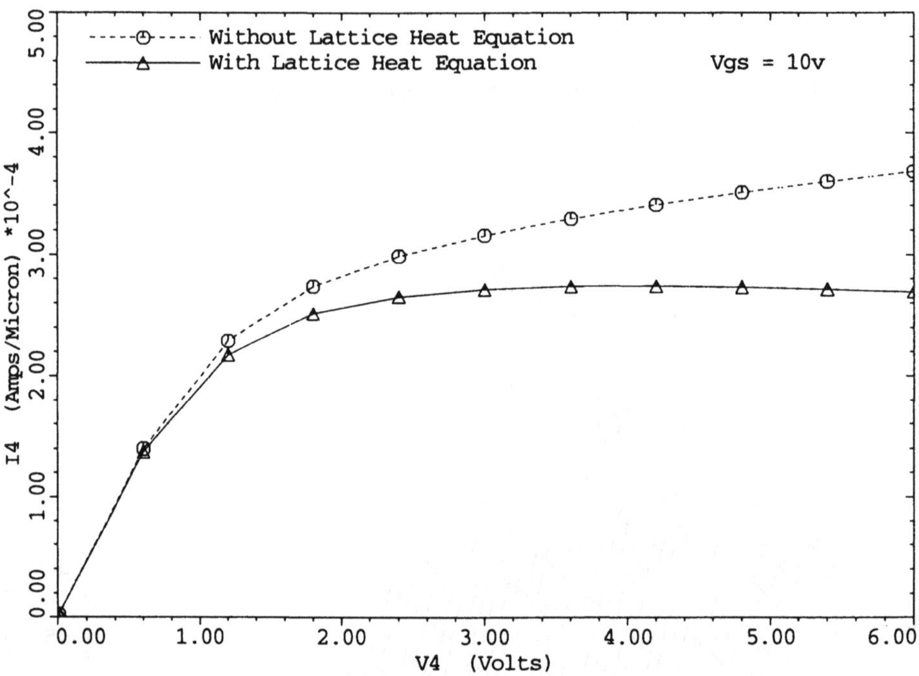

Figure 2.30 :The drain current vs drain voltage characteristics simulated with and without the lattice-heating effect (Source: [10]. Reprinted with permission).

2.5.5 Simulation using energy balance equations

This section demonstrates applications of the self-consistent solution of the carrier energy balance equations in addition to the standard drift-diffusion equations used in the previous sections. The impact ionization mechanism will also be included in the example. The device considered here is a 0.5 μm lightly-doped-drain (LDD) MOSFET. After the creation of the mesh structure (see section 2.4.1), three sets of solution are performed in a loop, as specified in Fig. 2.31 and elaborated below:

1. A conventional drift-diffusion solution. Here the energy balance equations are not solved and the impact ionization rate is calculated based on the local electric field. Since parameter IMPACT is specified on the **SOLVE** statement but not on the **MODEL** statement, it will be calculated as a post-processing analysis.

2. An approximate energy balance solution. In this mode, the energy balance equations are solved to determine the mean carrier temperature, but the carrier temperature is never fed back to the drift-diffusion equations (i.e., the energy balance equations are only solved once per bias point). The coupling of energy balance equations to the drift-diffusion equations is disabled by turning off the temperature dependent mobility model **TMPMOB** and the temperature enhanced diffusion model **TMPDIF**. However, since **II.TEMP** is specified, the calculations will be based on carrier temperature instead of electric field.
3. The full energy balance model. In this case, the electron temperature is fed back into the continuity equations through the **TMPMOB** and **TMPDIF** models. Impact ionization is again computed as a post processing step.

```
COMMENT   Apply drain bias (2.0v)
SYMBOL    CARR=0
METHOD    ICCG  DAMPED
SOLVE     V1=2.0  OUT.FILE=TEMPSOL

COMMENT   Perform three sets of solutions:
+         1. Drift-diffusion soln.   : mobility(E), impact ioniz(E)
+         2. Approx energy bal. soln: mobility(E), impact ioniz(T),
+                                      no temp. enhanced diffusion
+         3. Full energy bal. soln:    mobility(T), impact ioniz(T)

LOOP   STEPS=3
  ASSIGN    NAME=ELETEMP  L.VALUE=(F,T,T)
  ASSIGN    NAME=FULLEB   L.VALUE=(F,F,T)
  ASSIGN    NAME=LOGFIL  C1="MDE8AID"  C2="MDE8AIE"  C3="MDE8AIF"
  ASSIGN    NAME=SOLFIL  C1="MDE8ASD"  C2="MDE8ASE"  C3="MDE8ASF"

  LOAD      IN.FILE=TEMPSOL

  IF COND=@ELETEMP

    IF COND=@FULLEB
      MODELS     TMPDIF  TMPMOB   II.TEMP
    ELSE
      MODELS     ^TMPDIF ^TMPMOB  II.TEMP
    IF.END

    SYMBOL    CARR=1  NEWTON  ELECTRON  ELE.TEMP
    METHOD    ETX.TOL=0.10
  ELSE
    SYMBOL    CARR=1  NEWTON  ELECTRON
  IF.END

  COMMENT   Sweep gate bias from 0v to 6v.
  LOG       IVFILE=@LOGFIL
  SOLVE     V2=0.00  IMPACT.I
  SOLVE     V2=0.25  ELEC=2  VSTEP=0.25  NSTEP=4  IMPACT.I
  SOLVE     V2=1.50  ELEC=2  VSTEP=0.50  NSTEP=8  IMPACT.I
  SOLVE     V2=6.00  IMPACT.I  OUT.FILE=@SOLFIL
L.END
```

Figure 2.31 : MEDICI output file including the options of energy balance equations and impact ionization (Source: [10]. Reprinted with permission).

The resulting two-dimensional contours for the potential and electric field, impact ionization rate, and electron temperature for the MOSFET are plotted in Figs. 2.32(a), (b), and (c), respectively. The highest electron temperature (i.e., exceeds 1000 K) and impact ionization rate occur near the drain junction due to the high electric field in the region. Figure 2.33 compares the substrate currents simulated using the drift-diffusion (DD) model, the approximated energy balance equations (approx-EB), and the full energy balance equations (full-EB). It can be seen that the discrepancy between the DD model and full-EB model increases with increasing gate voltage.

More detailed descriptions for the physical mechanisms implemented, input/output statements, and simulation examples can be found in MEDICI manual [10].

2.6 Three-dimensional device simulation

Here we briefly discuss the capability of the three-dimensional device simulator DAVINCI [11]. DAVINCI is a powerful program that can be used in simulating three-dimensional distribution of potential and carrier concentrations in a semiconductor device and in predicting its electrical characteristics for any bias condition. The physical models and numerical algorithm implemented in DAVINCI are very similar to those in MEDICI, only that the three-dimensional grid structures are used and three-dimensional Poisson equation, continuity equations, and current equations are solved.

Due to the nature of three-dimensional structure, the number of nodes and the memory required for DAVINCI will be more than those for MEDICI. The maximum number nodes available in DAVINCI is dependent on how the executable is configured. Current DAVINCI version [11] allows maximum node counts of between 10,000 and 60,000. The amount of virtual memory required by DAVINCI is linearly dependent on N^{max}_p, the maximum number nodes available. The following equation gives a good approximation for the virtual memory requirement:

$$M_v = 3.7N_p^{max} + 2.4 \tag{2.42}$$

where M_v is the virtual memory required in megabytes and N^{max}_p is the maximum number of nodes in thousands. On the other hand, the amount of physical memory, or RAM, needed by DAVINCI is linearly dependent on N_p, the number of nodes used in a given simulation:

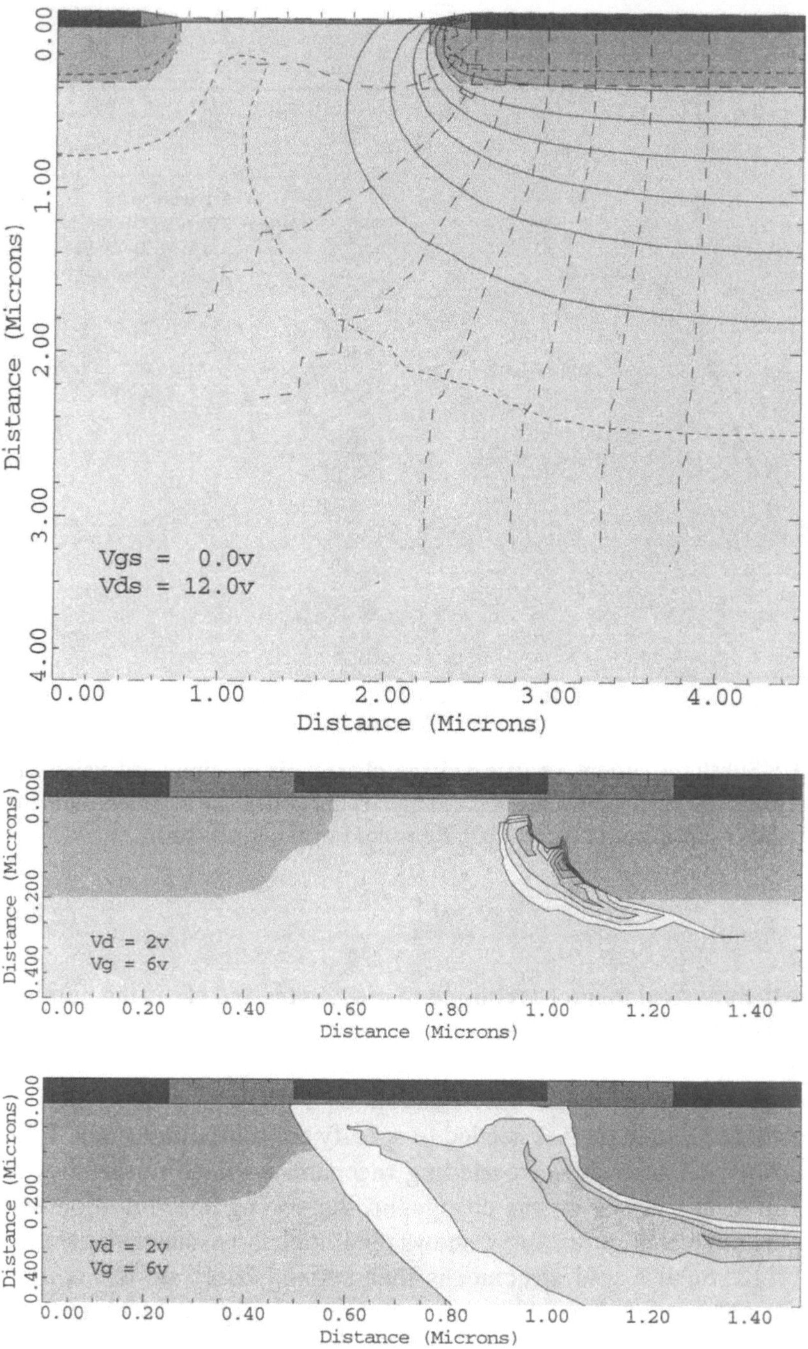

Figure 2.32 : Two-dimensional contours of (a) potential (solid lines) and electric field (dashed lines), (b) impact ionization rate, and (c) electron temperature near the drain junction of the MOSFET simulated using the output file in Fig. 2.30 (Source: [10]. Reprinted with permission).

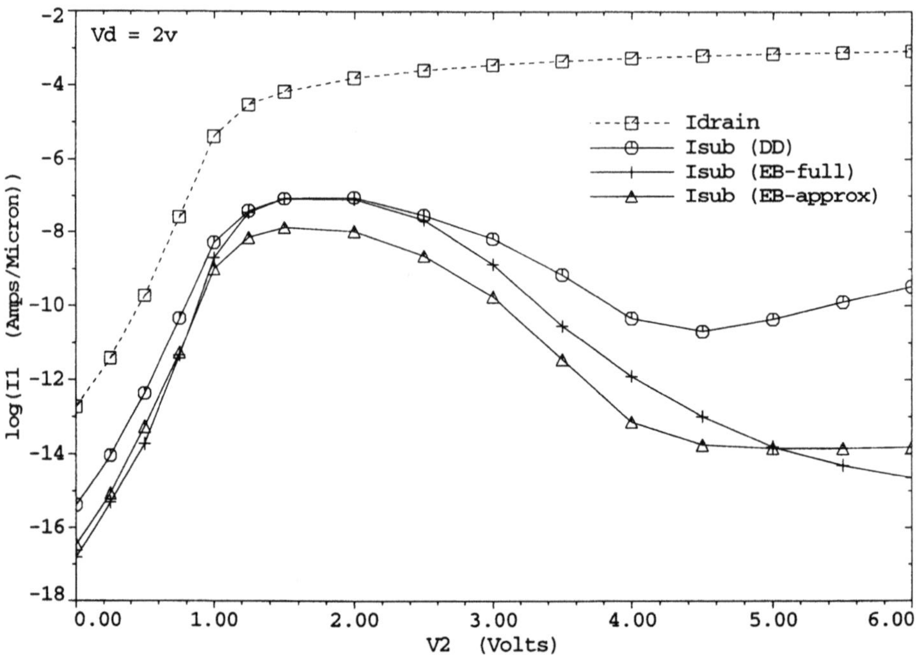

Figure 2.33 : Substrate current vs gate voltage characteristics simulated using the conventional diffusion-drift model, the approximated energy balance equations, and the full energy balance equations (Source: [10]. Reprinted with permission).

$$M_p = 0.7\, N_p\, N_{eq} \tag{2.43}$$

where M_p is the physical memory required in megabytes and N_{eq} is the number of equations being solved.

DAVINCI has the same mesh specification as MEDICI, except that an additional **Z.MESH** statement is needed to specify the third dimension. Like MEDICI, DAVINCI provides a regridding mechanism which automatically refines an initial grid based on the changes in the doping concentration and potential in the device. Figure 2.34(a) shows the initial, three-dimensional grid of a MOSFET. Such a grid structure is then refined based on the doping concentration and potential requirements, as shown in Figs. 2.34(b) and (c), respectively. The three-dimensional doping concentration contours and potential contours, among other physical insight, can be simulated, which are illustrated in Figs. 2.35(a) and (b), respectively.

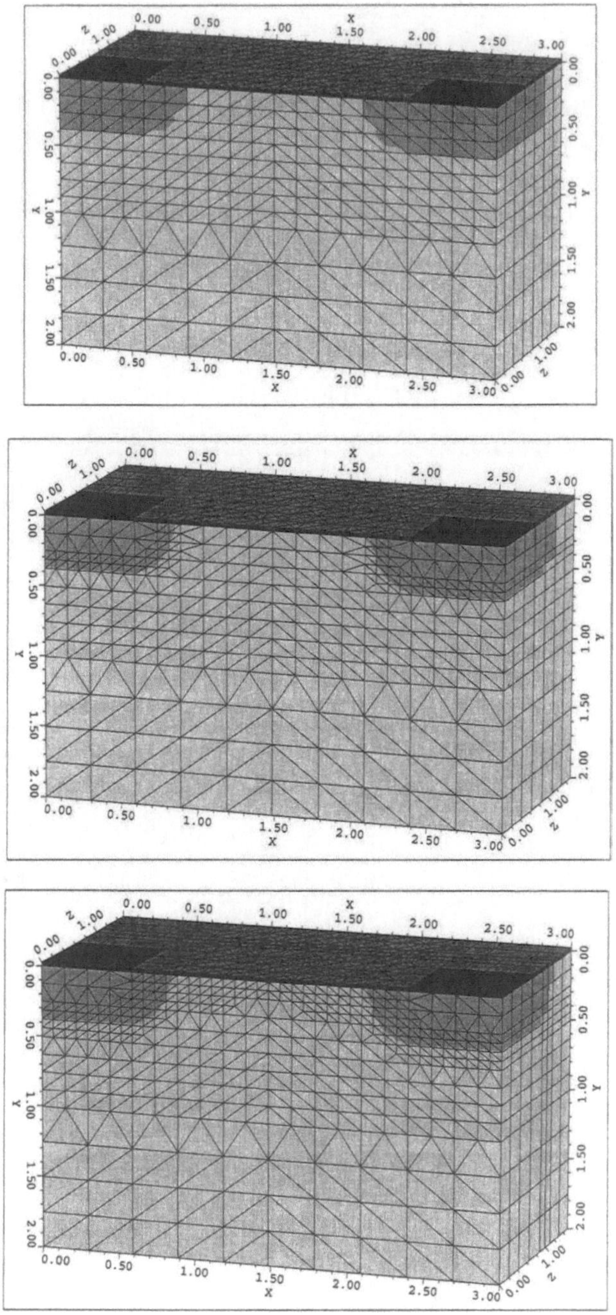

Figure 2.34 : Schematic of three-dimensional MOSFET (a) initial grid structure, (b) refined grid based on doping concentration, and © refined grid based on potential generated by the three-dimensional device simulator DAVINCI (Source: [11]. Reprinted with permission).

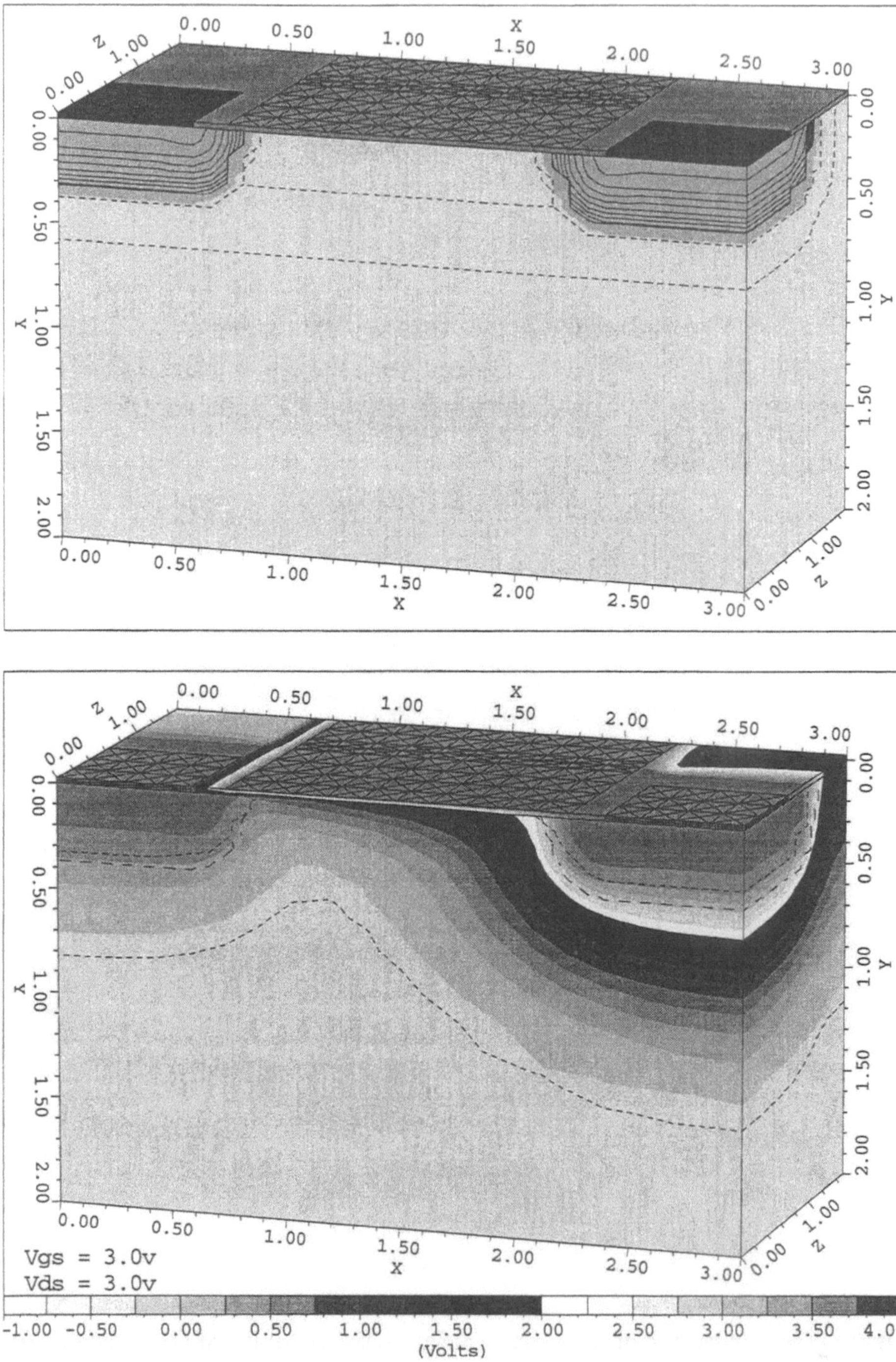

Figure 2.35 : Three-dimensional MOSFET (a) doping concentration contours and (b) potential contours generated by the three-dimensional device simulator DAVINCI (Source: [11]. Reprinted with permission).

The usefulness and applications of device simulation to MOSFET parameter extraction will be demonstrated in Chapters 3, 4, 5, and 6.

REFERENCES

[1] M. Sheu, *Physics of Semiconductor Devices*, Englewood Cliffs, NJ: Prentice Hall, 1990.

[2] J. J. Liou, *Advanced Semiconductor Device Physics and Modeling*, Boston: Artech House, 1994.

[3] D. P. Kennedy and R. R. O'Brien, "Two-dimensional mathematical analysis of a planar type junction field-effect transistor," IBM J. Res. Dev., vol. 13, p. 662, 1969.

[4] D. Vandrope and N. H. Xuong, "Mathematical two-dimensional model of semiconductor devices," Electron. Lett., vol. 7, p. 47, 1971.

[5] M. S. Mock, "A time-dependent numerical model of the insulated-gate field-effect transistor," Solid-St. Electron., vol. 24, p. 959, 1973.

[6] SEDAN-2: One-Dimensional Device Analysis Program, User's Manual, Technology Modeling Associates, Inc., 1984.

[7] BIPOLE3: Bipolar Semiconductor Device Simulation, User's Manual, Technology Modeling Associates, Inc., 1993.

[8] A. F. Franz, G. A. Granz, W. Kausel, G. Nanx, P. Dickinger, BAMBI 2.1 User's Guide, Institute for Microelectronics, Technical University Vienna, 1989.

[9] C. Fischer, P. Habas, O. Heinreichsberger, H. Kosina, P. Lindorfer, P. Pichler, H. Potzl, C. Sala, A. Schutz, S. Selberherr, M. Stiftinger, and M. Thurner, MINIMOS 6.0 User's Guide, Institute for Microelectronics, Technical University Vienna, 1994.

[10] MEDICI: Two-Dimensional Semiconductor Device Simulation, User's Manual, Technology Modeling Associates, Inc., 1993.

[11] DAVINCI: Three-Dimensional Semiconductor Device Simulation, User's Manual, Technology Modeling Associates, Inc., 1993.

[12] ATLAS: Device Simulation Software, User's Manual, Silvaco International Inc., 1995.

[13] MICROTEC: Software Package for Two-Dimensional Process and Device Simulation, User's Manual, Siborg Systems, Inc., 1996.

[14] N. D. Arora, J. R. Hauser, and D. J. Roulston, "Electron and hole mobilities in silicon as a function of concentration and temperature," IEEE Trans. Electron Devices, vol. ED-29, p. 292, 1982.

[15] D. M. Caughey and R. E. Thomas, "Carrier mobilities in silicon empirically related to doping and field," Proc. IEEE, vol. 55, p. 2192, 1967.

[16] D. Chen, E. C. Kan, U. Ravaioli, C. -W. Shu, and R. W. Dutton, "An improved energy transport model including nonparabolocity and non-Maxwellian distribution effects," IEEE Electron Device Lett., vol. EDL-13, 1992.

[17] B. Meinerzhagen and W. L. Engl, "The influence of the thermal equilibrium approximation on the accuracy of classical two-dimensional numerical modeling

of silicon submicron MOS transistors," IEEE Trans. Electron Devices, vol. 35, p. 689, 1988.

[18] C. W. Gear, *Numerical Initial Value Problems in Ordinary Differential Equations*, Englewood Cliffs, NJ: Prentice Hall, 1971.

[19] S. E. Laux, "Techniques for small-signal analysis of semiconductor devices," IEEE Trans. Electron Devices, vol. ED-32, p. 2028, 1985.

[20] R. S. Varga, *Matrix Iterative Analysis*, Englewood Cliffs, NJ: Prentice Hall, 1962.

Chapter 3

Extraction of the threshold voltage of MOSFETs

The threshold voltage V_T is an important parameter for MOSFET modeling, simulation and characterization [1-2], as can be seen clearly from the MOSFET models developed and discussed in Chapter 1. Such a voltage is conventionally defined as the gate voltage that causes the onset of strong inversion in the channel of MOSFETs. Figure 3.1 shows a qualitative plot of the inversion layer charge Q'_I vs. the applied gate voltage V_g, which includes the characteristics of depletion, weak inversion, moderate, and strong inversion regions. The point where the Q'_I plot becomes a straight-line is the onset of strong inversion. As seen in Fig. 3.1, the transition of the exact point toward the straight-line region is very gradual, and no clear point can be identified that could conventionally be taken as the onset of strong inversion. Thus a possible definition of the onset point, which yields V_T, can be the V_g value for which Q'_I is within an acceptable value not much larger than zero. The voltage V_{M0} shown in Fig. 3.1 fits into this definition. Alternatively, V_T can be determined from the extrapolation of the slope of the linear region, and V_{T0} shown in Fig. 3.1 would be the appropriate threshold voltage. The last possible definition of V_T is taken at the point where the curve becomes straight-line, which yields $V_T = V_{H0}$, as indicated in Fig. 3.1. It turns out that V_{H0} is about 0.6 V above V_{M0} at room temperature and for practical fabrication processes [3]. Clearly, the first two definitions have defined the voltage for onset of moderate inversion, whereas the last one has defined the onset for strong inversion. In fact, often no distinction is made between V_{M0}, V_{T0}, and V_{H0} in the literature, and all three are taken to be one and the same point called threshold.

This chapter will first review the existing methods for extracting V_T, and their advantages and disadvantages discussed. It turns out that the majority of the V_T extraction methods is influenced strongly by the presence of the source and drain series resistances of the MOSFET. This is highly undesirable because the value of V_T extracted should be independent of the parasitic components. To evade such a difficulty for extracting V_T, a new and improved extraction method will then be developed in the chapter, and the results of this method will be compared to the existing methods. The next topic covered is a unique phenomenon called the reverse short-channel effect (RSCE) (i.e., V_T increases with decreasing channel length), which contradicts the classical theory that V_T decreases monotonically with deceasing channel length. The RSCE can be found in some short-channel MOSFETs having a high degree of nonuniform doping contribution in the channel due to certain processing conditions. The last part of the chapter presents a method to determine the threshold voltage shift due to quantum mechanical effects associated with the decreased gate oxide thickness and increased substrate doping density in modern MOSFETs. Whenever applicable, device simulation will be used to aid parameter extraction process and give physical insight of MOSFET operation.

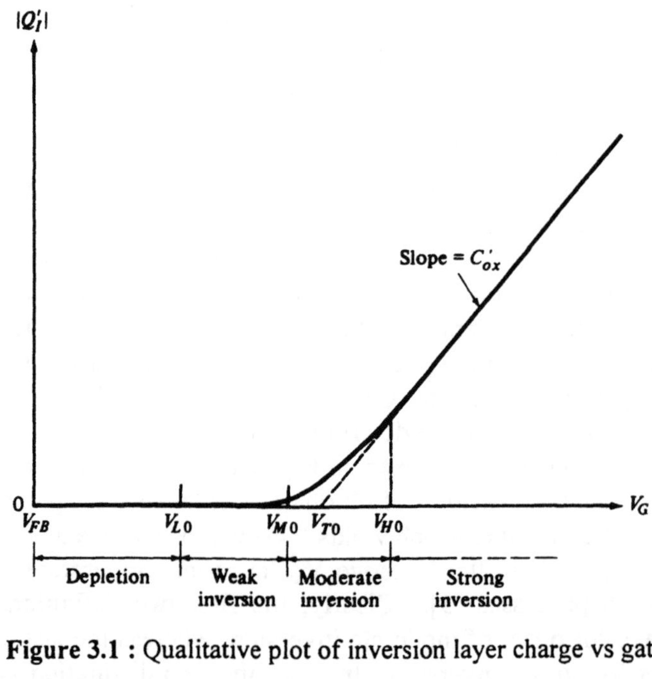

Figure 3.1 : Qualitative plot of inversion layer charge vs gate voltage.

3.1 Existing methods for extracting the threshold voltage

Several methods [1,4-6] have been developed to extract the threshold voltage V_T of a MOSFET biased in the linear region. They include 1) the constant-current method in which V_T is determined from the gate voltage at a given constant drain current [1]; 2) the linear-extrapolation method which determines V_T by extrapolating the point of maximum slope on the current-voltage characteristics to the gate-voltage axis [1]; 3) the second-derivative method in which V_T is determined by taking the second derivative of the drain current I_D with respect to gate voltage V_g (i.e., d^2I_D/dV_g^2) [4]; 4) the ratio method which calculates the ratio of the conductance $g = dI_D/dV_D$ to the transconductance $g_m = dI_D/dV_g$ and extrapolates such a ratio linearly to the gate bias axis [5]; and 5) the quasi-constant-current method which determines V_T from the subthreshold current behavior [6].

The constant-current method is widely used in industry because of its simplicity; the threshold voltage can be determined quickly with only one voltage measurement necessary, as shown in Fig. 3.2. However, it has the disadvantage of being strongly dependent of the arbitrary choice of the drain current level. This is evident by the results in Fig. 3.3, where different gate voltages at different drain currents can be taken to be the threshold voltages.

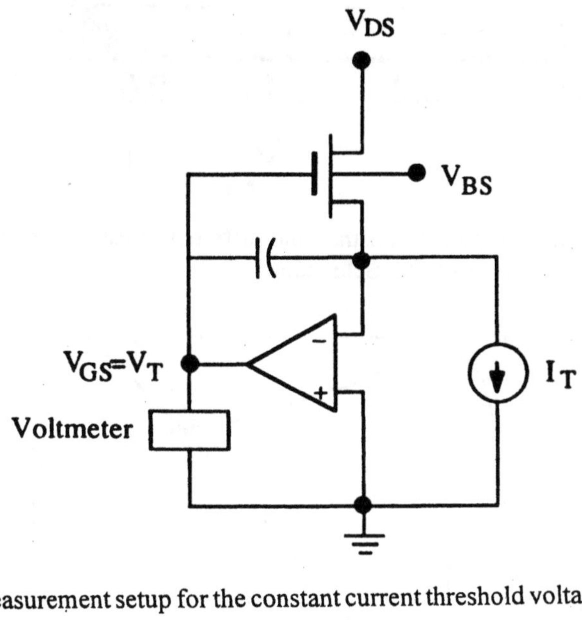

Figure 3.2 : Measurement setup for the constant current threshold voltage method (after Schroeder [1]).

The linear-extrapolation method is another popular threshold-voltage extraction method. Noting the fact that I_D vs V_g curve deviates from a straight line at gate voltage above V_T due to series resistance and mobility degradation effects, such a method first finds the point of maximum slope on I_D vs V_g curve (i.e., the point of maximum transconductance g_m), fits a straight line to the curve at that point, extrapolates to the gate-voltage axis (i.e., $I_D = 0$), and the intercept is the threshold voltage, as illustrated in Fig. 3.4. The main drawback is that the point of maximum slope is a function of the drain and source series resistances, and thus the threshold voltage value extracted from this method is influenced strongly by these resistances.

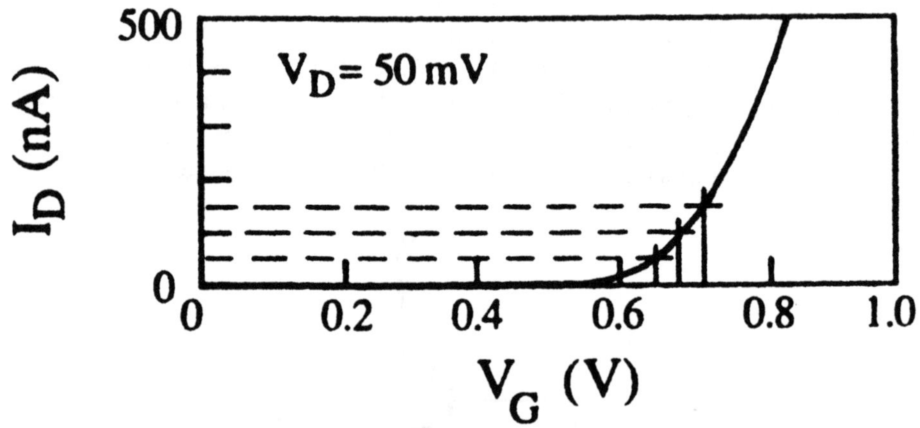

Figure 3.3 : Qualitative plot to illustrate the different threshold voltages determined from different drain currents (after Schroeder [1]).

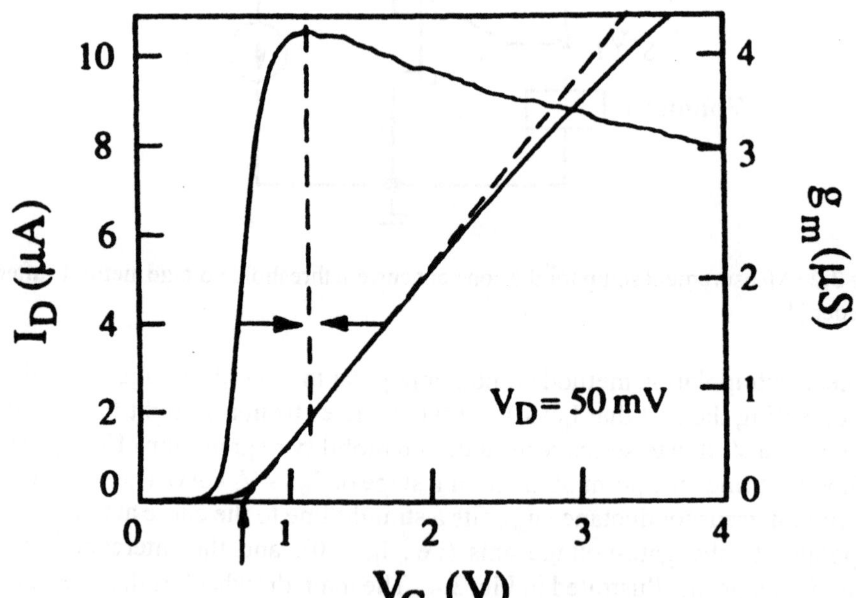

Figure 3.4 : Plots of drain current and transconductance to illustrate the linear extrapolation threshold voltage method (after Schroeder [1]).

The second-derivative method [4], developed to avoid the dependence on the series resistances, determines V_T from the gate voltage at which the derivative of the transconductance (i.e., $dg_m/dV_g = d^2I_D/dV_g^2$) is maximum, as denoted by V_T(TC) in Fig. 3.5. Also shown in the figure are the threshold voltages extracted from the linear extrapolation (i.e., V_T(LE)) and from the classical definition of surface potential equals twice the bulk potential (i.e., $V_T(2\phi_F)$). Such a method is highly sensitive to noise in the measurements, however, because the approach of second derivative is equivalent to using a high-pass filter in measurements.

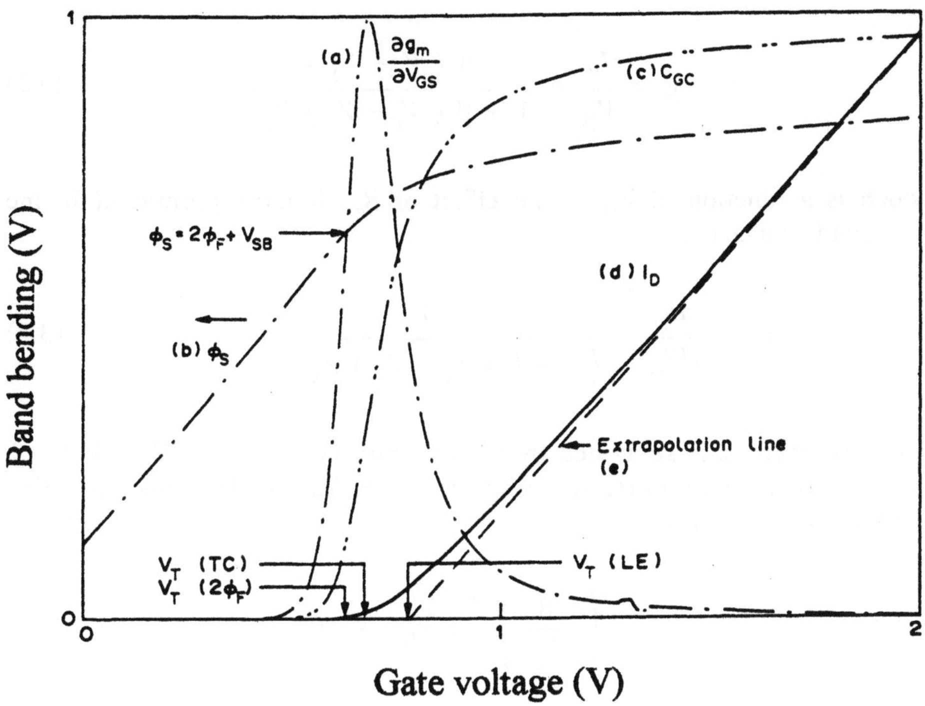

Figure 3.5 : Schematic illustration of determining the threshold voltage from the second-derivative method (i.e., V_T(TC)), from the line extrapolation method (i.e., V_T(LE)), and from the classical method of surface potential equals twice the bulk potential (i.e., $V_T(2\phi_F)$) (after Wong et al. [4]).

The ratio method [5], developed also to avoid the dependence on the series resistances, suggests that the ratio $g/g_m^{0.5}$ is a linear function of gate bias, whose intercept at the gate-voltage axis equals the threshold voltage. This can be demonstrated using the following simple theory. At a small drain voltage V_D (i.e., linear region), the drain current is related to the gate voltage V_G, the total series resistance R_{ext} (i.e., drain and source series resistances), and V_T as

$$I_D \equiv \beta\,(V_G - V_T)\,(V_D - I_D R_{ext}) \tag{3.1}$$

$$\beta \equiv \frac{A}{L_m - \Delta L} = \frac{A}{L_{eff}} \tag{3.2}$$

where A is a constant associated with the MOSFET structure, L_m is the mask channel length, L_{eff} is the effective channel length, and $\Delta L = L_m - L_{eff}$. The output conductance g is

$$g = \frac{I_D}{V_D} = \frac{\beta\,(V_G - V_T)}{1 + \beta\,(V_G - V_T)\,R_{ext}} \tag{3.3}$$

which is a function of R_{ext}. The effect of R_{ext} is more prominent in the transconductance g_m:

$$g_m \equiv \frac{dI_D}{dV_G} = \frac{\beta\,V_D}{\left(1 + \beta\,(V_G - V_T)\,R_{ext}\right)^2} \tag{3.4}$$

The ratio method is suggested by the common factor $[1 + \beta(V_G - V_T)R_{ext}]$, through which the series resistance appears in (3.3) and (3.4). Thus, according to (3.3) and (3.4), the ratio

$$\frac{g}{g_m^{0.5}} = \left(\frac{\beta}{V_D}\right)^{0.5}(V_G - V_T) \tag{3.5}$$

is independent of R_{ext} and is a linear function of V_G, whose intercept at the V_G-axis is equal to V_T. Figure 3.6 shows the measured $g/g_m^{0.5}$ versus the gate voltage for several n-channel MOSFETs. According to the method, the interception of $g/g_m^{0.5}$ lines of MOSFETs with different mask channel lengths (i.e., 0.75, 0.875, 1.25 μm shown in Fig. 3.6) on the gate-voltage axis yields the threshold voltage (i.e., about 0.81 V indicated in Fig. 3.6). The main drawback of such a method is the fact that the lines do not necessarily intercept at one point, due to the second-order effects not accounted for in the simple MOSFET model given in (3.1)-(3.2), as evidenced by the slight divergence of the intercepts of the different lines (i.e., for different channel lengths) at the gate-voltage axis shown in Fig. 3.6. The method also requires extra steps of finding the conductance and transconductance of the MOSFET.

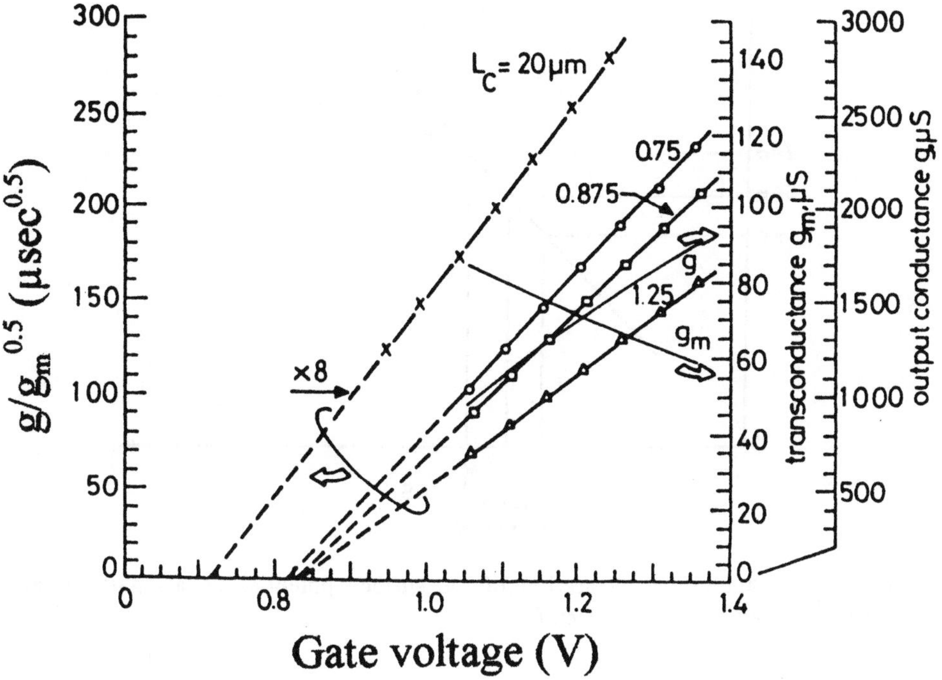

Figure 3.6 : Measured $g/g_m^{0.5}$ as a function of the gate voltage. The interception of the lines at the x-axis yields the threshold voltage (after Jain [5]).

The quasi-constant-current method [6] was derived based on the theory of drain current in the subthreshold region. It defines V_T as the gate voltage required for the surface band bending equals twice the bulk potential. In addition to its complexity in extracting V_T, such a method is valid only for the subthreshold region, where the electrical characteristics are not as well defined as the strong inversion region. Figure 3.7 shows the typical variation of the drain current as a function of gate voltage for MOSFETs operated in the subthreshold region. It is easy to see that there are two distinct features for subthreshold conduction of the MOSFET: 1) For gate biases up to the weak inversion region, the drain current is a pure diffusion current. The upper limit of this region is determined by the surface potential equals twice the bulk potential (i.e., $2\phi_F$); and 2) At higher gate biases in the strong inversion region, the drain current is predominantly a drift current, and the moderate inversion region exists between the weak and strong inversion regions. Based on this concept, an expression relating the threshold voltage and the measured subthreshold current can be derived [6]. Figure 3.8 shows V_T extracted from this method as a function of the channel length and different body-to-source voltage V_{BS}.

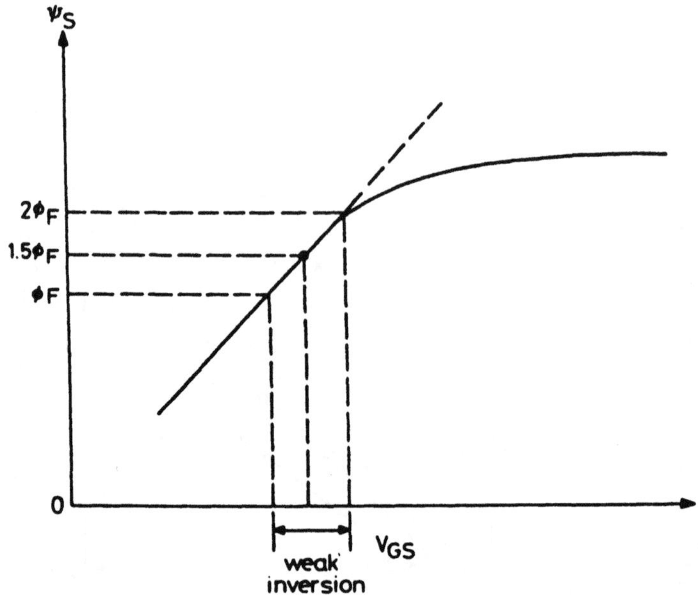

Figure 3.7 : Qualitative illustration of the surface potential versus gate voltage characteristics (after Yan and Deen [6]).

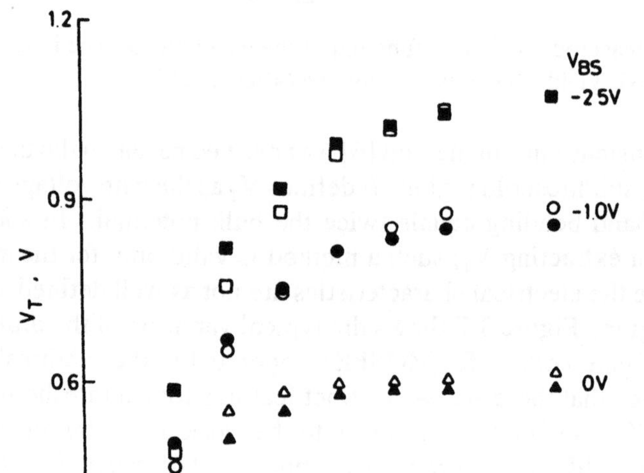

Figure 3.8 : Threshold voltage versus the effective channel length as a parameter of the body-to- source voltage extracted from the quasi-constant-current method (after Yan and Deen [6]).

3.2 Improved threshold voltage extraction method

An improved extraction method has been reported recently [7-9] to eliminate the effect of the series resistances on the extracted parameters in an arbitrary two-terminal device which contains linear (i.e., resistors) and non-linear elements. It was developed based on a generalized function called the integral function, which contains only integrations of current and voltage. The integration acts as a low-pass filter, thus reducing the possibility of measurement errors associated with the extraction procedure. Using the same concept, we present in this section a new method to extract the threshold voltage of the four-terminal MOSFET. The method will be tested in the environments of circuit simulator AIM-SPICE [10], device simulator MEDICI [11], and measurements. The results obtained from the new method will also be compared with those obtained from the existing methods.

3.2.1 Review of the integral function for two-terminal devices

For the purpose of extracting the model parameters of a two-terminal device (i.e., p/n junction diode), it would be convenient to use a function that, in addition to being easy to calculate from the device's experimentally measured I-V characteristics, does not depend on the parasitic series resistances. A possible such function can be defined by an integral function D [8-9]:

$$D(V, I) \equiv \int_0^I V\,dI - \int_0^V I\,dV \tag{3.6}$$

where the first and second terms on the right-hand side are the device's content and the co-content, respectively, and V and I are the device terminal voltage and current, respectively. The integral function D can also be considered as a measure of the device non-linearity, which would be zero for a device with only linear elements because the content and co-content are identical for such a device. The derivation and general theory of the integral function are discussed in more detail in Appendix C.

An arbitrary two-terminal device with both linear and nonlinear elements has the following properties [9]: (a) the summation of the contents over all the branches (i.e., all the parasitic resistances and other nonlinear components associated with the device's equivalent circuit) is zero; and (b) the summation of the co-contents over all the branches is zero. Therefore, the summation of the function D over all the branches is zero. In others words, the function D eliminates the effect of the linear elements, such as series resistances, and can be used for extracting intrinsic parameters of semiconductor devices. The proof of such properties for an arbitrary network with any number of generalized elements is given in Appendix C.

Let us now consider a simple case of p/n junction diode, which is frequently modeled by an ideal diode in series with a parasitic resistance R:

$$I = I_S \left(\exp\left(\frac{V - I R}{n V_{th}} \right) - 1 \right) \tag{3.7}$$

where I the current passing through the p/n junction, v is the voltage applied to the junction, I_S is the saturation current, n is the ideality factor, and $V_{th} = kT/q$ is the thermal voltage. It is well known that the presence of R can significantly obscure the linear characteristic of the ln(I) vs. v plot to such extent that the extraction of I_S and n from the plot becomes unreliable. The use of the integral function D can eliminate such a difficulty. Substituting (3.7) into (3.6) and performing the integrals yield the following equation not containing R [9]:

$$D \approx I n V_{th} \left(\ln\left(\frac{I}{I_S} \right) - 2 \right) \tag{3.8}$$

The values of I_S and n can be determined as follows. First the integral function D is calculated from (3.6) using the measured device's I-V data. Putting this into (3.8), and noting that D/I should produce a straight line whose slope and intercept allow the direct determination of n and I_S, respectively.

To illustrate the approach, we have applied the method to the source-body p/n junction of a silicon MOSFET at room temperature. The value of D/I were first calculated using (3.6) and experimental data, and the results are shown in Fig. 3.9. Although this function is not entirely linear, it does exhibit a wide range of linearity. The parameters n and I_S can then be calculated from D/I values, which are plotted in Fig. 3.10. Figure 3.11 presents the comparison between the I-V characteristics obtained from measurements and calculated from the model given in (3.7) and the extracted values of n and i_S shown in Fig. 3.10. Very good agreement is obtained.

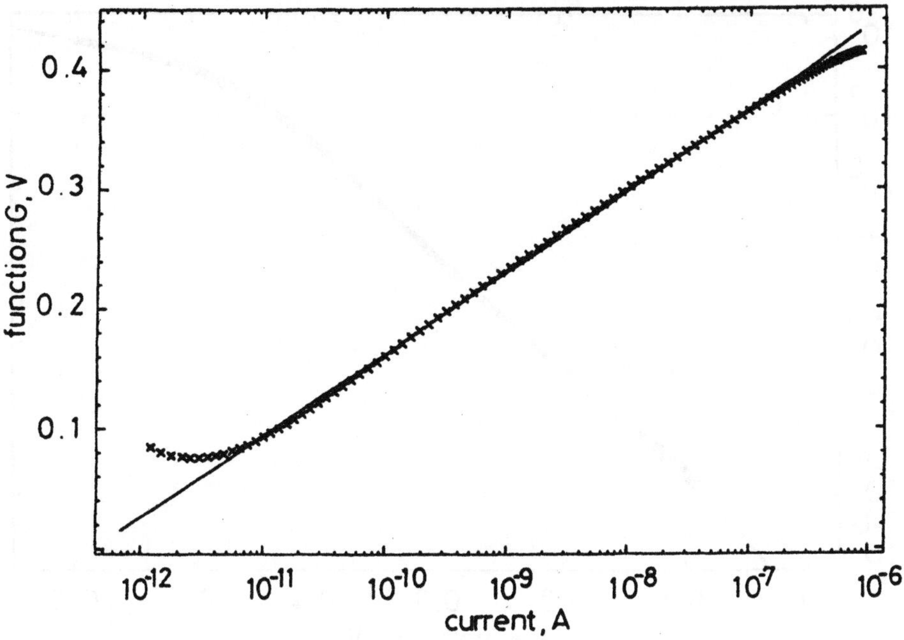

Figure 3.9 : Function G = D/I versus the logarithm of the current of a p/n junction (symbols) and its straight-line fit (line).

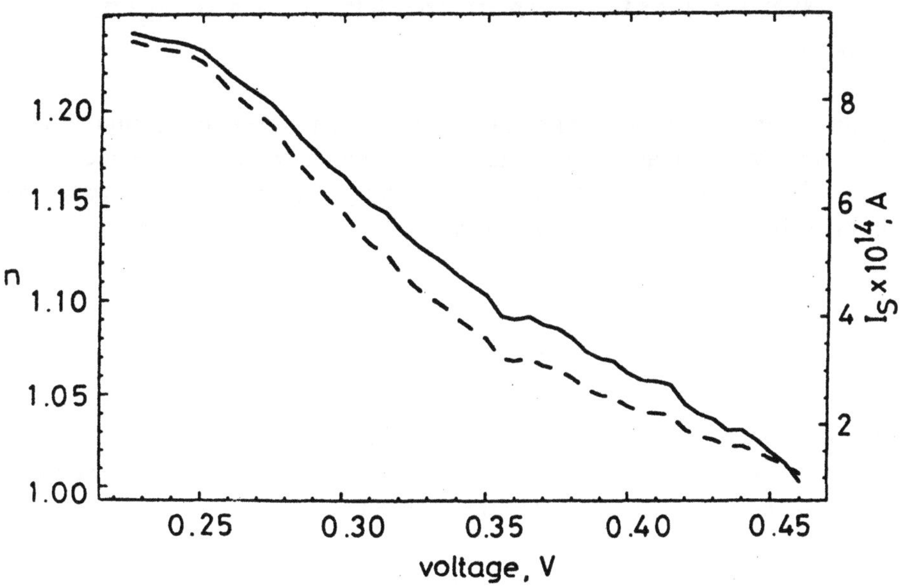

Figure 3.10 : Extracted n (solid line) and I_S (dashed line) versus the applied voltage.

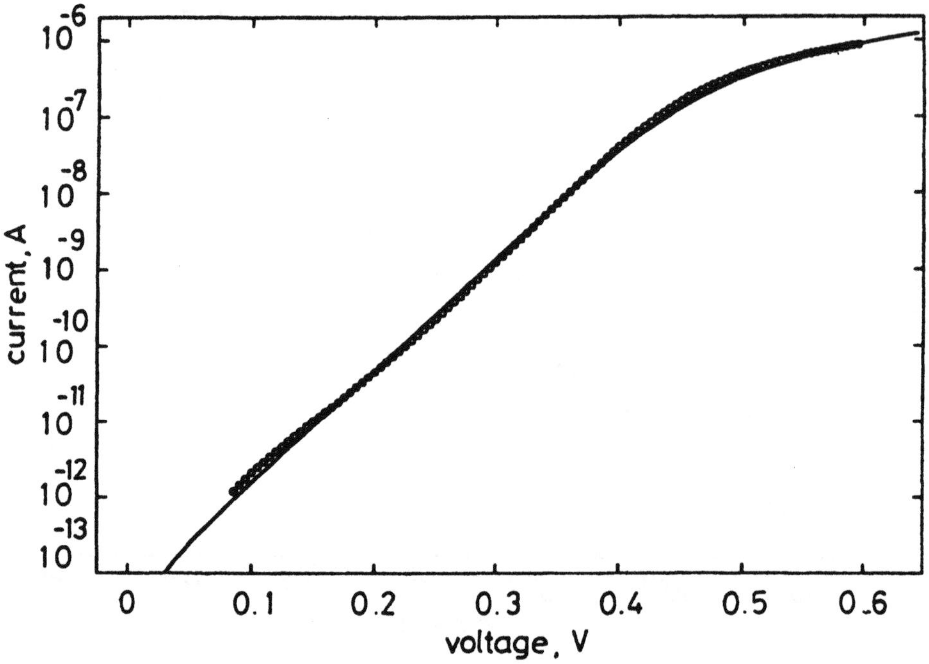

Figure 3.11 : I-V characteristics obtained from measurements (symbols) and from model (line) using the extracted values of I_S and n.

3.2.2 MOSFET extraction method based on integral function

The concept of the integral function discussed above for the two-terminal device parameter extraction is also applicable for three- and four-terminal devices. This section will develop a MOSFET threshold voltage extraction method based on the integral function. For the case of a MOSFET biased in the linear region, the drain current can be expressed as:

$$I_D = f(V_{GS} - V_T)\, V_{DS} \tag{3.9}$$

where f is a function defined by a particular MOSFET model, and V_{GS} and V_{DS} are the intrinsic gate-source and drain-source voltages, respectively. The two voltages can be related to the external voltages as

$$V_{GS} = V_g - I_D R_s \tag{3.10}$$

and

$$V_{DS} = V_d - I_D (R_s + R_d) \quad . \tag{3.11}$$

Here V_g and V_d are the extrinsic gate-source and drain-source voltages, respectively, and R_s and R_d are the source and drain series resistances, respectively, which are assumed independent of the gate voltage.

The total measured resistance R_m from drain to source is

$$R_m \equiv \frac{V_d}{I_D} = R_d + R_{ch} + R_s \tag{3.12}$$

where R_{ch} is the channel resistance, which can be obtained from (3.9):

$$R_{ch} \equiv \frac{V_{DS}}{I_D} = \frac{1}{f(V_{GS} - V_T)} \tag{3.13}$$

For the strong inversion, the extrinsic gate voltage is high (i.e., $V_{GS} >> V_T$). Since the extrinsic drain-source voltage is small in the linear region under study, we can approximate $(V_{GS} - V_T) \approx (V_g - V_T)$. On the other hand, for the weak inversion, $(V_{GS} < V_T)$, the drain current is very small, and the voltage drops in R_s and R_d are much smaller than V_{GS}. As a result, we can again approximate $(V_{GS} - V_T) \approx (V_g - V_T)$. Thus for either the strong or weak inversion, (3.13) can be simplified to

$$R_{ch} \approx \frac{1}{f(V_g - V_T)} \quad . \tag{3.14}$$

Next, we introduce a change of variable from V_g to V_{gb}, and the reason of using such a new variable will be given later. It is defined as

$$V_{gb} \equiv V_{max} - V_g \tag{3.15}$$

where V_{max} is a constant parameter, which physically represents the maximum gate voltage under consideration (V_{max} = 5 V in our case). In deriving the integral function D_{MOS} for the MOSFET, the variable I used in (3.6) will be changed to a new variable (R_mV_{gb}). This is because (3.6) was derived based on the property that the total series resistance R of a two-terminal device, such as a junction diode, depends linearly on I. For the MOSFET, however, R_d and R_s are assumed constant with respect to I_D, but they depend linearly on the variable (R_mV_{gb}). As a result, D_{MOS} can be expressed in terms of V_{gb} and (R_mV_{gb}) as

$$D_{MOS}(V_{gb}, R_mV_{gb}) = \int_0^{(R_mV_{gb})} V_{gb}\, d(R_mV_{gb}) - \int_0^{V_{gb}} (R_mV_{gb})\, dV_{gb} \tag{3.16}$$

Using the method of integration by parts [8-9], we have

$$D_{MOS}(V_{gb}, R_mV_{gb}) = R_mV_{gb}^2 - 2\int_0^{V_{gb}} (R_mV_{gb})\, dV_{gb} \quad . \tag{3.17}$$

It should be pointed out that R_m is a measurable parameter, and D_{MOS} is independent of R_s and R_d.

The effective electron mobility μ_{eff} is a function of V_g and can be expressed by [10]:

$$\mu_{eff} = \frac{\mu_o}{1 + \theta(V_g - V_T)} \tag{3.18}$$

where μ_0 is the electron mobility in the bulk and the parameter θ accounts for the mobility degradation at the oxide-semiconductor interface. Combining (3.9) and (3.18) yields

$$I_D = \frac{K(V_g - V_T)}{1 + \theta(V_g - V_T)} V_{DS} \tag{3.19}$$

where K is a parameter related to the MOSFET channel width, channel length, oxide capacitance, and free-carrier mobility. Putting (3.19) into (3.17), and after some algebraic manipulations, we obtain

$$D_{MOS}(V_{gb}, R_m V_{gb}) = \frac{2V_{gb}}{K} + \frac{V_{gb}^2}{K(V_{\max} - V_{gb} - V_T)} + \frac{2(V_{\max} - V_T)}{K} \ln\left[1 - \frac{V_{gb}}{(V_{\max} - V_T)}\right]. \tag{3.20}$$

The value of D_{MOS} as a function of V_{gb} can be found by putting the measured or simulated current-voltage data into (3.17). This, together with V_{max}, allows one to determine K and V_T from (3.20) by fitting the right-hand side of (3.20) to the known D_{MOS} on the left-hand side of equation. This procedure is the same as that used for extracting the ideality factor and saturation current of a junction diode discussed in Sec. 3.2.1 and in more detailed in Appendix C. It should be mentioned that choosing different V_{max} values does not alter the outcome of K and V_T extracted.

From the mathematical point of view, we can use V_g instead of V_{gb} in deriving (3.20), but R_m, and thus D_{MOS}, is very large for small V_g (i.e., weak inversion). Consequently, in the D_{MOS} versus V_g plot, D_{MOS} is very large in a narrow range of small V_g and is small in the remaining range of large V_g (i.e., strong inversion), which makes the fitting to the D_{MOS} versus V_g plot very difficult. This problem can be eliminated by using the new variable V_{gb}, which results in a more desirable plot with small D_{MOS} in the weak inversion region and large D_{MOS} in the wider region of strong inversion.

It is well known that accurate MOSFET modeling requires two different V_T; one for strong inversion and one for the subthreshold region. Like the majority of the existing threshold-voltage extraction methods, the current integral-function method extracts the threshold voltage for strong inversion. The technique should be applicable for the determination of V_T for the subthreshold region, but the drain-current model for strong inversion (Eq. (3.9)) will need to be changed to that for weak inversion, and the extraction will be more complicated and laborious.

In the following, the threshold voltage extraction method developed is tested in the environments of circuit simulation, device simulation, and measurements.

3.2.3 Circuit simulation results

We first simulate an n-channel MOSFET using AIM-SPICE simulator [10] and extract V_T from the SPICE simulation results. The level-3 MOSFET model in SPICE is used in simulation, and the following parameters are used: a threshold voltage of 0.5 V, a mask channel width of 10 μm, a mask channel length of 0.5 μm, a junction depth of 0.10 μm with a lateral extent of 0.075 μm, an oxide thickness of 100 Å, a substrate doping of 10^{17} cm^{-3}, a mobility of 331.5 cm^2/V-s, a bulk threshold parameter of 0.53 $V^{1/2}$, a mobility degradation factor of 0.8 V^{-1}, and a maximum drift velocity for carriers of 10^5 m/s. An increment of 50 mV for the gate voltage and V_d = 50 mV were used in the simulation.

Figure 3.12(a) shows the drain current as a function of the gate bias simulated for various cases of $R_d = R_s$, and Fig. 3.12(b) shows the corresponding D_{MOS} as a function of V_{gb}. Clearly, the same D_{MOS} is obtained for all cases considered and thus is independent of the drain and source series resistances. For demonstration purposes, excessive cases of $R_d = R_s$ = 0 and 100 KΩ were used in Fig. 3.12, but the other values of 0.1 KΩ to 10 KΩ considered represent a typical range for the series resistances of conventional and LDD MOSFETs.

Figure 3.13 shows the threshold voltages as a function of $R_d = R_s$ extracted from the different methods based on the SPICE simulation results. It can be seen that (a) the present method yields the best result of V_T = 0.44 V (i.e., closest to V_T = 0.5 V specified in SPICE simulation) among all the methods considered for a wide range of drain and source series resistances; (b) the linear-extrapolation method is highly dependent of the series resistances; (c) the second-derivative method becomes insensitive to the series resistances for $R_d = R_s$ < 10 KΩ; and (d) the ratio method depends only slightly on the series resistances but has the largest error among all methods considered.

Figure 3.14 shows the value of K^{-1} versus the mask channel length L_m extracted from the present method based on AIM-SPICE simulation using both level-3 and level-10 MOSFET models. The results indicate that K decreases with increasing L_m and that the extracted K value depends slightly on the level of the MOSFET model used. The latter is due to the fact that K is a function of the free-carrier mobility and different mobility models are used in different levels of MOSFET model. The procedure of extracting K, when applied to measurement data and given the geometry of MOSFET such as the channel length and channel width, can be used to determine the value of free-carrier mobility in the channel region of MOSFETs.

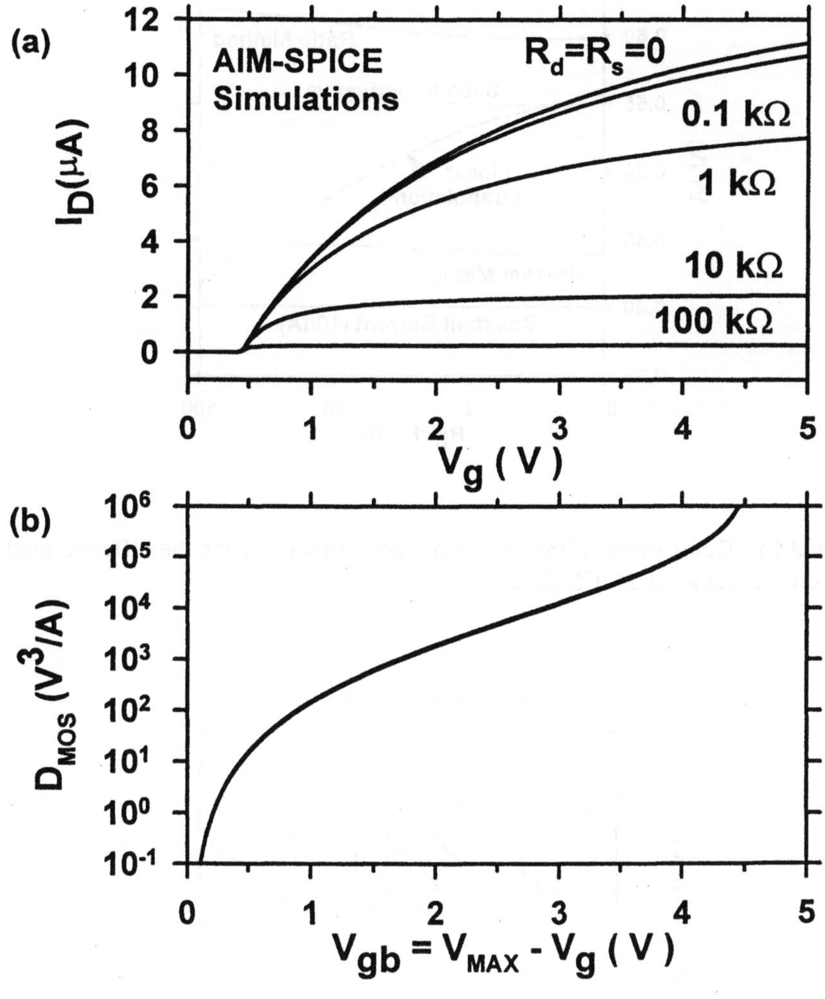

Figure 3.12 : (a) Drain current versus gate voltage characteristics and (b) corresponding D_{MOS} function simulated using AIM-SPICE circuit simulator for different drain and source series resistances. An increment of 50 mV for the gate bias and V_d = 50 mV are used. Note that the same D_{MOS} is obtained for all cases.

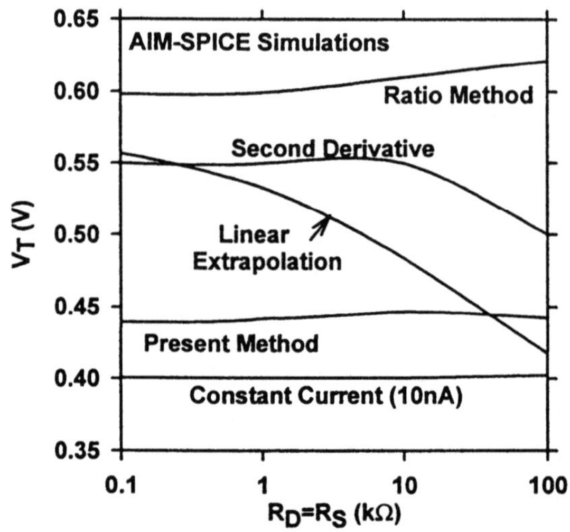

Figure 3.13 : Comparison of threshold voltages extracted using the different methods based on the results of SPICE simulation.

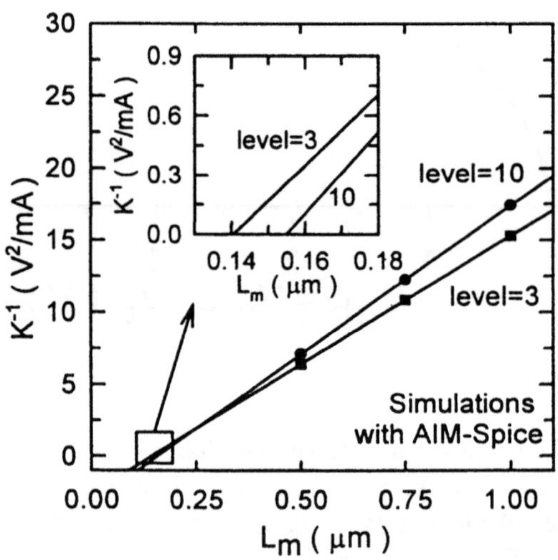

Figure 3.14 : The value of K^{-1} versus the mask channel length extracted using the new extraction method and two different levels of MOSFET model in SPICE.

3.2.4 Device simulation results

Here, a two-dimensional device simulator MEDICI [11] is used, and p-channel LDD MOSFETs with a mask channel length $L_m = 0.75$ μm and typical device make-up [12] are simulated.

Figure 3.15(a) presents the drain current as a function of the gate bias simulated for a fixed drain contact resistance $R_{cd} = 1$ KΩ-μm and several different source contact resistances R_{cs}. The reason for using R_{cd} and R_{cs} here, rather than R_d and R_s, is because only the drain and source contact resistances, not the drain and source series resistances, can be specified in device simulation. Also note that R_{cd} and R_{cs} are part of R_d and R_s, respectively. Figure 3.15(b) shows D_{MOS} as a function of V_{gb} calculated from the simulation results in Figure 3.15(a). Again, the same D_{MOS} is obtained for different contact resistances.

Figure 3.16 shows the threshold voltages as a function of R_{cs} extracted from the various methods based on the MEDICI simulation results. The trends are similar to those obtained from SPICE simulation shown in Fig. 3.13. First, the constant-current method is insensitive to R_{cs} and yields the smallest V_T for a wide range of R_{cs} (i.e., $R_{cs} < 5$ KΩ-μm). Second, both the linear-extrapolation and second-derivative methods depend strongly on R_{cs}. Third, among all methods considered, the present method is the least sensitive to R_{cs}. However, some discrepancies between Figs. 3.13 and 3.16 can be found, such as the different tendencies of the second-derivative and ratio methods obtained from SPICE and MEDICI simulations. This is due to the different types of resistances used in Figs. 3.13 and 3.16 and, to a less extent, the different free-carrier mobility models used in SPICE and MEDICI simulations.

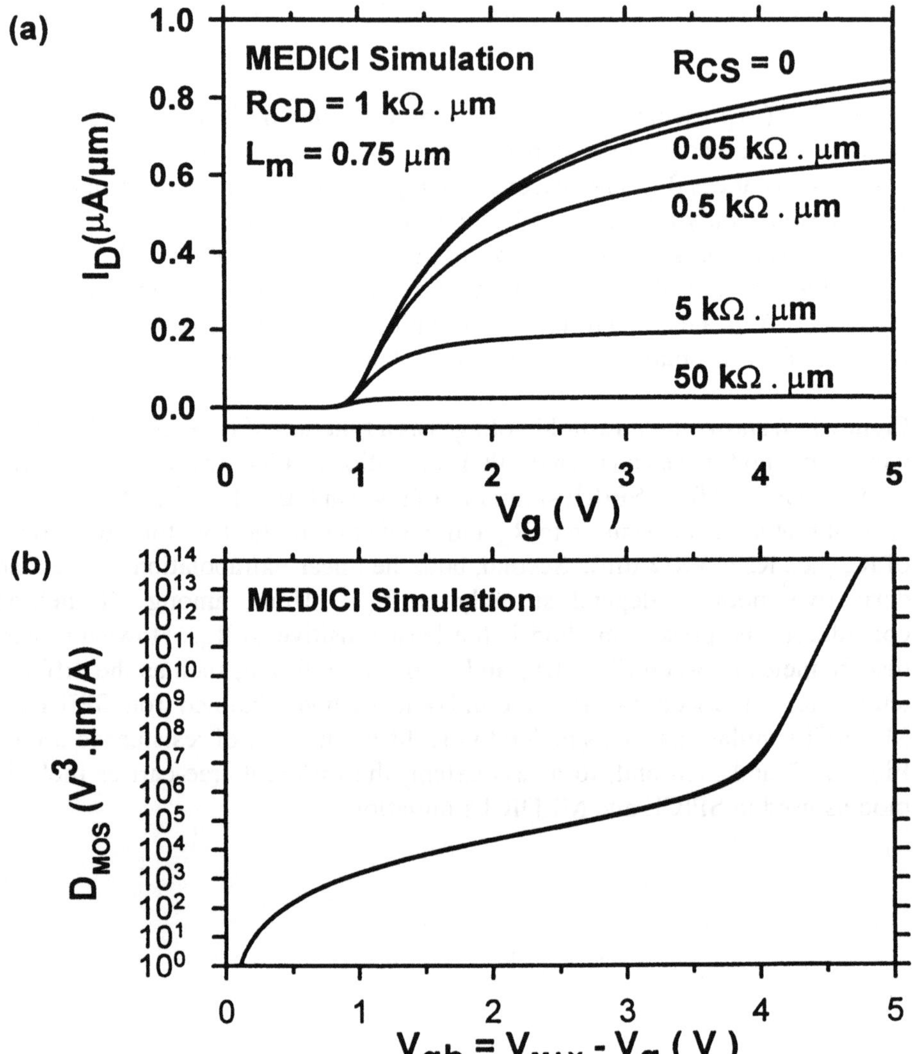

Figure 3.15 : (a) Drain current versus gate voltage characteristics and (b) corresponding D_{MOS} function simulated using MEDICI device simulator for different contact resistances. An increment of 50 mV for the gate bias and V_d = 50 mV are used.

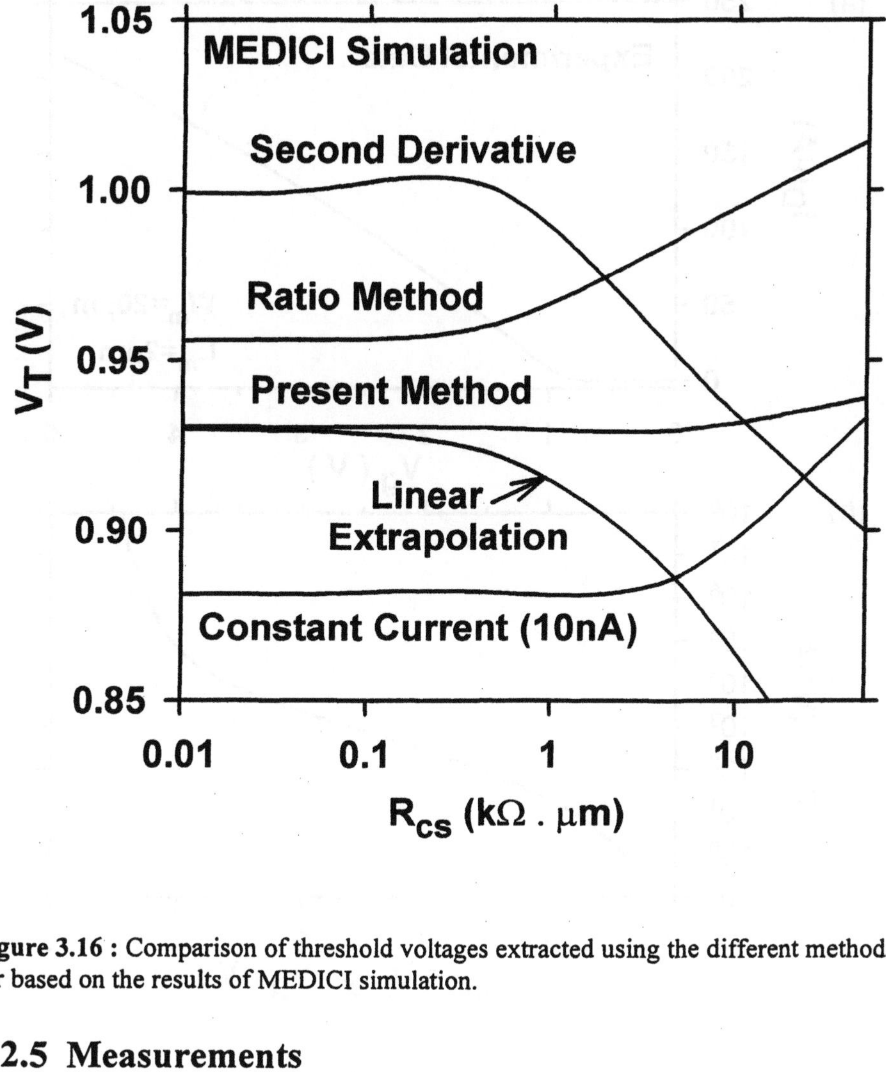

Figure 3.16 : Comparison of threshold voltages extracted using the different methods for based on the results of MEDICI simulation.

3.2.5 Measurements

The drain current versus the gate voltage characteristics measured from a 2-μm MOSFET are presented in Fig. 3.17(a). An increment of 100 mV for the gate voltage and V_d = 100 mV have been used in the measurement. Figure 3.17(b) shows the corresponding D_{MOS} calculated as a function of V_{gb}.

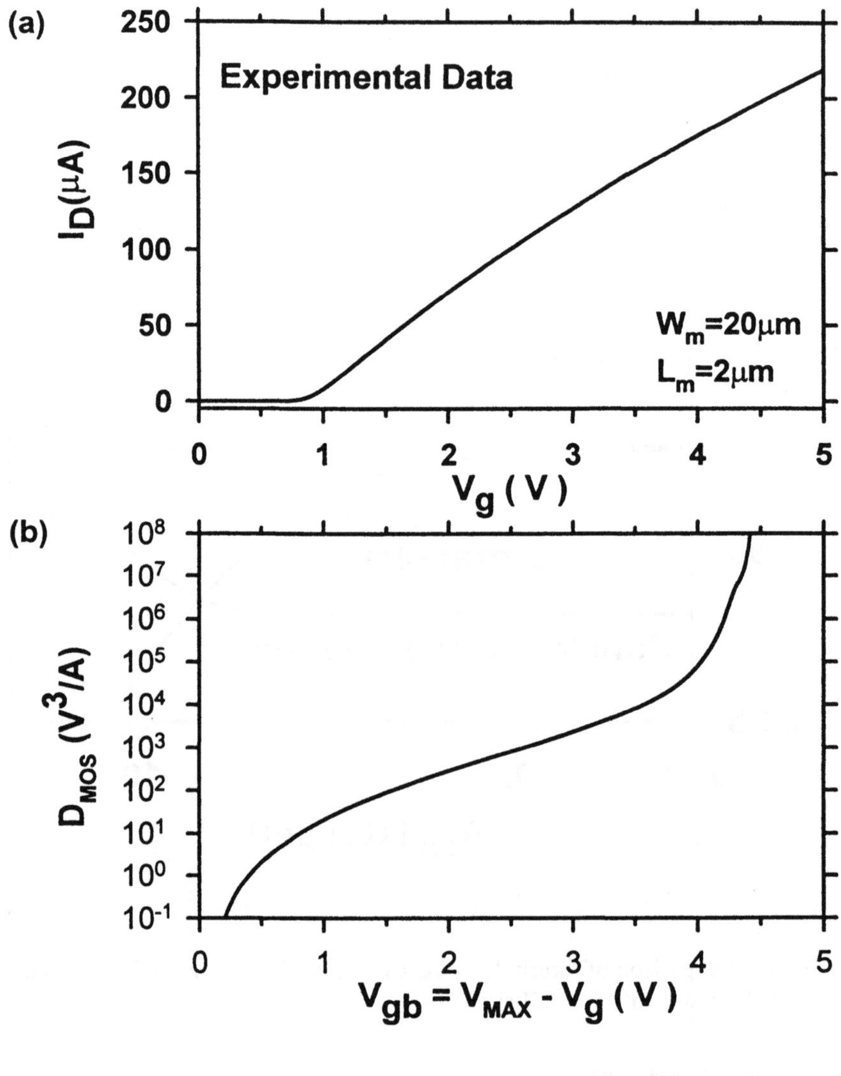

Figure 3.17 : (a) Drain current versus gate voltage characteristics and (b) corresponding D_{MOS} function obtained from measurements. An increment of 100 mV for the gate bias and V_d = 100 mV are used.

Table 3.1 shows the threshold voltages extracted from the various methods based on experimental data. The results show that the constant-current and second-derivative methods give the smallest and largest V_T, respectively, and that the present and linear-extrapolation methods yield comparable V_T. This trend is in good agreement with that obtained from MEDICI simulation shown in Fig. 3.16.

Method	Threshold Voltage (V) Device with width to length ratio W/L= 20/2
Constant current @ 100 nA	0.657
Linear extrapolation [1]	0.889
Second derivative [4]	1.000
Ratio of conductance and transconductance [5]	0.873
Integral function (new method)	0.909

Table 3.1 : Extracted values of the threshold voltage using different methods based on experimental data.

3.3 Threshold voltage shift reversal in short-channel MOSFETs

It has been reported that the threshold voltage V_T of a short-channel MOSFET can increase with decreasing mask channel length L_m [13-18]. This phenomenon is called the reverse short-channel effect (RSCE) because it is opposite to the conventional short-channel theory which suggests that V_T decreases with decreasing L_m [10]. Figure 3.18 shows the threshold voltage versus the mask channel length measured from – and p-channel MOSFETs fabricated using a 0.6 μm technology. The threshold voltage was obtained using the linear extrapolation at the point of maximum slope on the gate voltage V_G versus drain current I_D curve in the linear region. The devices have a channel width of 25 μm and a mask channel length ranging from 0.5 to 25 μm. The RSCE is found in both devices in the L_m range of about 2 μm to 1 μm.

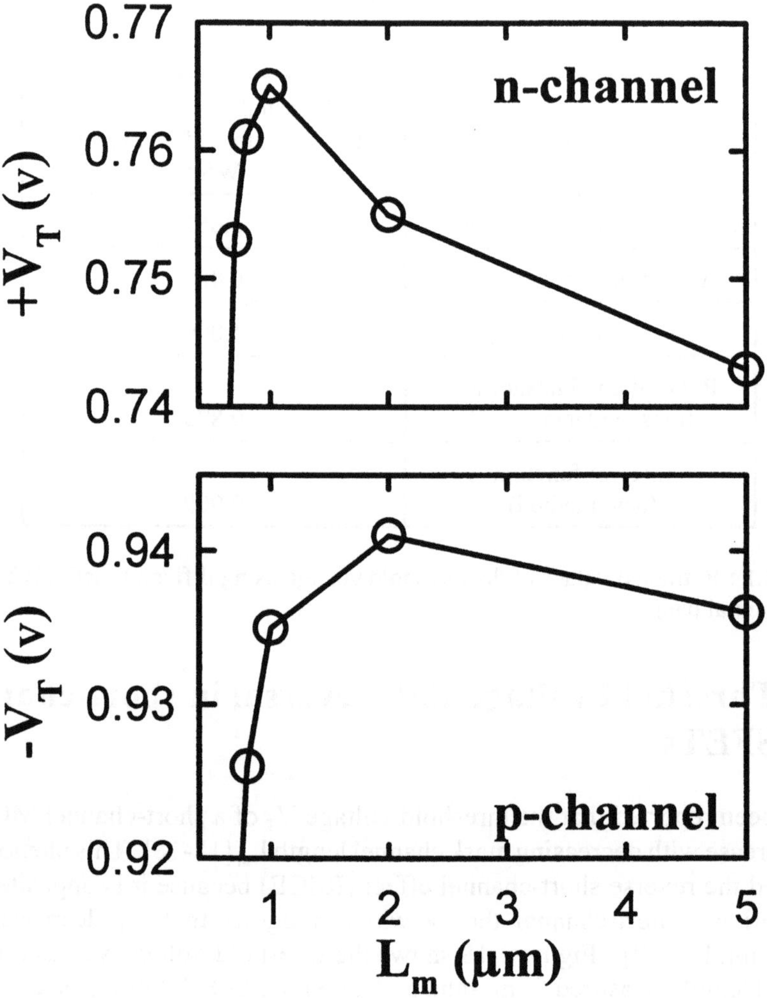

Figure 3.18 : Threshold voltages as a function of the mask channel length measured from n- and p- channel MOSFETs fabricated using a 0.6 µm technology.

Several different explanations to the origin of the RSCE have been presented. They include: 1) nonuniform lateral distribution of channel dopant due to diffusion of Frenkel pairs [17]; 2) oxidation-enhanced impurity diffusion during the poly-gate sidewall reoxidation [14]; 3) vacancies injections during silicide formation [13]; and 4) damage in the drain and source created by ion implantation [16]. It should be pointed out that the nonuniform lateral channel doping profile mentioned in 1) is related the effective channel length L_{eff}, as indicated by Gutierrez [19]. This is due to the fact that the doping concentrations in the drain and source decreases gradually from these regions into the channel, thus resulting in that L_{eff} is larger than the metallurgical channel length defined by the drain and source metallurgical junctions [20-21] and that the doping concentration in the "effective channel" is not uniform.

While the above explanations for the origins of the RSCE appear to differ from each other, they all implicate a physical mechanism of nonuniform lateral and vertical doping distributions in the channel region of MOSFET. As the mask channel is decreased, the source and drain regions are closer, and the degree of nonuniformity in the channel doping is increased. Consequently, the RSCE occurs when the channel is sufficiently short and a sufficiently large doping nonuniformity is reached. Specifically, the necessary conditions for the occurrence of the RSCE are [14]: 1) sufficient large concentration decrease towards the Si-SiO_2 interface of the channel doping; 2) laterally nonhomogeneous enhancement of the diffusivity of the channel dopant either by injection of interstitials or vacancies from outside the gate region; and 3) the minimum distance between the point-defect injection next to the gate and the metallurgical channel to drain junction is smaller than the characteristic lateral decay length of the point defects. These conditions imply that the RSCE is influenced by the oxide layer thickness t_{ox}, as well as by the lateral extent y_{sub} of the drain/source region underneath the gate. Figure 3.19 shows V_T versus channel length measured from two MOSFETs with different t_{ox}, indicating that the RSCE is less prominent if t_{ox} is reduced. Figure 3.20(a) shows the MOSFET structure with two different y_{sub} (i.e., y_{sub1} and y_{sub2}), and the threshold enhancement $\Delta V_T = V_T - V_T$(long channel) measured from MOSFETs with two different y_{sub} are given in Fig. 3.20(b). Clearly, ΔV_T is reduced as y_{sub} is increased.

An empirical model has been developed to describe the observed RSCE in submicron MOSFETs [15]. It was derived by assuming substrate bias-independent fixed charge at the source and drain ends, which are induced due to the various processing steps. The model can be expresses as

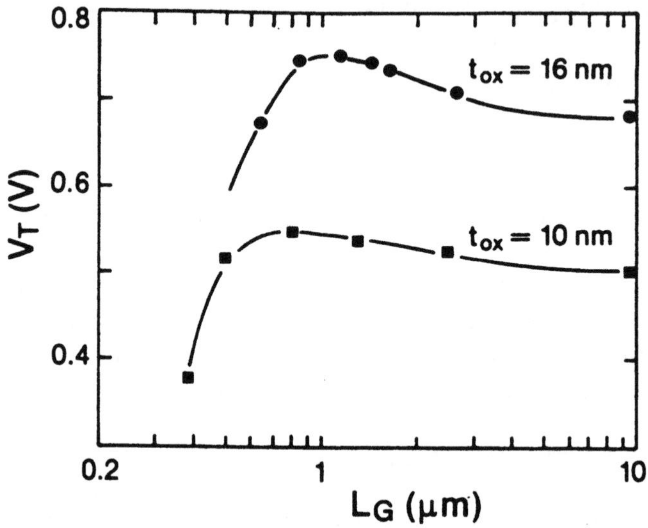

Figure 3.19 : Threshold voltage as a function of the gate length measured from MOSFETs with oxide layer thicknesses of 10 and 16 nm (Source Mazure and Orlowski [14]. Reprinted with permission).

$$V_T = V_{FB} + 2\,\phi_f + \frac{\left(2\,\epsilon\,q\,N_{eq}\,\phi_s\right)^{0.5}}{C_{ox}} - \frac{K_1\,\phi_s}{L_{eff}} \tag{3.21}$$

where

$$N_{eq} = N_s \quad for \quad \phi_s \le \phi_i;$$

$$N_{eq} = N_s\,\frac{\phi_i}{\phi_s}\left(1 - \frac{N_b}{N_s} + \frac{N_b}{N_s}\left(1 + \frac{N_s}{N_b}\left(\frac{\phi_s}{\phi_i} - 1\right)\right)^{0.5}\right)^2 \tag{3.22}$$

$$for \quad \phi_s > \phi_i;$$

and

$$\phi_i = \frac{q\,N_s\,W^2}{2\,\epsilon}\left(1 - \frac{K_2}{L_{eff}}\right) \tag{3.23}$$

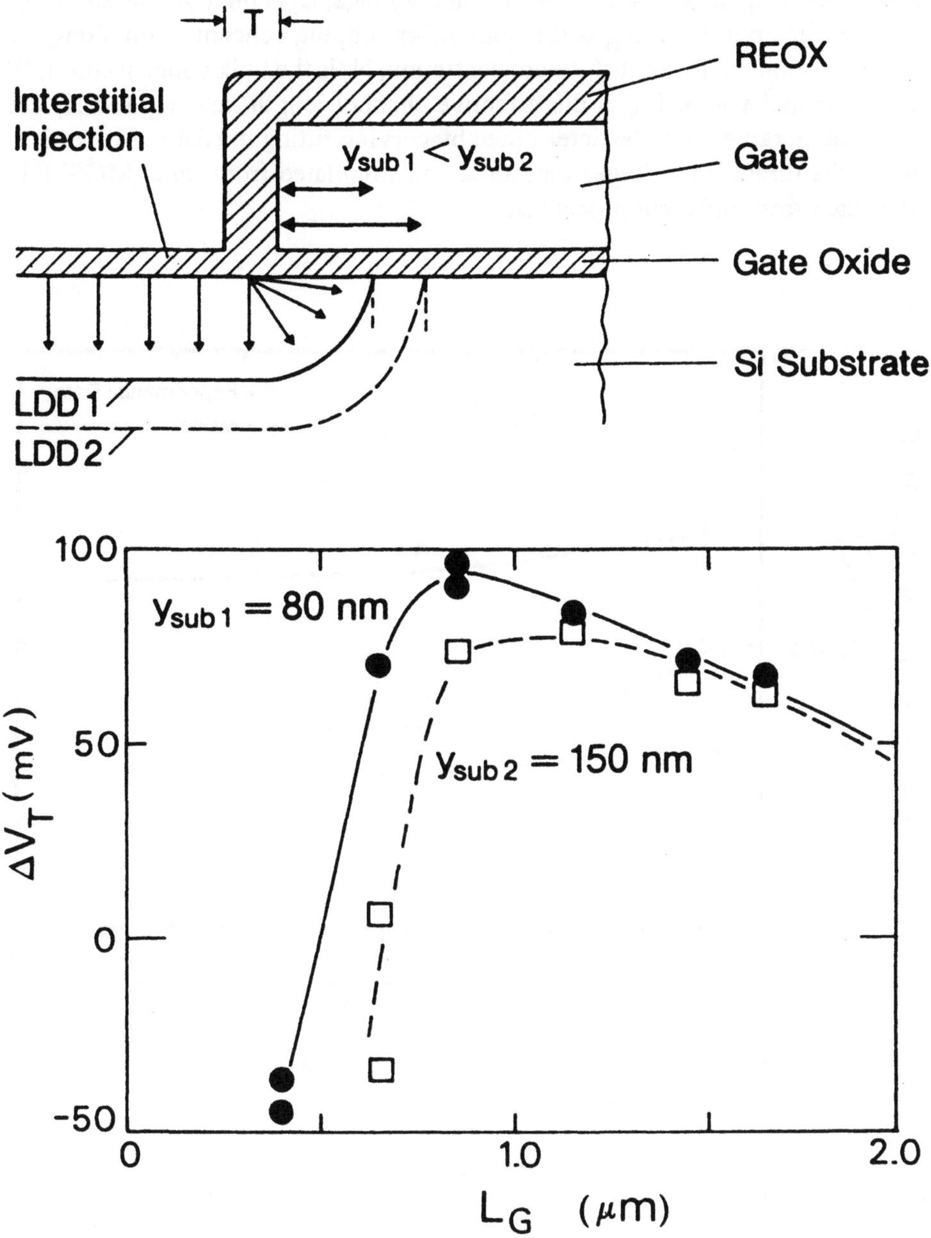

Figure 3.20 : (a) Schematic diagram showing the MOSFET proximity with two different drain/source lateral extent (y_{sub1} and y_{sub2}) underneath the gate, and (b) Threshold voltage enhancement measured from two MOSFETs with different y_{sub} (Source: Mazure and Orlowski [14]. Reprinted with permission).

In the above equations, V_{FB} is the flatband voltage, ϕ_f is the bulk potential, ϕ_s is the surface potential, N_{eq} is the nonuniform doping concentration along the channel region, N_s is the surface concentration, N_b is the bulk concentration, W is the channel width, L_{eff} is the effective channel length, and K_1 and K_2 are empirical parameters to be determined by curving fitting the data. Figure 3.21 shows the threshold voltages measured and calculated from three MOSFETs fabricated from different processes.

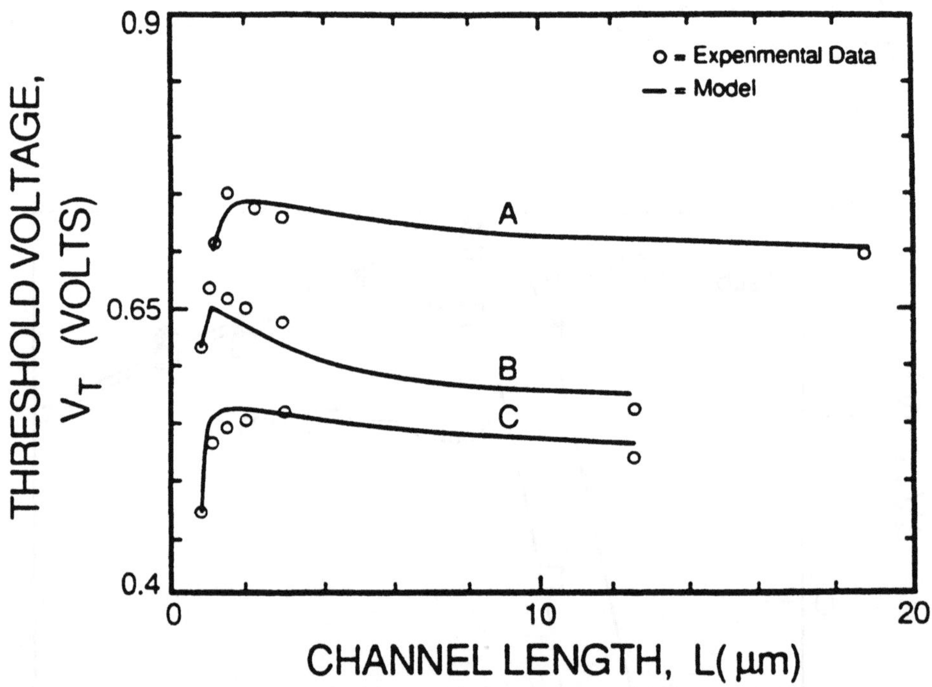

Figure 3.21 : Measured and modeled threshold voltages for three MOSFETs fabricated from different processes (Source: Arora and Sharma [15]. Reprinted with permission).

In the following, we will investigate, using device simulation, the effects of nonuniform doping profile in the channel on the threshold voltage of MOSFET and thus will verify whether such a mechanism can be attributed to the experimentally observed RSCE in MOSFETs. To this end, p-channel MOSFETs will be considered, in which the drain and source are surrounded by n-regions to emulate the nonuniform lateral and vertical doping profile in the channel, as shown in Fig. 3.22. The threshold voltage of such a structure is then simulated using a two-dimensional device simulator MEDICI [10]. All MOSFETs simulated have the following device make-up: a substrate doping concentration

N_{sub} of 10^{17} cm^{-3}, a gate oxide thickness of 25 nm, a mask channel length varying from 0.5 to 5 μm, a p-type drain and source regions with Gaussian profile and peak doping density $N_A = 10^{20}$ cm^{-3} inside an n-type Gaussian profile (n-well) with peak doping density $N_D = 10^{18}$ cm^{-3}. The junction depth for the p-type region is 0.34 μm with a lateral extent of 0.25 μm (i.e., 75% of vertical extent). Two different n-well junction depths X_j are considered: 0.50 and 0.38 μm with a lateral extent of 75% of vertical extent.

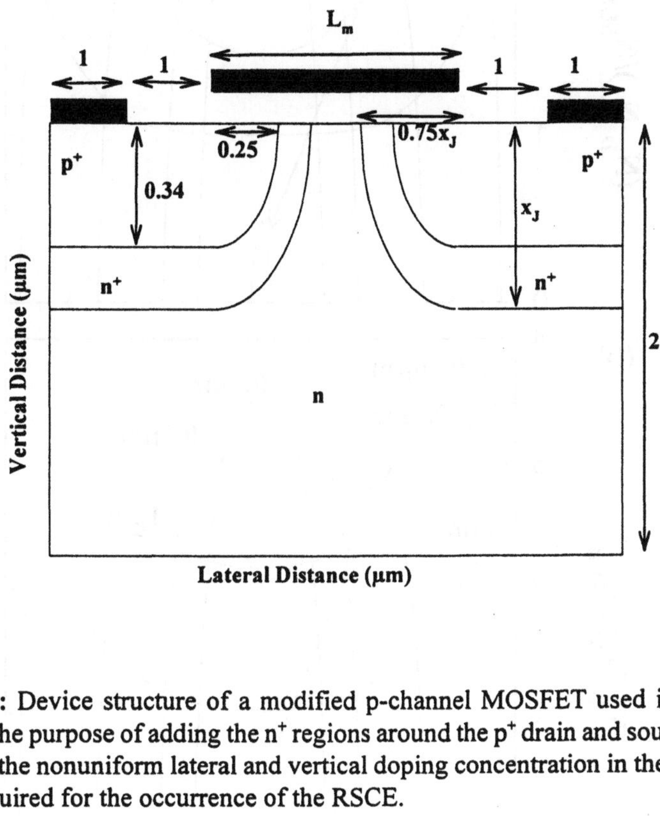

Figure 3.22 : Device structure of a modified p-channel MOSFET used in MEDICI simulation. The purpose of adding the n^+ regions around the p^+ drain and source regions is to emulate the nonuniform lateral and vertical doping concentration in the channel, a condition required for the occurrence of the RSCE.

Depending upon the values of L_m and X_j, the structure shown in Fig. 3.22 can result in various nonuniform doping profiles in the channel, as evidenced by the normalized doping concentrations $(N_D - N_A)/N_{sub}$ at the surface of the channel (i.e., oxide-Si interface) versus the normalized channel distance shown in Fig. 3.23. For a relatively long channel (i.e., $L_m = 2$ μm), the doping concentration in the channel is higher near the drain and source junctions and approaches to N_{sub} in the middle of the channel. As L_m is decreased, however, the lateral channel doping profile becomes narrower and highly nonuniform.

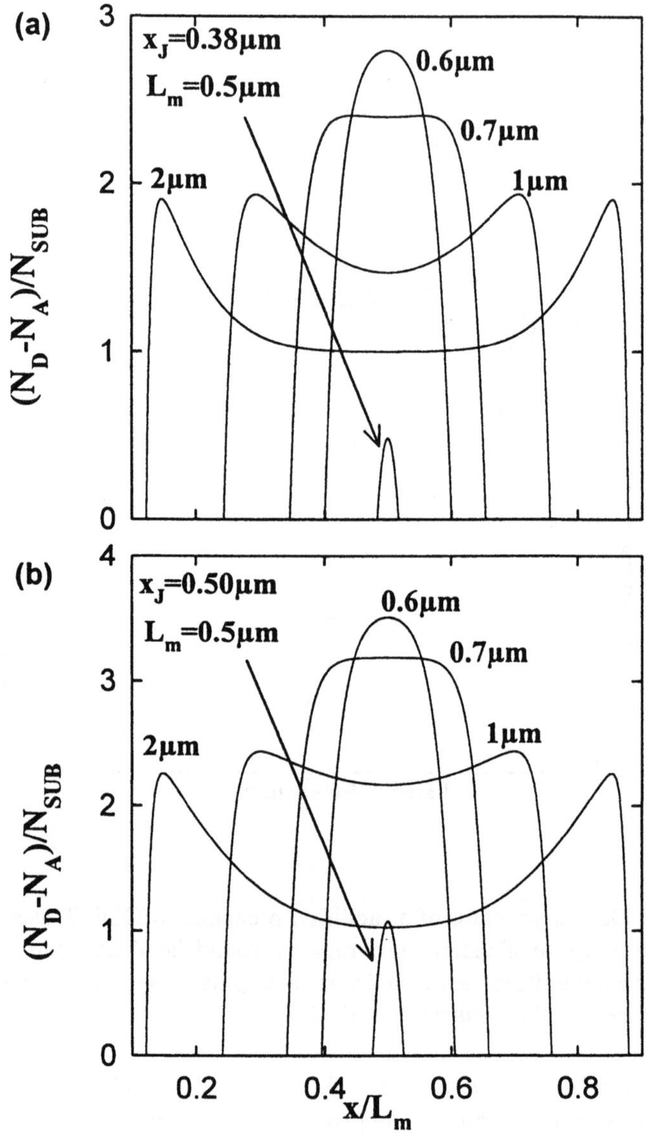

Figure 3.23 : Normalized doping concentration at the surface of the channel versus the normalized channel distance simulated for various L_m and (a) 0.38 µm and (b) 0.50 µm junction depths.

Figure 3.24 shows the V_T versus L_m characteristics simulated for three different cases: 1) without the n-well (i.e., uniform lateral channel doping profile); 2) with an $X_j = 0.38$ μm n-well; and 3) with an $X_j = 0.5$ μm n-well. For the case without n-well, the threshold voltage exhibits the classical short-channel behavior with V_T decreases with decreasing L_m. A small RSCE is observed in the MOSFET with $X_j = 0.38$ μm, and RSCE becomes more prominent as X_j is increased to 0.5 μm. This is due to the fact that a larger X_j gives rise to a higher degree of doping nonuniformity in the channel, a trend demonstrated in Fig. 3.23.

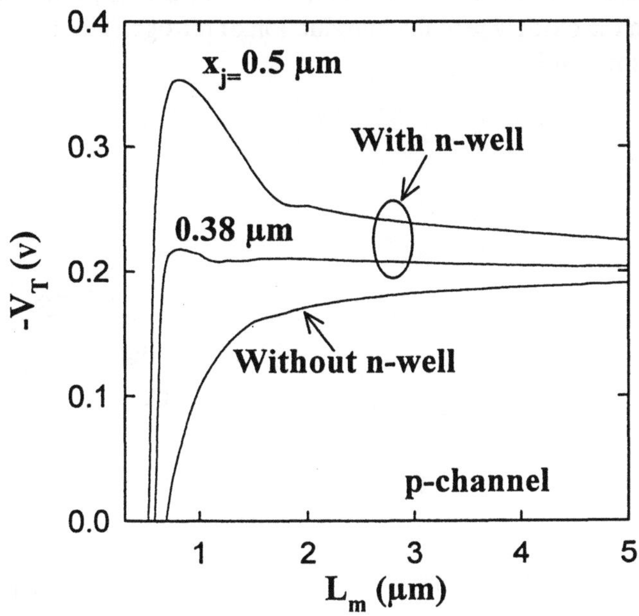

Figure 3.24 : Threshold voltages as a function of the mask channel length simulated for various L_m and two different n-well junction depths.

The preceding simulation results have confirmed the commonly used theory that the anomalous threshold voltage reversal in the short-channel MOSFETs is originated mainly from the nonuniform doping concentration in the channel region. Furthermore, it was suggested that the RSCE becomes more prominent in the direction of shorter channel length due to the fact that the reduced channel length increases the degree of doping nonuniformity in the channel.

It was found experimentally that nitrogen implantation into the channel region of MOSFET can be used to suppress RSCE [22]. The nitrogen atoms behave as dopant species for silicon and thus retard the redistribution of the channel doping concentration, a mechanism gives rise to the nonuniform doping concentration in the channel and thus the RSCE. Figure 3.25 compares the RSCE measured from a MOSFET without nitrogen implantation (denoted by reference wafer in Fig. 3.25) and a MOSFET with nitrogen implantation with a dose of 10^{15} cm^{-2}. The symbols from left to right in the figure for both devices represent the measured V_T for decreasing channel length. There is clearly a large reduction in the peak of the RSCE from 45 mV to 15 mV when the nitrogen implant is used. This approach, however, comes with the expenses of a reduced free-carrier mobility in the channel and a change in the dielectric permittivity in the oxide due to the introduction of nitrogen atoms in the channel and oxide, respectively.

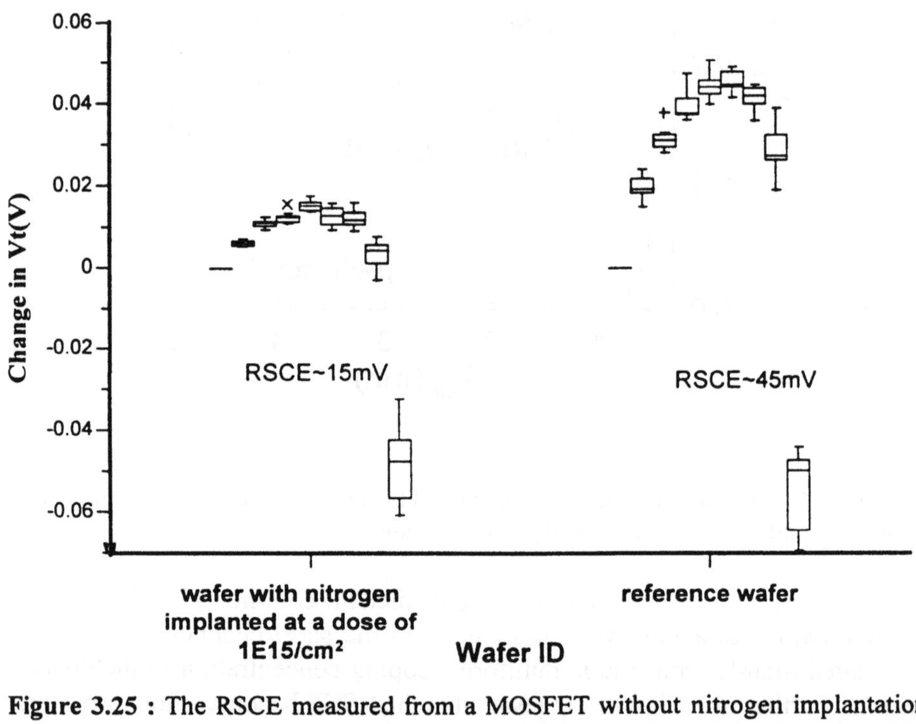

Figure 3.25 : The RSCE measured from a MOSFET without nitrogen implantation (denoted by reference wafer) and from a MOSFET with nitrogen implantation. The symbols for both devices from left to right represent the measured threshold voltages with decreasing channel length (after Lee et al. [22]).

Another way to reduce the RSCE is the use of MOS devices made in the silicon-on-insulator (SOI) material [23]. The reduction of the RSCE depends on the thickness of the Si overlayer; the thinner the Si film, the less the threshold voltage roll-on and thus the less RSCE. These findings are explained by a decrease of the lateral distribution of silicon interstitials generated at the source and drain regions and are related with their high recombination velocity at the buried oxide. Figure 3.26 shows the measured threshold voltage as a function of gate length for different silicon thicknesses (2 to 10 μm), as well as the case of bulk silicon wafer. Moreover, the influence of voltage V_{sub} applied to the substrate is illustrated in the same figure. Clearly, the roll-on of the threshold voltage is reduced as the silicon thickness is decreased. Figure 3.27 presents the dependence of threshold voltage on the silicon thickness for two SOI MOSFETs with gate lengths of 12 and 2 μm. It is observed from the figure that the influence of the Si film on V_T is more prominent in 2-μm than in 12-μm SOI devices. These results clearly support the idea that using thick SOI material is an effective way to reduce the RSCE in short-channel MOSFETs.

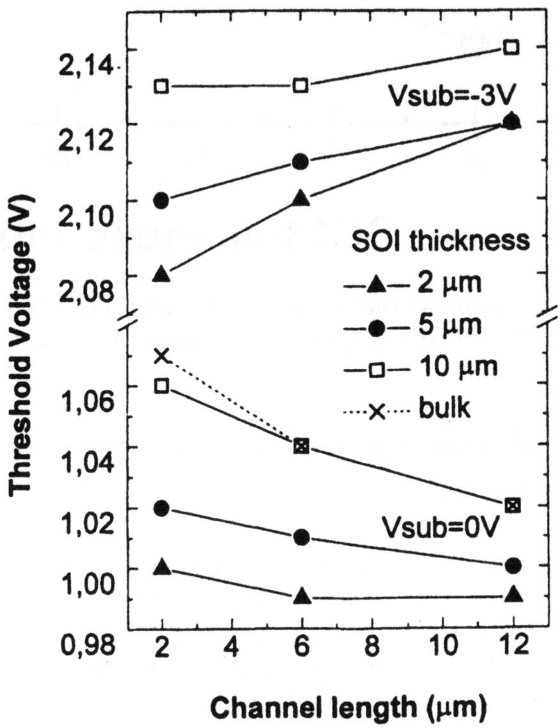

Figure 3.26 : Measured threshold voltage as a function of channel length for SOI MOSFETs with different silicon thicknesses (Source: Tsoukalas et al. [23]. Reprinted with permission).

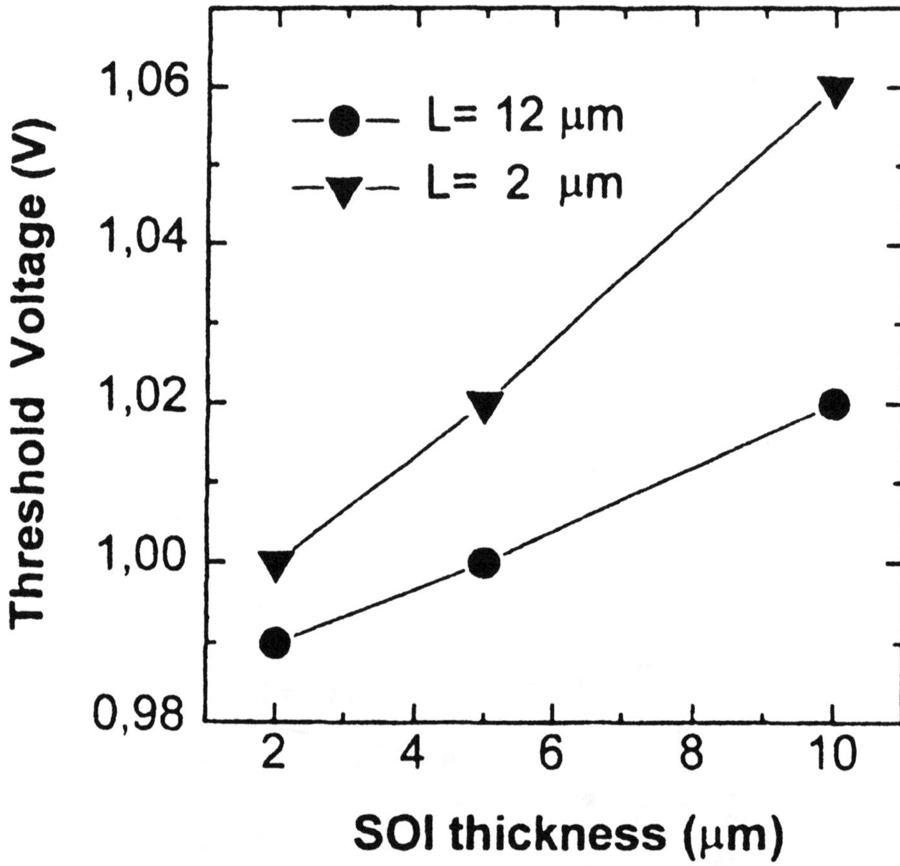

Figure 3.27 : Measured threshold voltage variation as a function of silicon thickness for 2-µm and 12-µm SOI MOSFETs (Source: Tsoukalas et al. [23]. Reprinted with permission).

3.4 Threshold voltage shift due to quantum mechanical effects

The continued scaling of MOS device gate length requires decreased gate oxide thickness t_{ox} and increased substrate doping density N_A. For gate length less than 0.25 µm, the decreased tox and increased NA result in large enough transverse electric fields to cause significant quantization of the carriers, ever near threshold, in the potential well at the Si/SiO_2 interface [24]. The consequence of the quantization is to delay the onset of inversion and cause the redistribution of accumulated or inverted carriers at the interface. Specifically, as discussed in Chapter 1 (see Sec. 1.8), the quantum mechanical (QM) effects give rise to the quantization of energy in the conduction band, which consequently increases the effective bandgap and causes a displacement of inversion layer charge away from the interface. Such a displacement of charge is equivalent to an increase in the effective oxide layer thickness, hence an increase in the threshold voltage and a reduction in the current level. Thus, using the classical MOS analysis, without including the QM effects, may lead to substantial errors in the prediction of the performance of modern deep-submicron MOSFETs. In this section, an experimental procedure developed recently [25] for the determination of threshold voltage shift due to the QM effects is presented.

Let us consider a large area MOS capacitor (i.e., MOS structure without the source and drain). Using such a structure eliminates the hot-carrier effect, and capacitors with large area increase measurement accuracy. The substrate doping density ranges from 5×10^{15} to 10^{18} cm^{-3}, and the electrical characterization involved the capacitance-voltage (C-V) measurements performed with a very slow sweep rate of 10 mV/sec.

The extraction methodology for the threshold voltage shift ΔV_T due to the QM effects is as follows [25]. Two-dimensional MEDICI device simulations are first carried out to illustrate how different physical parameters cause changes in the C-V curve. As shown in Figs. 3.28(a) and (b), the oxide thickness variations mainly affect the flatband and accumulation regions of the C-V curve, while the doping variation affect the depletion region of the C-V curve. This mutually exclusive influence of t_{ox} and N_A on the C-V curves has been exploited in order to extract their values accurately. A classical C-V curve was generated using MEDICI, with an initial estimate for t_{ox} based on the ellipsometer measurements, and using the doping profile generated from a process simulator. The oxide thickness was then adjusted slightly to match the flatband and accumulation region of the experimental and simulated C-V curves. This process was repeated until a very good match was obtained between the experimental and simulated C-V curves in accumulation, flatband, and depletion regions. Based on this technique, the gate oxide thickness and the doping density in the substrate can be extracted accurately. The experimental C-V curves corresponding to both the lightly and heavily doped substrate region on the same wafer were then analyzed using the technique described above. In the low doping case, where the QM effects are expected to be minimal, very good agreement was obtained in all bias regimes between the measured and simulated C-V curves. On the other hand, in the case of high substrate doping, despite a very good match between the two C-V curves in the flatband and depletion regions, a deviation, or a shift, due to the QM effects is seen near the threshold region. Since both the lightly and heavily doped MOS devices are on the same wafer, such a shift can be concluded not caused by any other physical mechanism but the QM effects.

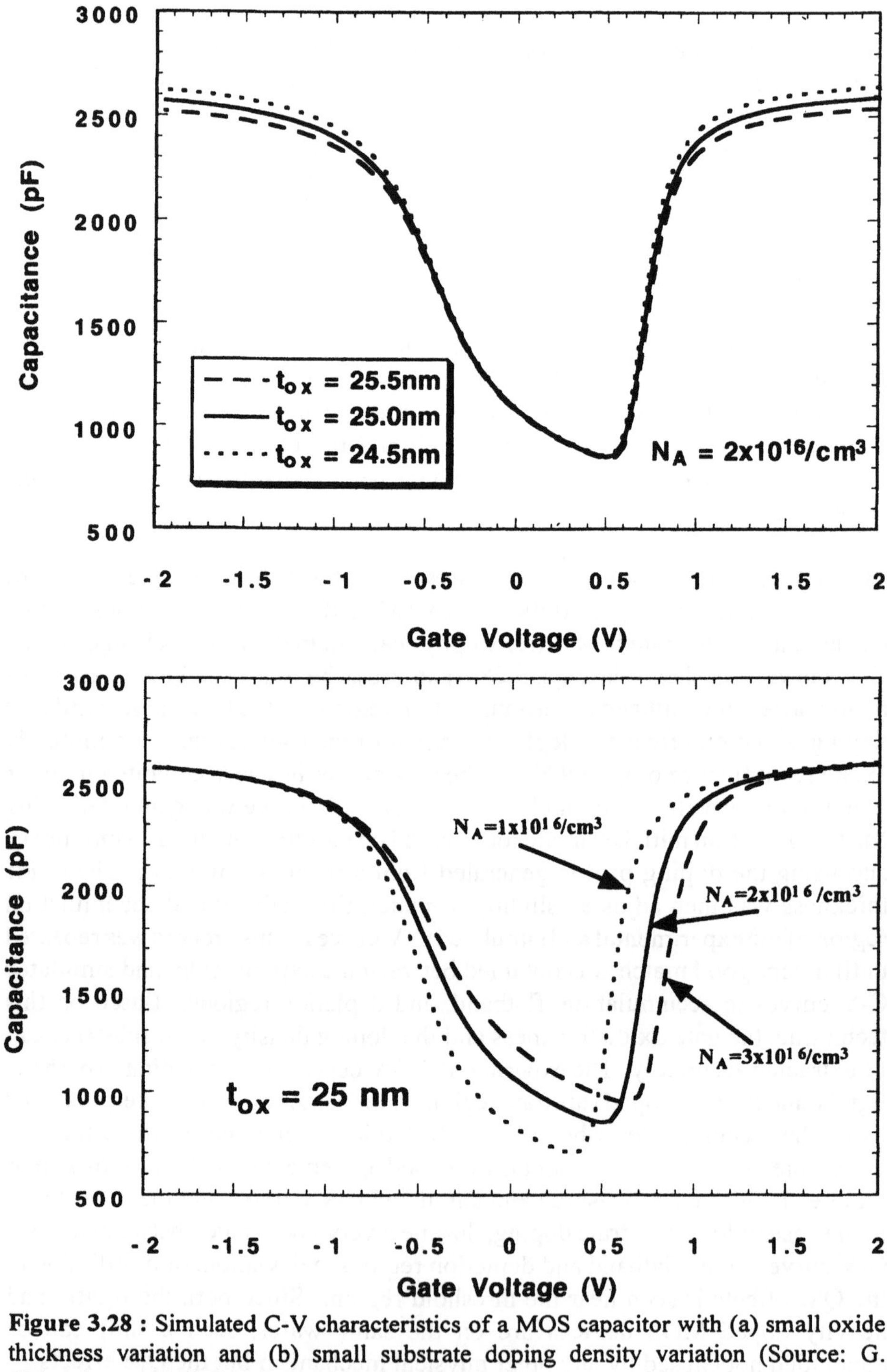

Figure 3.28 : Simulated C-V characteristics of a MOS capacitor with (a) small oxide thickness variation and (b) small substrate doping density variation (Source: G. Chindalore et al. [25]. Reprinted with permission).

Figure 3.29 summarizes the results of ΔV_T as a function of substrate doping density for both electron and hole inversion layers. First, the ΔV_T values are found to increase with increasing doping density and increasing oxide thickness. This is because, for a given oxide thickness, the potential well at the interface becomes steeper as the doping density is increased, leading to an increasing separation between the subbands. The effective "bandgap widening" has the effect of decreasing the inversion layer charge density for a given bias condition, thereby increasing the threshold voltage. On the other hand, the increase in ΔV_T with increasing oxide thickness is because the surface potential becomes a weaker function of gate voltage with increasing oxide thickness. Hence higher voltages are required to achieve a given inversion layer density for thicker oxides. Finally, as can be seen in Fig. 3.29, the measurement of [26], three-subband model [27], fullband model [28], and simplified model [29] all agree well with the values extracted from this method (i.e., denoted experiment in Fig. 3.29).

The extraction method discussed above can be useful in two different ways. First, in determining the threshold voltage of a deep submicron MOSFET, one can predict V_T based on the classical model (discussed in Chapter 1) and then add to it the threshold voltage shift due to the quantum mechanical effects extracted from this method. The other useful application would be when V_T of a deep submicron MOSFET is extracted from the conventional extraction methods (discuused in Secs. 3.1 and 3.2) based on device simulation results. Since quantum mechanical effects are not incorporated in device simulators, V_T extracted based on these results would be its classical value. The more accurate V_T can be obtained by adding ΔV_T extracted from the method discussed to such a value. It should be pointed out that if the extraction method is performed based on measurements, then quantum mechanical effects are automatically imbedded in the data, hence extraction of ΔV_T to account for the quantum mechanical effects is not necessary.

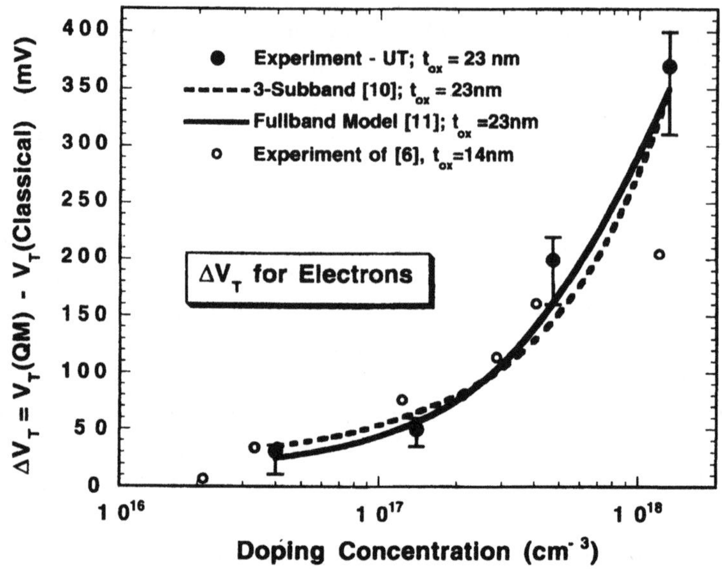

(a)

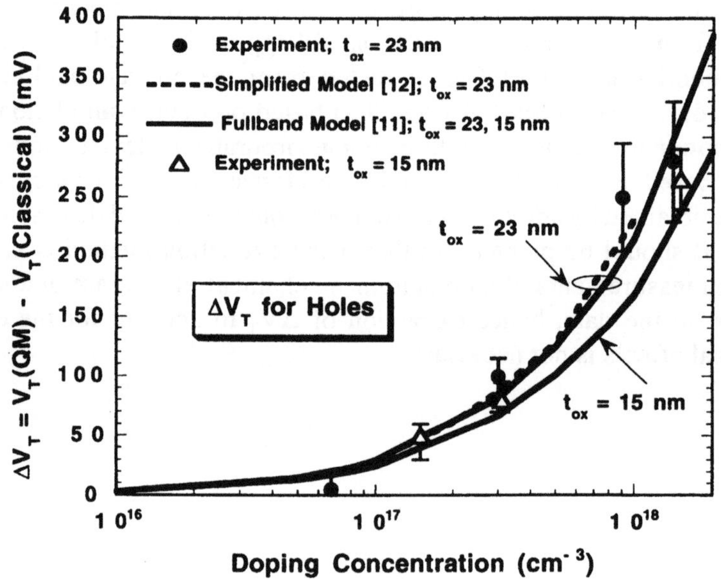

(b)

Figure 3.29 : Threshold voltage shift due to the QM effects obtained from the present method, method of [6] (Ref. [26] in this chapter), 3-subband model [10] (Ref. [27] in this chapter), fullband model [11] (Ref. [28] in this chapter), and simplified model [12] (Ref. [29] in this chapter) for (a) electron inversion layer and (b) hole inversion layer (Source: G. Chindalore et al. [25]. Reprinted with permission).

REFERENCES

[1] D. K. Schroeder, *Semiconductor Material and Device Characterization*, Wiley, New York, 1990.

[2] J. J. Liou, *Advanced Semiconductor Device Physics and Modeling*, Artech House, Boston, 1994.

[3] Y. P. Tsividis, *Operation and Modeling of the MOS Transistor*, McGraw-Hill, New York, 1987.

[4] H. S. Wong, M. H. White, T. J. Krutsick and R. V. Booth, "Modeling of transconductance degradation and extraction of threshold voltage in thin oxide MOSFET's," Solid-St. Electron., vol. 30, p. 953, 1987.

[5] S. Jain, "Measurement of threshold voltage and channel length of submicron MOSFETs," IEE Proc. Cir. Dev. and Sys., vol. 135, p. 162, 1988.

[6] Z. X. Yan and M. J. Deen, "Physically-based method for measuring the threshold voltage of MOSFETs," IEE Proc. Cir. Dev. and Sys., vol. 138, p. 351, 1991.

[7] A. Ortiz-Conde, F. J. García Sánchez, J. J. Liou, J. Andrian, R. J. Laurence, and P. E. Schmidt, "A generalized model for a two-terminal device and its application to parameter extraction," Solid-St. Electron., vol. 38, p. 265, 1995.

[8] F. J. García Sánchez, A. Ortiz-Conde, and J. J. Liou, "A parasitic series resistance-independent method for device-model parameter extraction," IEE Proc. Cir. Dev. and Sys., vol. 143, p. 68, 1996.

[9] F. J. García Sánchez, A. Ortiz-Conde, G. De Mercato, J. J. Liou, and L. Recht, "Eliminating parasitic resistances in parameter extraction of semiconductor device models," Proc. of First IEEE Int. Caracas Conf. on Dev. Cir. and Sys., Caracas, Venezuela, 1995, p. 298.

[10] K. Lee, M. Shur, T. A. Fjeldly, and T. Ytterdal, *Semiconductor Device Modeling for VLSI*, Prentice-Hall, Englewood Cliffs, NJ, 1993.

[11] MEDICI Manual, Technology Modeling Associates, Inc., 1993.

[12] Md. Rofiqul Hassan, J. J. Liou, A. Ortiz-Conde, F. J. García Sánchez, and E. Gouveia Fernandes, "Drain and source resistances of short-channel LDD MOSFETs," Solid-St. Electron., vol. 41, p. 778, 1997.

[13] C. Y. Lu and J. M. Sung, "Reverse short-channel effects on threshold voltage in submicrometer silicide devices," IEEE Electron Device Lett., vol. 10, p. 446, 1989.

[14] C. Mazure and M. Orlowski, "Guidelines for reverse short-channel behavior," IEEE Electron Device Lett., vol. 10, p. 556, 1989.

[15] N. D. Arora and M. S. Sharma, "Modeling the anomalous threshold voltage behavior of submicrometer MOSFET's," IEEE Electron Device Lett., vol. 13, p. 92, 1992.

[16] H. I. Hanfi, W. P. Nobel, R. S. Bass, K. Varahramyan, Y. Li, and A. J. Dally, "A model for anomalous short-channel behavior in submicron MOSFET's," IEEE Electron Device Lett., vol. 14, p. 575, 1993.

[17] T. Kunikiyo, K. Mitsui, M. Fujinage, T. Uchida, and N. Kotani, "Reverse short-channel effects due to lateral diffusion of point-defect induced by source/drain ion implantation," IEEE Trans. CAD IC System, vol. 13, p. 507, 1994.

[18] C. -Y. Chang, C. -Y. Lin, J. W. Chou, C. C. -H. Hsu, H. -T. Pan, and J. Ko,

"Anomalous reverse short-channel effects in p^+ polysilicon gated p-channel MOSFET," IEEE Electron Device Lett., vol. 15, p. 437, 1994.

[19] E. Gutierrez, "The drain threshold voltage V_{Td} in submicrometer MOS transistors at 4.2 K," IEEE Electron Device Lett., vol. 16, p. 85, 1995.

[20] R. Narayanan, A. Ortiz-Conde, J. J. Liou, F. J. Garcia Sanchez, "Two-dimensional numerical analysis for extracting the effective channel length of short-channel MOSFETs," Solid-State Electron., vol. 38, p. 1155, 1995.

[21] Y. Taur, Y. -J. Mii, R. Logan, and H. -S. Wong, "On effective channel length in 0.1-µm MOSFET's," IEEE Electron Device Lett., vol. 16, p. 136, 1995.

[22] T. K. Lee, P. C. Liu, C. H. Gan, Y. Q. Zhang, and Y. A. Nga, "Suppression of reverse short channel effect by nitrogen implantation and its implications on nitrogen as a dopant species for applications in 0.25 µm technology," Proc. of IEEE Hong Kong Electron Device Meeting, Hong Kong, Aug. 30, 1997, p. 61.

[23] D. Tsoukalas, C. Tsamis, D. Kouvatsos, P. Revva, and E. Tsoi, "Reduction of the reverse short channel effect in thick SOI MOSFET's," IEEE Electron Device Lett., vol. 18, p. 90, 1997.

[24] T. Ando, A. B. Fowler, and F. Stern, "Electronic properties of two-dimensional systems," Rev. Mod. Phys., vol. 54, p. 437, 1982.

[25] G.Chindalore, S. A. Hareland, S. Jallepalli, A. F. Tasch, Jr., C. M. Maziar, V. K. F. Chia, and S. Smith, "Experimental determination of threshold voltage shifts due to quantum mechanical effects in MOS electron and hole inversion layers," IEEE Electron Device Lett., vol. 18, p. 206, 1997.

[26] M. J. Van Dort, P. H. Woerlee, A. J. Walker, C. A. H. Juffermans, and H. Lifka, "Influence of high substrate doping levels on the threshold voltage and mobility of deep-sbmicrometer MOSFETs," IEEE Trans. Electron Devices, vol. 39, p. 932, 1992.

[27] S. A. Hareland, S. Kirshnamurty, S. Jallepalli, C. Yeap, K. Hasnat, A. F. Tasch, Jr., and C. M. Maziar, "A computationally efficient model for inversion layer quantization effects in deep submicron n-channel MOSFETs," IEEE Trans. Electron Devices, vol. 43, p. 90, 1996.

[28] S. Jallepalli, J. Bude, W. K. Shih, M. R. Pinto, C. M. Maziar, and A. F. Tasch, Jr., "Effects of quantization on the electrical characteristics of deep submicron p- and n-MOSFETs," 1996 Symp. VLSI Tech. Dig. Tech., p. 138.

[29] S. A. Hareland, S. Jallepalli, G. Chindalore, A. F. Tasch, Jr., and C. M. Maziar, "A computationally efficient model for quantum mechanical effects in hole inversion layers in silicon PMOS devices," IEEE Trans. Electron Devices, to be published.

Chapter 4

Methods for extracting the effective channel length of MOSFETs

The channel length is a very critical parameter in CMOS technology for performance projection, device design, modeling and circuit simulation of MOSFETs [1], as evidenced by its numerous appearances in the MOSFET models developed and discussed in Chapter 1. The so-called channel length is a broad description of three different channel lengths in the MOSFET. One is the mask channel length L_m, which denotes the physical length of the gate mask. Another is the electrical effective channel length L_{eff}, which defines the length of a region near the S_i-S_iO_2 interface in which the inversion free-carrier density is controlled by the gate voltage. This channel length is given by

$$L_{eff} = L_m - \Delta L_{eff} \tag{4.1}$$

where ΔL_{eff} is the effective channel length reduction (see Fig. 4.1). The third channel length used frequently is the metallurgical channel length L_{met}, which is the distance between the source and drain metallurgical junctions at the S_i-S_iO_2 interface :

$$L_{met} = L_m - \Delta L_{met} \tag{4.2}$$

where $\Delta L_{met} = 2L_D$, and L_D is the length of the lateral diffusion of the source or drain region (Fig. 4.1).

Since the MOSFET electrical behavior is governed by the effective channel in which the inversion free carriers are controlled by the gate voltage, L_{eff} is the most important channel length for MOSFET modeling and characterization. This chapter will provide an in-depth and comprehensive coverage of the extraction of the effective channel length. The physics, methodology, and results of the existing extraction methods developed based on different approaches will be discussed and compared. Device simulation will be used to obtain various MOSFET characteristics and to aid the understanding of the extraction of MOSFET effective channel length

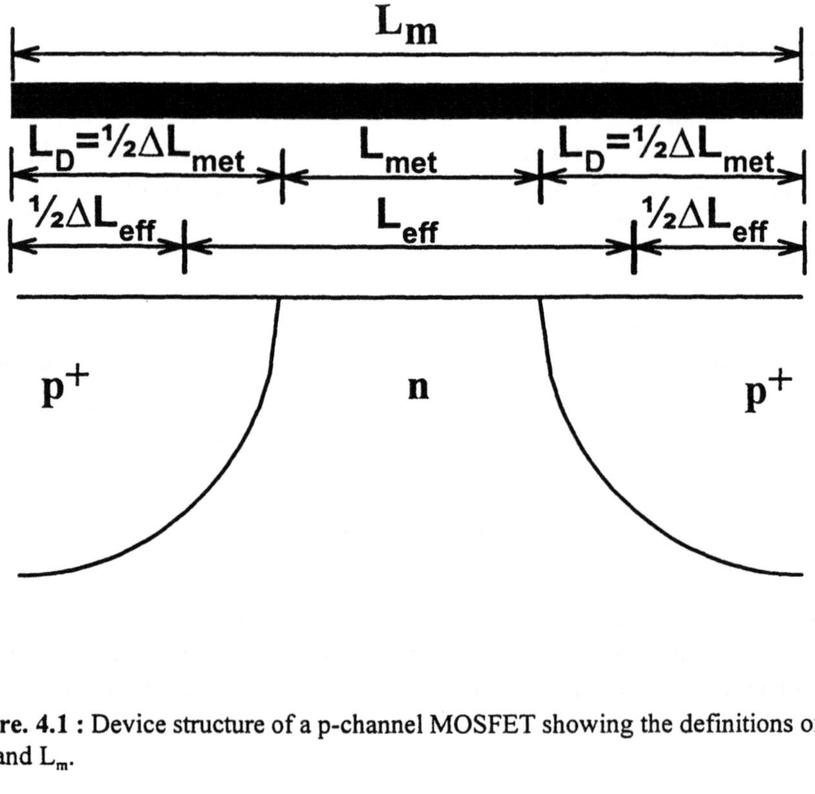

Figure. 4.1 : Device structure of a p-channel MOSFET showing the definitions of L_{eff}, L_{met} and L_m.

4.1 Introduction

The precise determination of the effective channel length is not straightforward due mainly to the fact that the gate mask length is larger than L_{met} and the gate extends over the drain and source regions in the vicinity of the metallurgical junctions (see Fig. 4.1). This gives rise to an uncertainty as to whether the portion of the drain and source regions underneath the gate should be considered as part of L_{eff} (i.e., $L_{eff} > L_{met}$) or as part of the drain and source series resistance and thus not part of L_{eff} (i.e., $L_{eff} = L_{met}$) [2-3]. Numerous studies have conducted in the past twenty years and have concluded [4-7] that the theory of $L_{eff} > L_{met}$ is more appropriate because the free-carrier density in the drain and source regions underneath the gate, like those in the channel between the drain and source metallurgical junctions, is influenced by the gate voltage.

Since ΔL_{eff}, and thus L_{eff}, cannot be measured directly, various methods have been developed in the literature to extract them from the current-voltage characteristics [8-15], capacitance-voltage characteristics [16-23], or physical insight provided by numerical simulation [4-5,24]. The main disadvantage of the methods based on current-voltage characteristics, called the I-V methods, is that they are often obscured by the presence of the parasitic drain and source series resistance. On the other hand, the main disadvantage of the capacitance-voltage (C-V) methods, is that equipments with high resolution are required to measure the small capacitances in the MOSFET (in the order of fento farads) and that it is somewhat difficult to correlate the C-V data and L_{eff}. Methods based on device physical insight require results simulated from device simulators, the accuracy of which depends on the proper selection of model parameters and may be questionable under certain conditions.

Ng and Brews [10] and Schroeder [25] published excellent reviews on analyzing the strength and weaknesses of the different methods to extract L_{eff} . Their papers pointed out the different assumptions used in developing various extraction methods. For example, one of the assumptions employed in the two most widely used I-V methods [8, 26-27] is that the drain and source series resistances are independent of the gate bias. Assumptions used in other methods include: 1) the free-carrier mobility in the channel is vertical-field dependent [28], lateral-field dependent [29], or constant [26, 30]; 2) the voltage drop across the drain and source series resistances is much smaller than the applied drain voltage [31]; 3) L_{eff} is independent of the substrate bias [32]; and 4) the drain and source series resistances are negligible [26]. Some works have been proposed to evade the use of some of these assumptions. For example, Peng et al. [31] developed an Leff extraction method in which the gate-bias dependent mobility was measured and included. On the other hand, the method by Hu et al. [33] attempted to incorporate the dependence of the drain and source series resistances with respect to the gate bias.

Because of its simplicity, the extraction method developed independently by Terada and Muta [8] and Chern et al. [9], but based on an identical approach, is the most popular one in the past twenty years. Many other methods have also been reported in the literature [11-15, 19-22, 34-37, 38-39]. Depending on the type of data used to extract L_{eff}, these methods can be categorized into the I-V method, C-V method, and simulation-based method. In the following sections, the development of the different extraction methods will be discussed, and L_{eff} extracted from these methods compared. Both measurements and a two-dimensional device simulator MEDICI [40] will be used to carry out the extraction process.

4.2 Current-voltage methods

4.2.1 Terada-Muta or Chern et al. method

The method by Terada-Muta [8] or Chern et al. [9] was derived based on a simple current-voltage relationship for MOSFET. Figure 4.2 shows the MOSFET equivalent circuit including the drain and source series resistances, and with the source and body terminals grounded. The drain current I_D in the linear region can be expressed as [1]

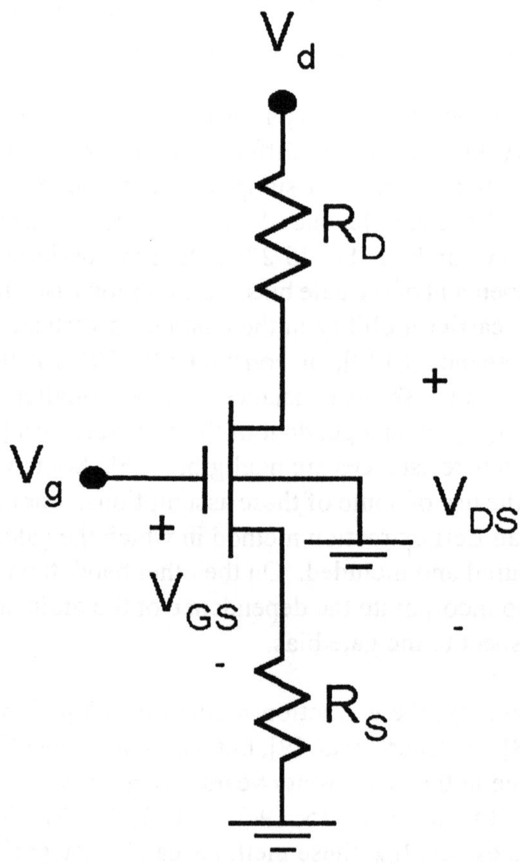

Figure 4.2 : MOSFET equivalent circuit including the source and drain series resistances (R_s and R_d) and having the body and source terminals grounded.

$$I_D = \frac{W}{L_{eff}} \mu C_o \left(V_{GS} - V_T \right) V_{DS} \tag{4.3}$$

where W is the channel width, C_o is the oxide capacitance per unit area, μ is the effective free-carrier mobility, V_T is the threshold voltage, and V_{GS} and V_{DS} are the intrinsic gate-source and drain-source voltages, respectively. The intrinsic voltages can be related to the external gate-source and drain-source voltages (V_g and V_d):

$$V_{GS} = V_g - I_D R_s \tag{4.4}$$

and

$$V_{DS} = V_d - I_D (R_S + R_D) \quad . \tag{4.5}$$

Here R_D and R_S are the drain and source series resistances, respectively. Combining equations (4.3) and (4.5), the total channel resistance, R_m, can be expressed by:

$$R_m \equiv \frac{V_d}{I_D} = R_{DS} + \frac{\left(L_m - \Delta L_{eff} \right)}{\mu C_o W \left(V_{GS} - V_T \right)} \tag{4.6}$$

where $R_{DS} \equiv (R_D + R_S)$ is the total drain and source resistance.

For the linear region under study, $(V_g - V_T)$ is much larger than $I_D R_{DS}$, and $V_g \approx V_{GS}$. This results in

$$R_m = R_{DS} + \frac{\left(L_m - \Delta L_{eff} \right)}{\mu C_o W \left(V_g - V_T \right)} \quad , \tag{4.7}$$

Then, according to (4.7), the plot of R_m versus L_m is a straight line for a given $(V_g - V_T)$, and the unique intersection of all the straight lines for different $(V_g - V_T)$ yields ΔL_{eff} on the L_m axis (i.e., x-axis) and R_{DS} on the R_m axis (i.e., y-axis). It is important to point out that the threshold voltage can be extracted from other methods (discussed in Chapter 3) and that the value of V_T is a function of L_m.

Although widely used, the Terada-Muta method has been found fail to yield accurate L_{eff} for MOSFETs operating at nitrogen liquid temperature [13-14, 41-42]. An example of this failure is illustrated in Figs. 4.3(a) and (b), which show the R_m versus L_m plots of p-channel devices at temperatures of 300 K and 77 K, respectively. At 300 K, the unique intersection of the straight lines yields $\Delta L_{eff} \approx 0.3$ μm on the x-axis and $R_{DS} \approx 60\ \Omega$ on the y-axis. On the other hand, the analogous procedure at 77 K yields no unique intersection of the straight lines, and even if the intersection of three of lines is used, a negative ΔL_{eff} is obtained, which is possible for the lightly-doped drain (LDD) MOSFET but is physically unsound for the conventional MOSFET under consideration [43]. The details of LDD MOSFET parameter extraction will be presented in Chapter 6.

The Terada method may also fail at room temperature under certain conditions. Recent numerical simulations [43-44], illustrated in Figs. 4.4(a) and (b) for MOSFETs with two different substrate doping concentration N_{Dsub}, have shown that such a method fails for MOSFETs having a relatively high doping concentration in the substrate. The simulated p-channel LDD MOSFETs had different N_{Dsub}, but the same heavily-doped source and drain p-type Gaussian profile with a peak doping concentration of N_A^+. In Fig. 4.4(a), $N_A^+ = 10^{20}$ cm^{-3} and $N_{Dsub} = 10^{17}$ cm^{-3} are considered, and a macroscopically unique intersection of the straight lines is obtained, which yields $R_{DS} = 1.8$ KΩ and $\Delta L_{eff} = -0.05$ μm. When N_{Dsub} is increased to 10^{18} cm^{-3}, however, more than one intersections exist (see Fig. 4.4(b)), and the precise value of ΔL_{eff} is not clear. As a result, it can be concluded that the Terada method becomes questionable for MOSFETs having a relatively low N_A^+ to N_{Dsub} ratio.

The failure of the Terada method can be attributed to the following assumptions used in developing the method: 1) the drain and source series resistances are independent of the gate bias; 2) V_T used in the method, and thus L_{eff} extracted, does not account for the effects of the series resistances; 3) $V_g \approx V_{GS}$; and 4) the free-carrier velocity saturation effect in the channel is negligible.

Recently, Terada and co-workers presented an improved extraction method [39], which proposed that ΔL_{eff} and R_{DS} extracted using their original method can be a function of the gate voltage due to the fact that the R_m versus L_m plot possesses several intersections of the straight lines. From these different intersections, a statistical approach is then used in their new method to determine the correct and unique ΔL_{eff} and R_{DS} based on the concept that the most accurate ΔL_{eff} and R_{DS} give rise to the least dependence of these two parameters on the gate bias. This method, which is particularly important for the lightly-doped drain (LDD) MOSFET, will be discussed in detail in Chapter 6.

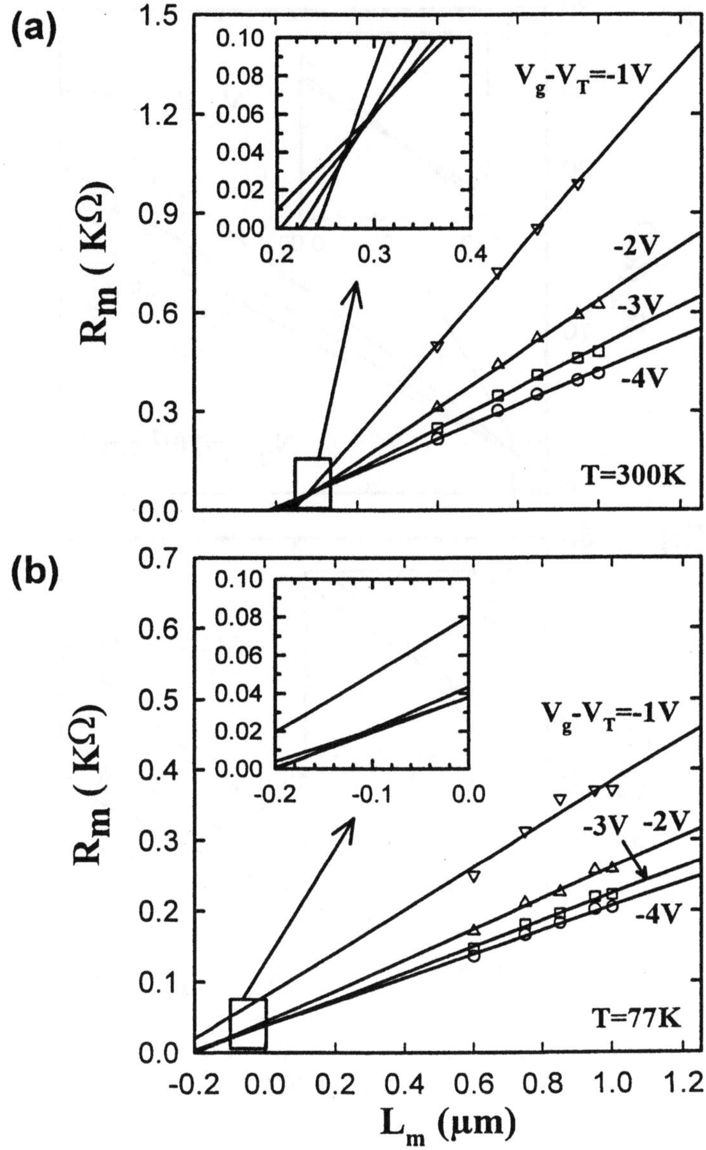

Figure 4.3 : The total channel resistance versus mask channel length for various gate voltages at (a) 300 K and (b) 77 K. The symbols are the measured data and the lines are the fittings to data using straight lines.

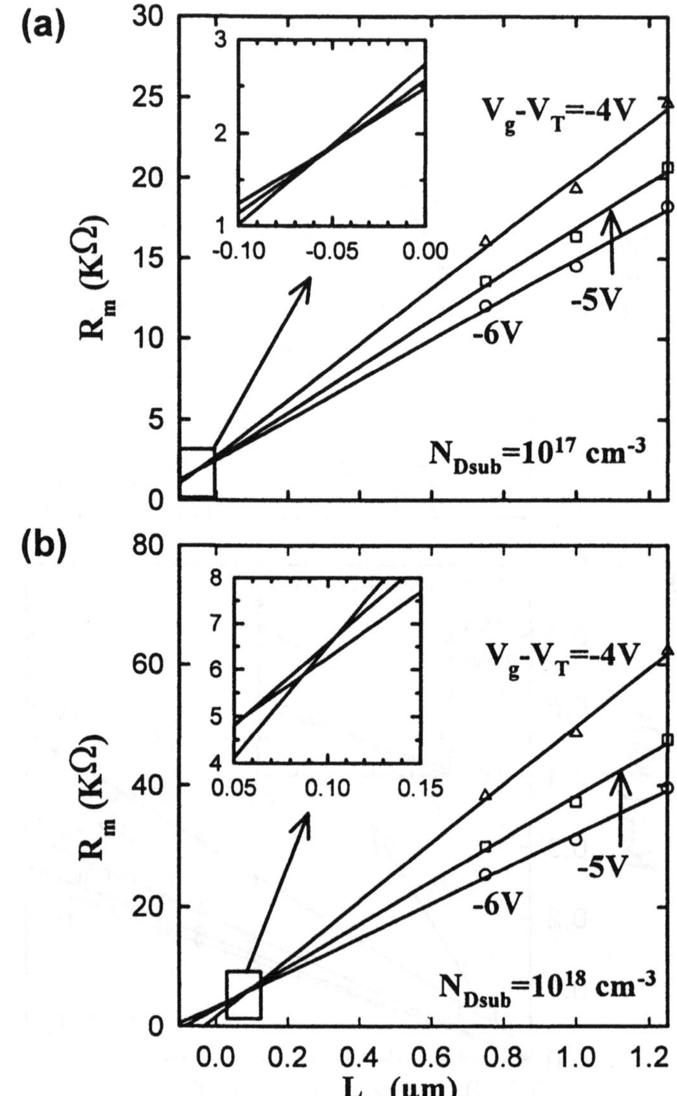

Figure 4.4 : Simulated total channel resistance versus mask channel length for two different MOSFETs with: (a) $N_A^+ = 10^{20}$ cm^{-3} and $N_{Dsub} = 10^{18}$ cm^{-3}; and (b) $N_A^+ = 10^{20}$ cm^{-3} and $N_{Dsub} = 10^{17}$ cm^{-3}.

4.2.2 Shift and ratio method

The shift and ratio (S&R) method developed by Taur et al. [11] uses an extraction concept that differs considerably from that of Terada counterpart. Their theory is centered around the total channel resistance, which was given in (4.7) and can be rewritten as

$$R_m = R_{DS} + \left(L_m - \Delta L_{eff} \right) f(V_g - V_T) \tag{4.8}$$

where $f(V_g - V_T)$ is a general function describing the MOSFET behavior. The S&R method extracts ΔL_{eff} using at least two devices (i.e., ith and jth devices) having different mask channel lengths (i.e., L_{mi} and L_{mj}, one of which needs to be long), and the following functions S_i and S_j:

$$S_i \equiv \frac{d R_{mi}}{d V_g} \approx \left(L_{mi} - \Delta L_{eff} \right) \frac{d f(V_g - V_{Ti})}{d V_g} \tag{4.9}$$

$$S_j \equiv \frac{d R_{mj}}{d V_g} \approx \left(L_{mj} - \Delta L_{eff} \right) \frac{d f(V_g - V_{Tj})}{d V_g} \quad , \tag{4.10}$$

where the assumption that R_{DS} and ΔL_{eff} are independent of V_g has been used. According to these equations, curves of S_i and S_j versus V_g can be constructed. To extract ΔL_{eff}, the S_i curve is first translated ("shift") horizontally in the V_g axis with respect to the S_j curve by the amount

$$\Delta V_{ij} \equiv \left(V_{Ti} - V_{Tj} \right) \quad , \tag{4.11}$$

because the threshold voltage is a function of the channel length. Also, the S_i curve is magnified ("ratio") in the S axis, with respect to the curve S_j, by a factor

$$r_{ij} \equiv \frac{L_{mi} - \Delta L_{eff}}{L_{mj} - \Delta L_{eff}} = \frac{S_i(V_g - \Delta V_{ij})}{S_j(V_g)} \quad . \tag{4.12}$$

The key here is to find the ΔV_{ij} value for which r_{ij} is a constant. Taur et al. [11]

solved ΔV_{ij} and r_{ij} using a statistical approach. Once the values of ΔV_{ij} and r_{ij} are found, L_{eff} can be calculated from (4.12).

The extraction can also be carried out more effectively using two new functions T_i and T_j, where

$$T_i \equiv \ln\left(|S_i|\right) = \ln\left(L_{mi} - \Delta L_{eff}\right) + \ln\left(\left|\frac{d\,f(V_g - V_{Ti})}{d\,V_g}\right|\right) \tag{4.13}$$

$$T_j \equiv \ln\left(|S_j|\right) = \ln\left(L_{mj} - \Delta L_{eff}\right) + \ln\left(\left|\frac{d\,f(V_g - V_{Tj})}{d\,V_g}\right|\right)\quad , \tag{4.14}$$

These new functions permit a more straightforward extraction of ΔL_{eff}. According to (4.13) and (4.14), the T_i curve (i.e., the curve of T function for the short-channel device) is, with respect to the T_j curve (i.e., the curve of T function for the long-channel device), 1) translated horizontally in the V_g-axis by the amount given in (4.11), and 2) translated vertically in the T-axis by the amount

$$\Delta T_{ij} \equiv \ln\left(\frac{L_{mi} - \Delta L_{eff}}{L_{mj} - \Delta L_{eff}}\right)\quad . \tag{4.15}$$

Now the key here is to find the correct values of ΔV_{ij} (i.e., horizontal translation between the two T curves) and ΔT_{ij} (i.e., vertical translation of the two T curves) from the plots of T_i and T_j versus V_g. This can be done graphically as follows: 1) assume a value for ΔV_{ij}; 2) choose a point on the T_j curve; 3) calculate ΔT_{ij} between the current T_i and T_j curves; 4) translate T_i horizontally by ΔV_{ij} and vertically by ΔT_{ij}; 5) calculate the error defined by the difference between T_j and translated T_i; and 6) the correct solution is the values of ΔV_{ij} and ΔT_{ij} that produces the smallest error. Putting the correct ΔT_{ij} into (4.15), together with the information of L_{mi} and L_{mj}, yields ΔL_{eff}. Note that now the problem of a translation and a magnification in the original S&R method has been changed to a more straightforward dual-translation problem. While it is possible to eliminate the translation in the T-axis by differentiating T_i with respect to V_g, it is better not to do so because such a mathematical manipulation would reduce the signal level and increase the effect of the noise on the experimental data.

We have applied this method to n-channel MOSFETs with a channel width of 20 μm and mask channel lengths of 1.75, 2.00 and 20 μm. An increment of 100 mV for the gate voltage and drain voltage of 100 mV were used in the measurements. Figure 4.5 shows the S function versus V_g characteristics, and Figure 4.6(a) shows the T function versus V_g characteristics obtained from the experimental data. Then, using the plots for $L_m = 2$ and 20 μm and different values of ΔV_{ij}, we calculated the corresponding ΔT_{ij} by shifting the plot and carrying out a numerical fit for the range $2V < V_g < 5V$. The range $V_g < 2V$ was not included in order to avoid moderate and weak inversion. Figure 4.6(b) presents ΔT_{ij} and the corresponding error versus ΔV_{ij} using the 2- and 20-μm MOSFETs. Since the error is minimal at about $\Delta V_{ij} = 0.07$ V, the solution is $\Delta T_{ij} = 2.58$. Putting this into (4.15), we obtained $\Delta L_{eff} = 0.53$ μm. If 1.75- and 20-μm MOSFETs are used, then $\Delta L_{eff} = 0.48$ μm is obtained. We also show in

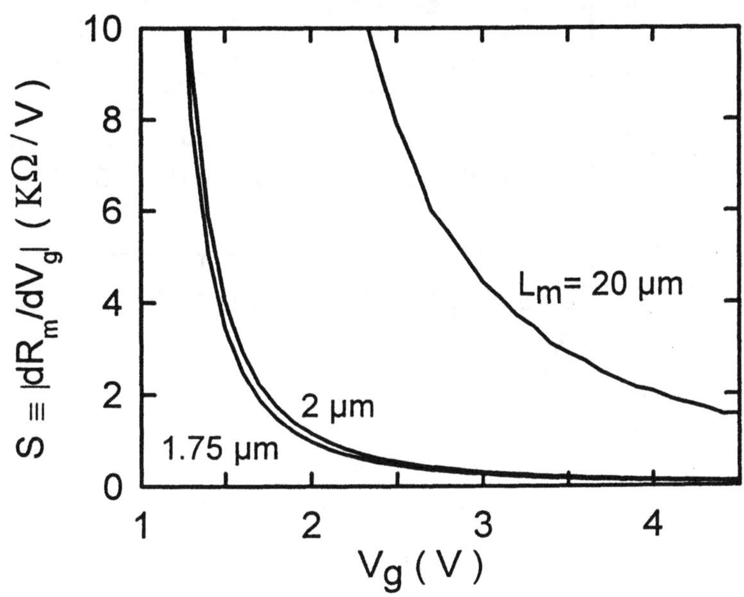

Fig. 4.5 : The function S versus the gate voltage for MOSFETs with three different mask channel lengths.

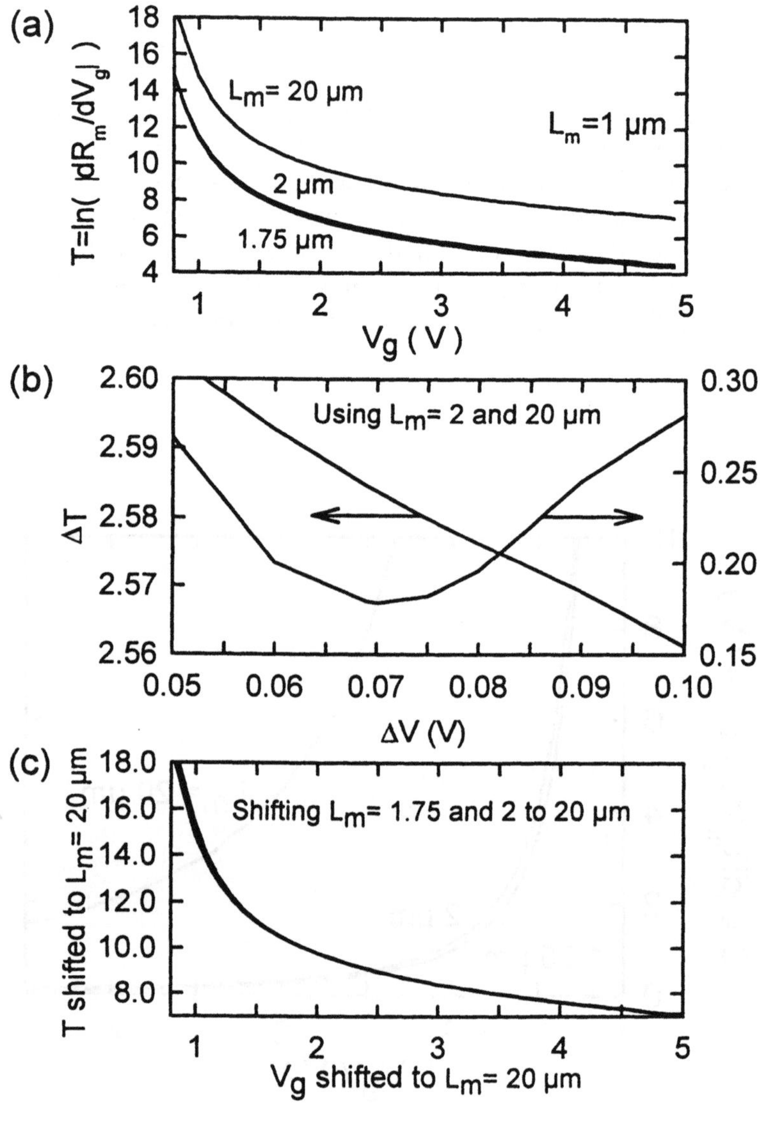

Figure 4.6 : (a) The function T versus the gate voltage for MOSFETs with three different mask channel lengths. (b) Shift in T versus the shift in gate bias and the corresponding error. (c) The plots for L_m= 1.75 and 2 μm shifted to the plot for L_m = 20 μm.

It is evident that the S&R method is more complex in extracting L_{eff} than the Terada method In addition, such a method may not be accurate in some cases due to the use of following assumptions: 1) the series resistances are assumed independent of the gate bias; 2) $V_g \approx V_{GS}$; and 3) the effect of drift velocity saturation along the channel is assumed negligible.

4.2.3 Conductance method

The effective channel length can also be extracted based on the MOSFET conductance characteristics [13-14, 42]. This method, called the conductance method, accounts for the carrier drift velocity saturation effects [45-46] and has been used to extract the parameters at both room and liquid nitrogen temperatures.

The benefits of operating CMOS devices at low temperature over traditional operation at room temperature have been widely reported in the literature. Some of these benefits are [47] high carrier mobility, improved latch-up immunity, and low subthreshold current. Modeling and parameter extraction of the devices at liquid nitrogen temperature (77 K) differ considerably from those at room temperature. For example, as mentioned earlier, the Terada-Muta method [8-10] to extract ΔL_{eff} can give an unrealistic negative value for ΔL_{eff} at 77 K. Also, while some conventional methods have been successfully used [48] at 77 K for long-channel devices (i.e., 2 μm ≤ L_m ≤ 6 μm) operating at relatively low drain voltages (i.e., V_d < 40 mV), there are other studies [12, 41] demonstrating that such methods failed for short-channel MOSFETs (i.e., 0.4 μm ≤ L_m ≤ 2 μm) at low temperatures.

Most conventional methods were developed based on the questionable assumptions that the drain and source series resistances are independent of the gate voltage, and that the drain current I_D is linearly proportional to the extrinsic drain voltage, because the drain voltage is kept very small and the device is biased in the so called "linear region". The latter assumption becomes invalid for MOSFETs having a sufficiently small channel length. For such devices, the electric field along the channel is very high, even if a small drain voltage is applied, which then causes the drift velocity to saturate. This is particularly true at low temperatures where the free-carrier mobility is increased. Consequently, the conventional methods have a greater difficulty in determining L_{eff} at low temperatures. The assumption about the voltage-independent series resistance is also critical to the extraction of L_{eff}, which has been analyzed theoretically by Ng and Lynch [2], experimentally by Hu and coworkers [33], and more comprehensively by Ng and Brews [10].

The conductance method presented in this section accounts for the carrier velocity saturation effects [45-46], thus allowing one to evaluate the dependence of the series resistance on the gate voltage and to extract L_{eff} of short channel MOSFETs at both room and liquid nitrogen temperatures.

Following the model proposed by Shur et al. [49-50] for p-channel MOSFETs, and using the strong inversion condition and the approximation $V_g \approx V_{GS}$, the drain current can be expressed as:

$$I_D = \frac{W \mu_{lf} C_o \left(V_g - V_T \right) V_{DS}}{L_{eff} \left(1 + \frac{V_{DS}}{V_{SATE}} \right)} \quad . \tag{4.16}$$

where μ_{lf} is the effective free-carrier mobility for low field and V_{SATE} is an effective voltage which accounts for the carrier velocity saturation effect. Combining (4.16) and (4.5), I_D can be expressed in terms of the extrinsic drain voltage V_d as

$$I_D = \frac{\beta V_{gT} V_{SATE}}{2} + \frac{V_{SATE} + V_d}{2 R_{DS}} - \frac{V_X}{2 R_{DS}} \tag{4.17}$$

where

$$V_{gT} \equiv (V_g - V_T) \tag{4.18}$$

$$\beta \equiv \frac{W \mu_{lf} C_o}{L_{eff}} \tag{4.19}$$

and

$$V_X \equiv [(\beta V_{gT} V_{SATE} R_{DS})^2 + 2 R_{DS} \beta V_{gT} V_{SATE} (V_{SATE} - V_d) + (V_{SATE} + V_d)^2]^{1/2} \tag{4.20}$$

Using the approach of Wen and coworkers [51], we now take the first and second derivatives of I_D with respect to V_d from equation (4.17):

$$G \equiv \frac{\partial I_D}{\partial V_d} = \frac{1}{2\,R_{DS}} + \frac{\beta\, V_{gT}\, V_{SATE}}{2\,V_X} - \frac{V_{SATE} + V_d}{2\,R_{DS}\,V_X} \tag{4.21}$$

and

$$\frac{\partial^2 I_D}{\partial V_d^2} = \frac{2\,\beta\,V_{gT}\,V_{SATE}^2}{V_X^3} \quad . \tag{4.22}$$

where G is the MOSFET conductance. Then, combining equations (4.21) and (4.22) yields the following expression:

$$G = \frac{1}{2\,R_{DS}} + C_1\,L_{eff}^{\ 1/3} + C_2\,L_{eff}^{\ -2/3} \tag{4.23}$$

where C_1 and C_2 are two constants governing by the following relationship:

$$\frac{C_2}{C_1} \approx -\,G_o\,R_{DS}\,L_{mo} \quad . \tag{4.24}$$

L_{mo} is the mean mask channel length of all the MOSFETs considered, and G_o is the mean conductance of these devices. Equation (4.23) allows one to determine R_{DS} and L_{eff} from the data of G as a function of V_g and L_m.

Figures 4.7(a) and (b) show the conductance versus mask channel length obtained from measurements (symbols) and from fitted model calculations (lines) for various gate voltages at 300 K and 77 K, respectively. The corresponding total resistance R_m obtained from measurements (symbols) and model calculations (lines) for 300 and 77 K are illustrated in Figs. 4.8(a) and (b). We wish to stress that although the results in Figs. 4.8(a) and (b) are not straight lines, their intersections give roughly ΔL_{eff} and R_{DS} on the L_m and R_m axises, respectively. Alternatively, the values of these two parameters can be extracted from the conductance method more precisely by fitting numerically the model to experimental data.

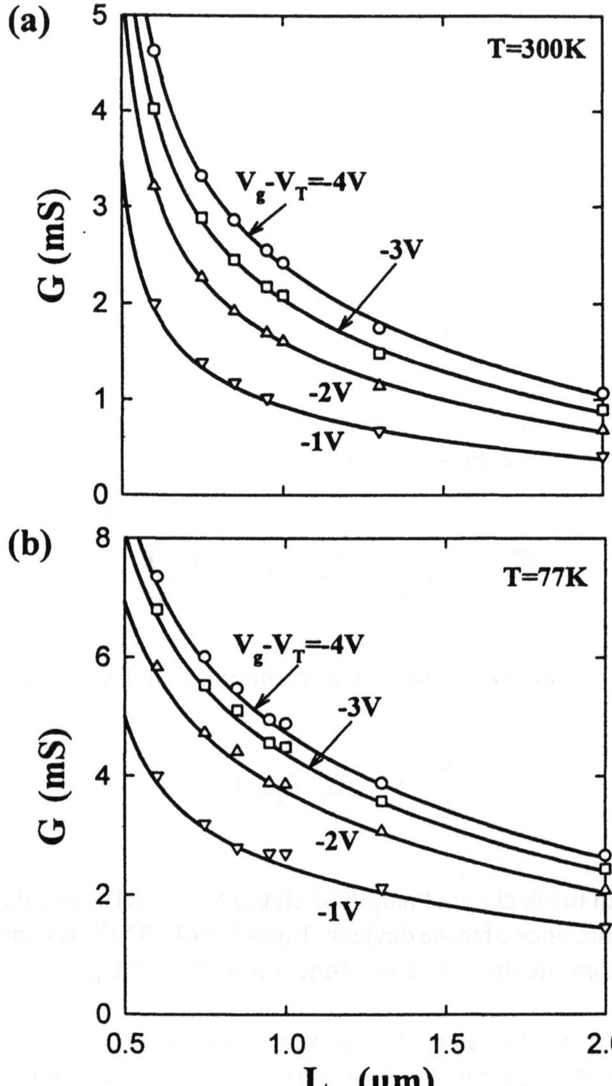

Figure 4.7 : Total channel conductance versus mask channel length for various gate voltages at (a) 300 K and (b) 77 K. The symbols are the measured data and the lines are the fittings to data using the conductance method.

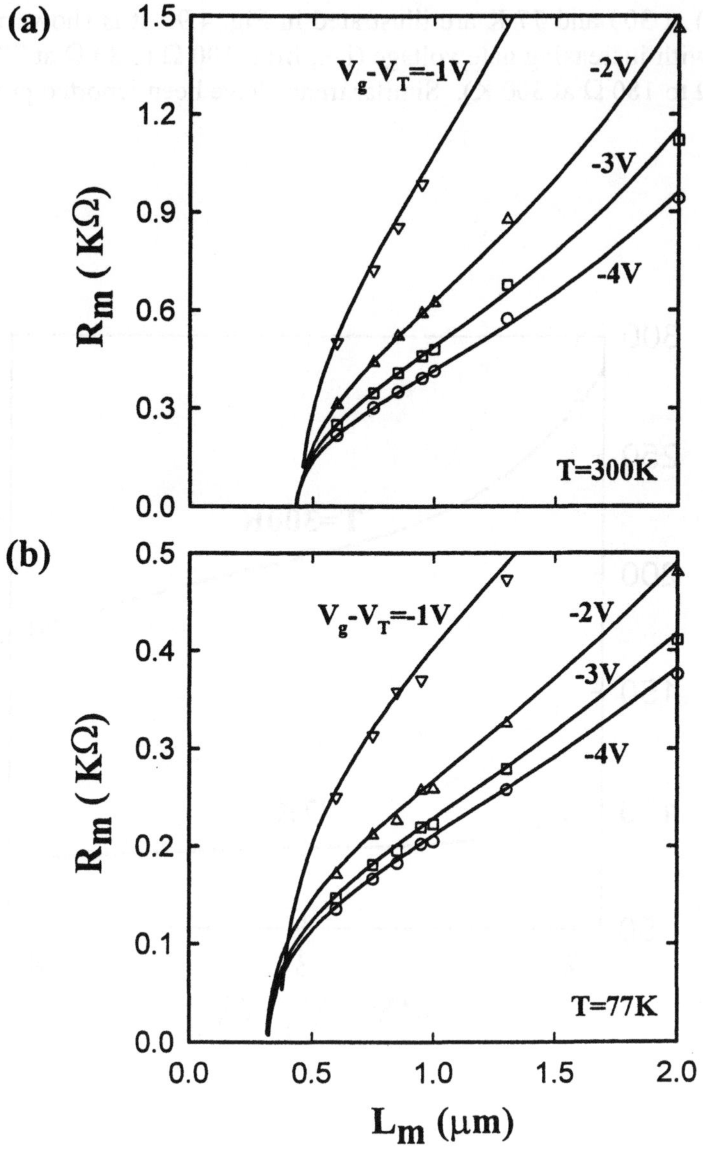

Figure 4.8 : Total channel resistance versus mask channel length for various gate voltages at (a) 300 K and (b) 77 K. The symbols are the measured data and the curves are the fittings to data using the conductance method.

The extracted values of the total series resistance (i.e., drain and source series resistances) at 300 and 77 K are illustrated in Fig. 4.9. It is shown that R_{DS} decreases with increasing gate voltage (i.e., from 100 Ω to 80 Ω at 77 K, and from 270 Ω to 180 Ω at 300 K). Similar trends have been reported previously [12].

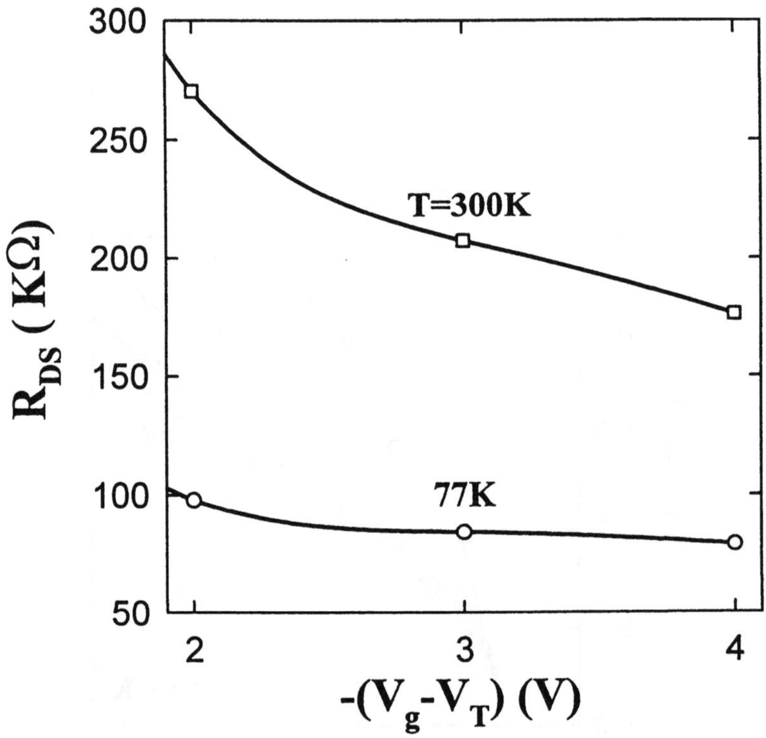

Figure 4.9 : Extracted values of the total drain and source series resistance versus gate voltages for two temperatures.

The extracted values of the effective channel length reduction, ΔL_{eff} (= L_m - L_{eff}), for the two temperatures are shown in Fig. 4.10. The results suggest that ΔL_{eff} depends weakly on V_g but strongly on temperature.

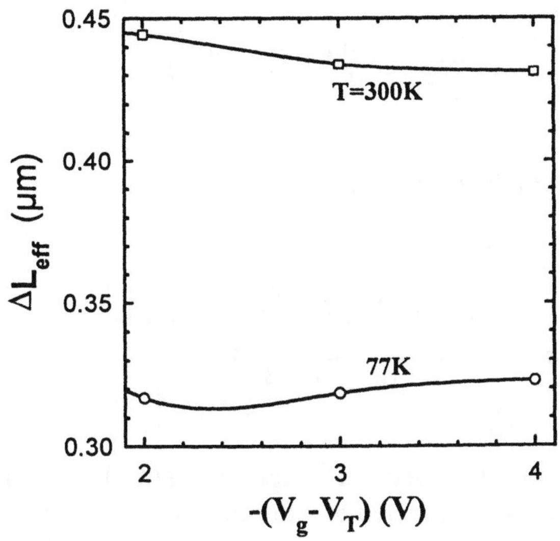

Figure 4.10 : Extracted values of the difference between the mask channel length and the effective channel length (i.e., $\Delta L_{eff} = L_m - L_{eff}$) for two different temperatures.

4.2.4 Fikry et al. method

The method by Fikry et al. [35] is similar to the conductance method discussed in the previous section in the sense that it accounts for the carrier velocity saturation effect in the channel and uses the assumption of $V_g \approx V_{GS}$. The two approaches to implement the saturation effect and to determine the effective channel length, however, are quite different. According to the Fikry et al. method [35], the velocity saturation effect is imbedded in the following free-carrier mobility model:

$$\mu = \frac{\mu_o}{\left(1 + \theta\left(V_g - V_T\right)\right)\left(1 + \frac{\mu_o V_d}{L_{eff}\, v_{sat}}\right)} \tag{4.25}$$

where μ_o is the low-field mobility, θ is the mobility degradation factor due to the vertical field, and v_{sat} is the saturation velocity of the carriers.

The following function is then used, which is analogous to the approach proposed in the threshold-voltage extraction method by Jain [52]:

$$\frac{I_D}{g_m^{1/2}} = s^{-1/2}\left(V_g - V_T\right) \tag{4.26}$$

where g_m is the transconductance and

$$s = \frac{L_m - \left(\Delta L_{eff} - \dfrac{\mu_o V_d}{v_{sat}}\right)}{W \mu_o C_o V_d} \tag{4.27}$$

The above two equations were derived by combining (4.3), (4.5) and $V_g \approx V_{GS}$. The values of V_T and s are extracted by plotting $I_D/g_m^{1/2}$ versus V_g. Then, the plot of s versus L_m allows one to obtain μ_o from its slope and ($\Delta L_{eff} - \mu_o V_d/v_{sat}$) from its intercept to the L_m axis. Thus, ΔL_{eff} can be determined from ($\Delta L_{eff} - \mu_o V_d/v_{sat}$), provided the value of v_{sat} is calculated from the following equation describing transconductance of the device biased in the saturation region:

$$g_m = W C_o v_{sat} \tag{4.28}$$

Alternatively, ΔL_{eff} can also be obtained from the extrapolation of ($\Delta L_{eff} - \mu_o V_d/v_{sat}$) versus V_d plot, which is a straight line, to the y axis where V_d is zero.

We have tested this method using MEDICI simulation results of p-channel LDD MOSFETs having mask channel lengths varying from 0.2 to 1.0 μm. The LDD structure, which is commonly used in modern MOSFETs, has a lightly doped region near the drain junction to reduce the electric field and thereby minimize the hot-carrier effects [1]. All simulated devices had drain and source metal contact width of 1 μm, a substrate doping concentration of 10^{17} cm^{-3}, and the same source and drain make-up as follows (see Fig. 4.11): (i) a heavily doped p-type Gaussian profile with a peak doping concentration of 10^{20} cm^{-3}, a junction depth of 0.12 μm with a lateral extent of 0.09 μm (0.75 of their vertical extent) with respect to the spacer; and (ii) a lightly doped p-type Gaussian profile with a peak doping concentration of 10^{19} cm^{-3}, a junction depth of 0.08 μm with a lateral extent of 0.06 μm (0.75 of their vertical extent) with respect to the gate. The gate oxide thickness is 10 nm (i.e., the oxide capacitance is 3.45x10^{-7}

F/cm²). The spacers are 60 nm width and the separation between the source- and drain-contact to the spacer is 1 μm. The carrier saturation velocities for electrons and holes are $v_{satn} = v_{satp} = 10^6$ cm/s.

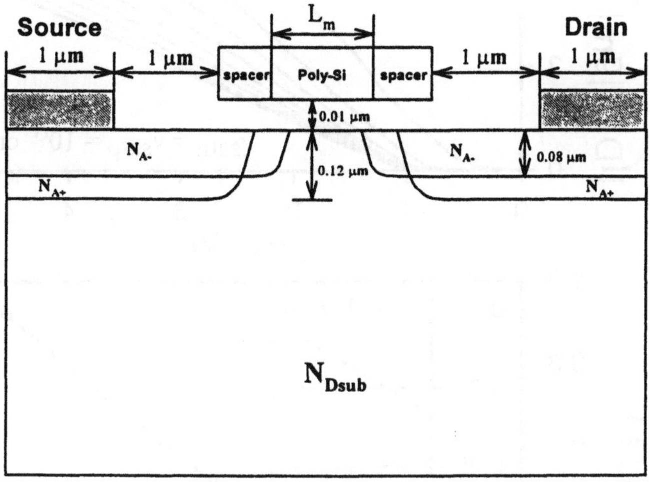

Figure 4.11 : Device structure of a p-channel lightly-doped drain (LDD) MOSFET used in the simulations.

Figure 4.12(a) shows the MEDICI simulation results of $I_D/g_m^{1/2}$ versus V_g for several mask channel lengths and V_d = -50 mV. The linear extrapolation of the curves to the V_g axis gives V_T. The corresponding plot of s versus L_m, illustrated in Fig. 4.12(b), yields $W\mu_o C_o V_d = 0.99 \times 10^{-6}$ μm/Ω from its slope and $(\Delta L_{eff} - \mu_o V_d/v_{sat}) = -0.0134$ μm from its intercept to the L_m axis. Then, using $C_o = 3.45\text{x}10^{-7}$ F/cm², $|V_d| = 0.05$ V, W = 1 μm and $v_{satn} = v_{satp} = 10^6$ cm/s, we obtain $\mu_o = 57$ cm²/V.s and $\Delta L_{eff} = 0.015$ μm from the Fikry method.

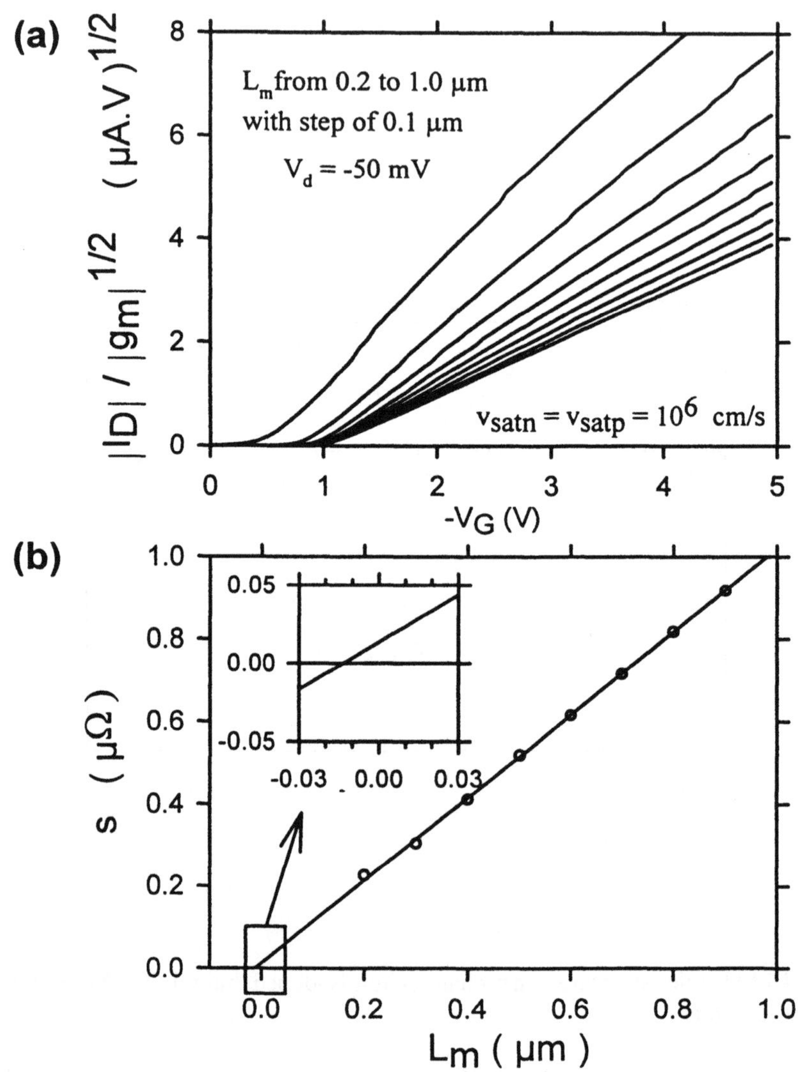

Figure 4.12 : (a) Calculated values of $I_D/g_m^{1/2}$ versus V_g for several mask channel length and V_d = -50 mV. The slopes of these approximate straight lines give the values of s. (b) Calculated values of s versus L_m. The slope of this approximate straight line yields $W\mu_o C_o^a V_d = 0.99 \times 10^{-6}$ μm/Ω and the intercept of the line at the L_m axis gives $(\Delta L_{eff} - \mu_o V_d/v_{sat}) = -0.0134$ μm.

If a larger bias condition of V_d = -100 mV is used in simulation, then $W\mu_o C_o V_d$ = 1.94 x 10^{-6} μm/Ω, $(\Delta L_{eff} - \mu_o V_d/v_{sat})$ = -0.053 μm, μ_o = 56 cm²/V.s and ΔL_{eff} = 0.042 μm. The fact that different V_d gives rise to different ΔL_{eff} suggests that the method is sensitive to the bias condition and that a small voltage should be used to make sure the MOSFET operated in the linear region. An alternative way to extract ΔL_{eff} is extrapolating the $(\Delta L_{eff} - \mu_o V_d/v_{sat})$ versus V_d plot to the point of V_d = 0 (i.e., y axis), as illustrated in Fig. 4.13, which gives ΔL_{eff} = 0.026 μm.

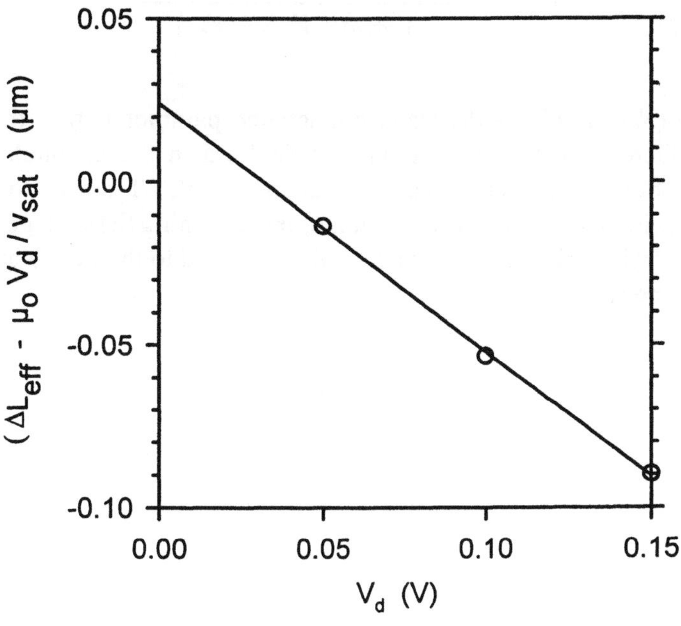

Figure 4.13 : Extracted values of $(\Delta L_{eff} - \mu_o V_d/v_{sat})$ (open circles) for three different V_d. The intercept of the straight line passing through these points at the vertical axis (i.e., V_d = 0) yields ΔL_{eff} = 0.026 μm.

4.2.5 Nonlinear optimization method

The nonlinear optimization method [34, 38] extracts ΔL_{eff} based on optimization techniques applied to the MOSFET current-voltage characteristics. It has two main advantages: (1) the consistent determination of all the parameters of the model because of the simultaneous extraction; and (2) the reduction of the effects of the noise on the experimental data due to the optimization techniques. There are two main disadvantages, however: (1) nonphysical parameters values can be obtained because of the pure fitting scheme, and (2) the requirement of a long computational process.

The outline of the development of such a method, proposed by Karlsson [38], is given below. The drain current for the MOSFET is

$$I_D = \frac{\beta \left(V_{GS} - V_T - \frac{V_{DS}}{2} \right) V_{DS}}{1 + \theta \left(V_{GS} - V_T \right)} \tag{4.29}$$

where $\beta = (W/L_{eff})\mu C_o$ is the transconductance parameter, θ is the mobility reduction factor due to the vertical electric field in the channel, and other parameters have their usual meaning. For the MOSFET biased in the strong inversion region with a small drain voltage, and assuming the voltage drop in the source and drain series resistances is small compared to the gate bias, the drain current can be rewritten as

$$I_D = a \frac{V_g - b}{V_g - c} V_d \tag{4.30}$$

where

$$a = \frac{\beta}{\theta + \beta R_{DS}} \tag{4.31}$$

$$b = V_T + \frac{V_d}{2} \tag{4.32}$$

$$c = V_T - \frac{1}{\theta + \beta R_{DS}} \tag{4.33}$$

Figure 4.14(a) shows the simulated I_D versus V_g characteristics (closed circles) for V_d = -50 mV using the same devices described in the previous section. The fits (lines) to the simulated results were obtained by optimizing the values of a, b and c such that the following parameter ϵ has the minimum value:

$$\epsilon \equiv \sum_{i=1}^{N} \left(I_D - a \frac{V_g - b}{V_g - c} V_d \right)^2 \tag{4.34}$$

Then, knowing the values of a, b and c, we can calculate the following three parameters:

$$\beta = \frac{a}{b - c \dfrac{V_d}{2}} \tag{4.35}$$

$$\theta + \beta R_{DS} = \frac{1}{b - c - \dfrac{V_d}{2}} \tag{4.36}$$

$$\beta^{-1} = (\mu C_o)^{-1} \frac{(L_m - \Delta L_{eff})}{W} \tag{4.37}$$

Figure 4.14(b) shows the straight line fitted to the ($\theta + \beta R_{DS}$) versus β simulation data (symbols). From the slope of this straight line we obtain R_{DS} = 1.72 KΩ, and from the intercept of the line at the vertical axis we get θ = 0.097 V^{-1}. Figure 4.14(c) presents the straight line fitted to the β^{-1} versus L_m simulation data. This allows us to extract $\mu C_o = 2.0 \times 10^{-5}$ A/V^2 from the slope (using W = 1 μm), and ΔL_{eff} = -0.02 μm from the intercept at the L_m axis. Finally, using $C_o = 3.45 \times 10^{-7}$ F/cm^2, we obtain μ = 57 cm^2/V.s.

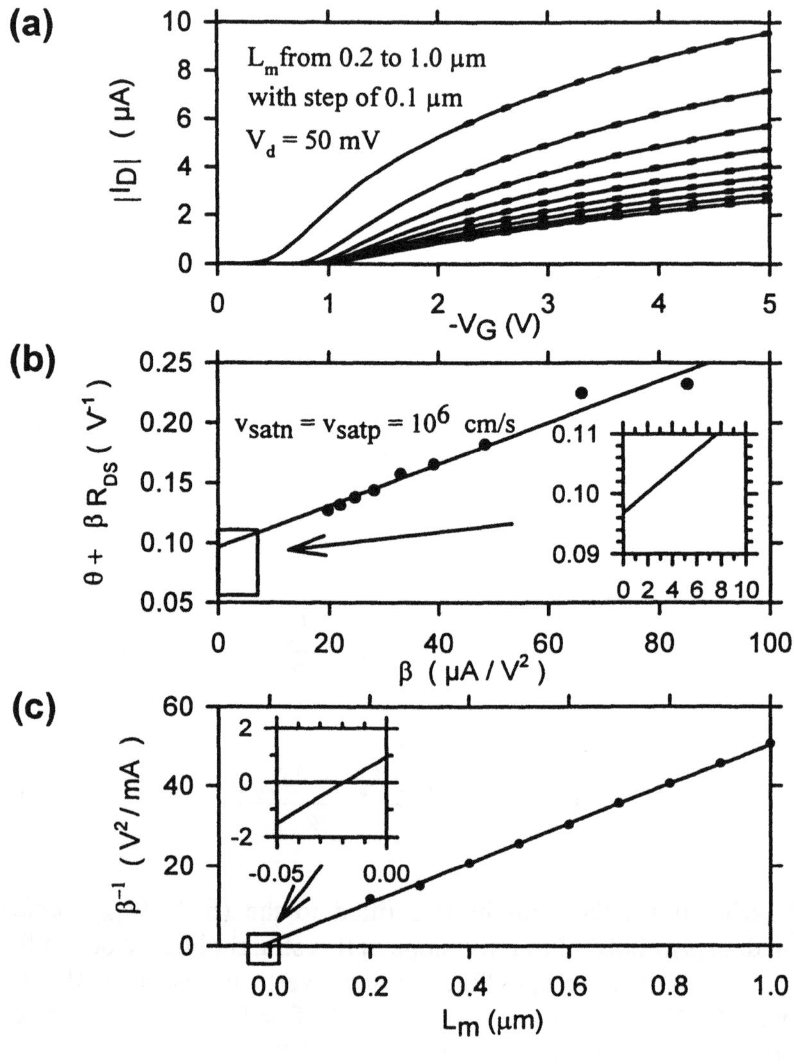

Figure 4.14 : (a) Simulated (closed circles) and fitted (lines) I_D versus V_g for V_d = -50 mV. The fits give the values of a, b and c. (b) Calculated ($\theta + \beta R_{DS}$) versus β using the values of a, b and c. The slope of ($\theta + \beta R_{DS}$) versus β gives R_{DS} = 1.72 KΩ, and the intercept at the vertical axis yields θ = 0.097 V^{-1}. (c) Extracted β^{-1} versus $L_{m,}$ which gives $\mu C_o^a = 2.0 \times 10^{-5}$ A/V^2 from the slope (using W = 1 μm), and ΔL_{eff} = -0.02 μm from the intercept at the L_m axis.

4.2.6 Jean and Wu method

The method by Jean and Wu [15] is intended to modify Terada's method so that the effects of gate-voltage dependence on ΔL_{eff} and R_{DS} can be accounted for. Assuming $V_g \approx V_{GS}$, equation (4.6) can be rewritten as:

$$R_m = Y_{cept} + S_{lope} L_m \tag{4.38}$$

where

$$S_{lope} = \frac{1}{\mu C_o W \left(V_g - V_T \right)} \tag{4.39}$$

and

$$Y_{cept} = R_{DS} - S_{lope} \Delta L_{eff} \quad . \tag{4.40}$$

Then ΔL_{eff} is obtained by differentiating Y_{cept} with respect to S_{lope} [15]:

$$\frac{d Y_{cept}}{d S_{lope}} = - \Delta L_{eff} \quad . \tag{4.41}$$

Unfortunately, such a method fails to deliver the promise of improving Terada's method, due to the fact that the above equations are valid only if ΔL_{eff} and R_{DS} are independent of V_g. This can be verified analytically by noting both Y_{cept} and S_{lope} are functions of V_g, as (4.41) can be rewritten in the form of

$$\frac{d Y_{cept}}{d S_{lope}} = \frac{\dfrac{d Y_{cept}}{d V_g}}{\dfrac{d S_{lope}}{d V_g}} \tag{4.42}$$

4.3 Capacitance-voltage method

To avoid the effect of the parasitic drain and source series resistances, which is a main mechanism causing the difficulty in the I-V methods, various methods have been developed to extract ΔL_{eff} from the capacitance-voltage characteristics (i.e., C-V methods) [16-23]. The main drawback of the C-V methods is the requirement of high resolution equipments to measure the small capacitances in MOSFETs, particularly in state-of-art MOSFETs having a very small geometry. Moreover, as will be shown later, it is somewhat difficult to correlate the C-V data and L_{eff}.

This section reviews the development of the C-V methods and investigates the validity these methods based on results simulated from MEDICI [40]. The effective channel length extracted from the C-V methods will be compared with those obtained from the widely used I-V methods [8-9] and the device simulation-based method [4-5]. It will be demonstrated later that L_{eff} obtained from the C-V method is much smaller than those obtained from the I-V methods and that such a discrepancy results from inconsistencies imbedded in the development of the C-V methods.

4.3.1 Device structure and C-V simulations

P-channel MOSFETs with mask channel lengths of 0.75, 1 and 1.25 μm are simulated using MEDICI. Figure 4.15 shows a conventional p-channel MOSFET with the following source and drain make-up: a p-type Gaussian profile with a peak doping concentration of 10^{20} cm^{-3}, a junction depth of 0.34 μm with a lateral extent of 0.25 μm (75% of their vertical extent), and a contact width of 1 μm. The gate oxide thickness is 25 nm (i.e., the oxide capacitance per unit area C_o^a is 1.37×10^{-15} F/μm^2) and the substrate doping concentration is 10^{17} cm^{-3}. The superscript "a" in the capacitances indicates that this is a capacitance per unit area.

Figure 4.16 shows the doping concentration at the Si-SiO_2 interface (i.e., zero vertical distance) along the channel (i.e., from the source to drain) for the device shown in Fig. 4.15. Note that the source and drain metallurgical junctions are located at 2.25 μm and 2.5 μm, respectively. For this particular MOSFET, L_m = 0.75 μm because the gate extends from 2.0 μm to 2.75 μm, the lateral diffusion length L_D of the source/drain region is 0.25 μm, and the metallurgical channel length $L_{met} = L_m - 2L_D$ = 0.25 μm.

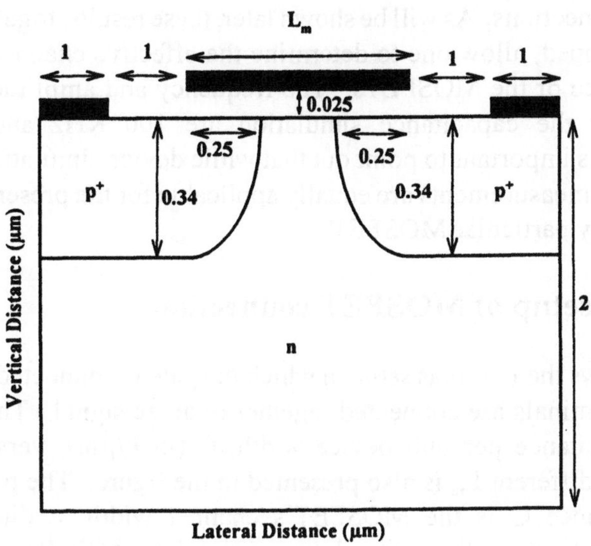

Figure 4.15 : Device structure of a conventional p-channel MOSFET used in simulations.

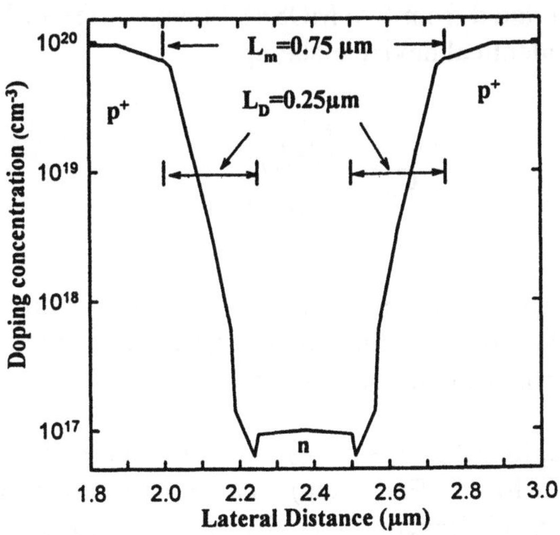

Figure 4.16 : Impurity doping concentration at the Si-SiO$_2$ interface along the channel for the conventional MOSFET.

We will simulate the C-V characteristics of the MOSFETs using three different small-signal connections. As will be shown later, these results, together with the equations developed, allow one to determine the effective channel length and oxide capacitance of the MOSFETs. The frequency and amplitude of the ac signal used for the capacitance simulation are 100 KHz and 100 mV, respectively. It is important to point out that while device simulations are used here, direct C-V measurements are equally applicable for the present method to extract L_{eff} of any particular MOSFET.

4.3.1.1 First setup of MOSFET connection

Figure 4.17 shows the first bias setup in which the gate is connected to a dc bias and all other terminals are connected together to an ac signal. The simulated MOSFET capacitance per unit device width, C (in F/μm), versus the gate voltage, V_g, for different L_m is also presented in the figure. The per μm in the unit of capacitance C is the MOSFET's channel width, a dimension not accounted for in the two-dimensional device simulator MEDICI. Let us first focus on the accumulation region (i.e., $V_g > 2$ V). It can be seen that this capacitance, called C_{A1} (subscript A1 denotes accumulation for the first setup), increases with increasing L_m. Since the channel is accumulated with majority carriers, the substrate does not contribute to the increase of C_{A1} with increasing L_m. Therefore, the increase of C_{A1} is due only to the increase of the oxide capacitance C_o^a. Thus, C_o^a can be calculated based on the difference in C_{A1} (i.e., ΔC_{A1}, in F/μm) simulated from two MOSFETs with different L_m (i.e., ΔL_m, in μm) but otherwise identical device make-up:

$$C_o^a = \left[\frac{\Delta C_{A1}}{\Delta L_m}\right] \quad . \tag{4.43}$$

From equation (4.43) and the results in Fig. 4.17, we obtain $C_o^a = 1.36\times10^{-15}$ F/μm^2, which is the correct value for the gate oxide thickness of 25 nm (1.37×10^{-15} F/μm^2 from calculation) considered.

For the inversion region (i.e., $V_g < -2$ V), the capacitance, called C_{I1} (subscript I1 denotes inversion for the first setup), is also proportional to L_m. Here, the channel is inverted and is linked electrically to the drain and source, which are connected to the ac signal. Since the same ac signal is applied to the body terminal, the channel has the same potential as the substrate, and the small-signal variation does not produce a change in the free-carrier charge in the channel. Therefore, the increase of the capacitance with increasing L_m in the inversion region is due only to the oxide capacitance increase, and C_o^a can also be calculated by

$$C_o^a = \left[\frac{\Delta C_{II}}{\Delta L_m}\right] \tag{4.44}$$

where ΔC_{II} (i.e., in F/μm) is the difference in C_{II} simulated from two MOSFETs with different L_m (i.e., ΔL_m, in μm). Using (4.44), we obtained $C_o^a = 1.37 \times 10^{-15}$ F/μm², which is almost the same as that calculated using (4.43).

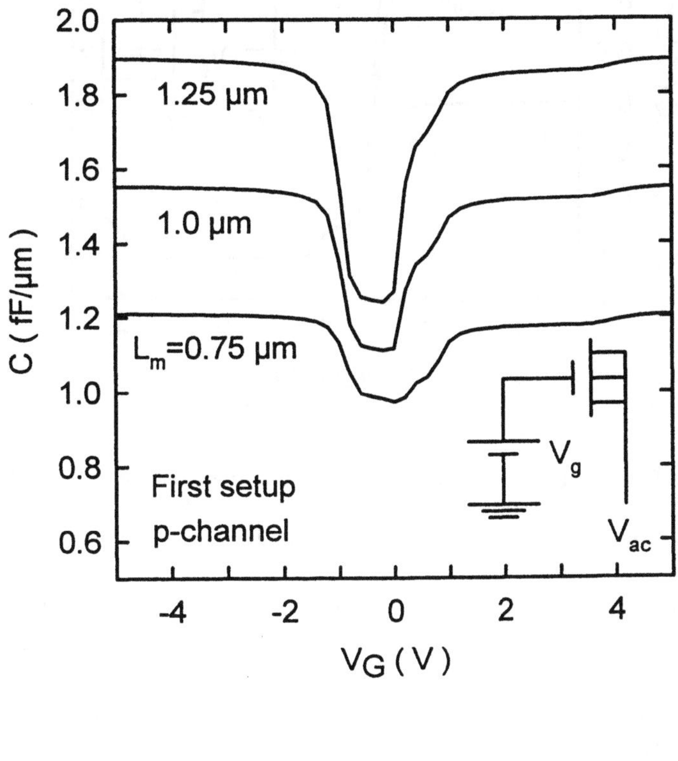

Figure 4.17 : The simulated capacitance per unit length versus the gate voltage for three different L_m obtained from the first setup with the gate connected to a dc bias and all other terminals connected to an ac signal.

4.3.1.2 Second setup of MOSFET connection

Figure 4.18 shows the second setup in which the gate is connected to a dc bias, the body is grounded, and the drain and source are connected to an ac signal. The simulated MOSFET capacitance per unit device width, C (F/μm), versus the gate bias, for different L_m is also presented.

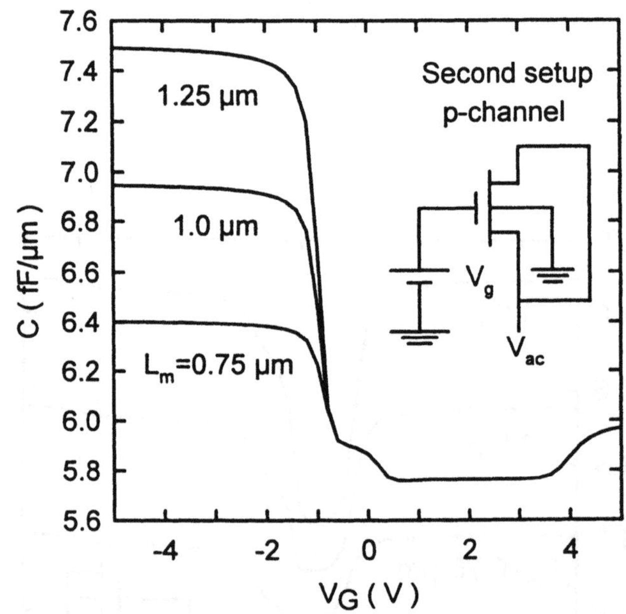

Figure 4.18 : The simulated capacitance per unit length versus the gate voltage for three different L_m obtained from obtained from the second setup having the gate connected to a dc bias, the body grounded, and the drain and source connected to an ac signal.

Let us first focus on the accumulation region (i.e., $V_g > 2$ V). It can be seen that this capacitance, called C_{A2} (subscript A2 denotes accumulation for the second setup), does not depend on L_m. Since the channel is accumulated and thus is not linked electrically to the drain and source, which are connected to the ac signal, this capacitance is associated only with the gate-source, gate-drain, body-source, and body-drain junction regions. It is important to point out that C starts to increase for $V_g > 4$ V. This is because a very large V_g (i.e., strong accumulation) extends the hole accumulation from the channel region to the curative drain and source p/n junctions. Since C_{A2} is proportional to the drain and source p/n junction peripheries, such an increase in hole accumulation increases C_{A2}. This does not happen, and C_{A2} remains nearly constant, however, for small V_g because hole accumulation is confined in the channel region.

Figure 4.18 also shows the capacitance in the inversion region, called C_{I2} (subscript I2 denotes inversion for the second setup), as a function of L_m. The channel is now inverted, which is linked to the ac signal via the source and drain, and both the substrate and gate are connected to the ground from the small-signal point of view. For this bias condition, C_{I2} is proportional to L_m because the channel capacitance is contributing to C_{I2}. Thus, the oxide capacitance C_o^a and channel region capacitance C_s^a contribute to C_{I2} in a parallel manner. Therefore, the gate-channel-substrate capacitance $C_{gcs}^a = (C_o^a + C_s^a)$ can be calculated by:

$$C_{gcs}^a = (C_o^a + C_s^a) = \left[\frac{\Delta C_{I2}}{\Delta L_m}\right] \tag{4.45}$$

where ΔC_{I2} (i.e., in F/μm) is the difference in C_{I2} simulated from two MOSFETs with different L_m (i.e., ΔL_m, in μm).

4.3.1.3 Third setup of MOSFET connection

Figure 4.19 shows the third setup in which the gate is connected to a dc bias, the drain and source are connected to ground, and the body is connected to an ac signal. The capacitance for the accumulation region (i.e., $V_g > 2$ V), called C_{A3} (subscript A3 denotes accumulation region for the third setup), increases with increasing L_m. For this bias condition, the channel does not contribute to the capacitance because of the accumulation condition, the gate-source and gate-drain capacitances are not affected by the ac signal, and the channel-source and channel-drain capacitances are independent of L_m. Therefore, C_o^a can also be extracted by

$$C_o^a = \left[\frac{\Delta C_{A3}}{\Delta L_m}\right] \tag{4.46}$$

where ΔC_{A3} (i.e., in F/μm) is the difference in C_{A3} simulated from two MOSFETs with different L_m (i.e., ΔL_m, in μm). We obtain again $C_o^a = 1.36\text{x}10^{-15}$ F/μm^2 using the values from this setup. An increase in C_{A2} for $V_G >$

4 V is also observed, and the reason is the same as that provided in the previous section.

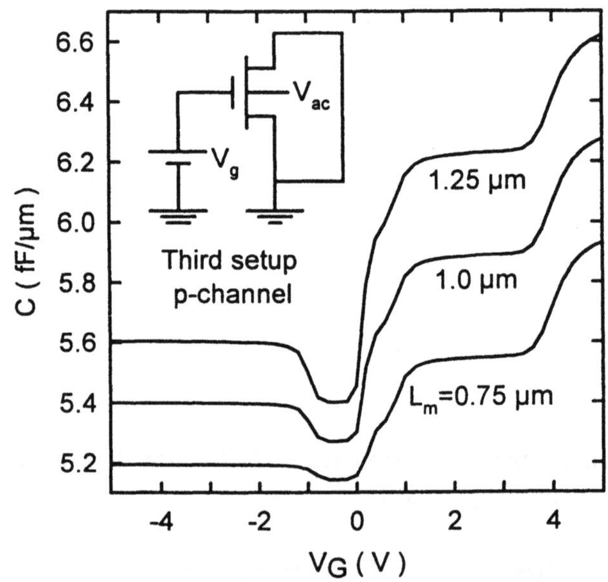

Figure 4.19 : The simulated capacitance per unit length versus the gate voltage for three different L_m obtained from the third setup having the gate connected to a dc bias, the drain and source grounded, and the body connected to an ac signal.

The capacitance C_{13} in the inversion region is also in direct proportion to L_m, as shown in Fig. 4.19. For this bias condition, the channel is inverted and is linked electrically to the source and drain, which are grounded. The grounding at the drain and source as well as at the gate creates a series combination of C_o^a and C_s^a. Since $C_o^a > C_s^a$ has been suggested from the second setup, the total capacitance simulated from this setup results mainly from the channel region capacitance, and

$$C_s^a \approx \left[\frac{\Delta C_{I3}}{\Delta L_m}\right] \text{(4.47)} \qquad (4.47)$$

where ΔC_{13} (i.e., in F/μm) is the difference in C_{13} simulated from two MOSFETs with different L_m (i.e., ΔL_m, in μm). Using the previous equation, we obtain $C_s^a = 0.82\text{x}10^{-15}$ F/μm², which is in reasonable agreement with C_s^a calculated using the second setup.

It is important to note that when the device is in strong inversion, the inverted channel provides a path that separates the capacitance associated with the oxide from that associated with silicon. Therefore, the capacitances for the different setups can be related by:

$$C_{I2} = C_{I1} + C_{I3} \tag{4.48}$$

because C_{I1} is only related to the oxide, C_{I3} is only related to the silicon region, and C_{I2} is the parallel combination of C_{I1} and C_{I3}.

Several extraction methods have been developed based on the C-V characteristics discussed above.

4.3.2 Sheu and Ko method

Sheu and Ko proposed an extraction method based on the second setup [17]. According to this method, all the significant capacitances in the MOSFET under strong inversion and accumulation are illustrated in Fig. 4.20. Subtracting the capacitance C_{I2} under inversion from the capacitance C_{A2} under accumulation yields the oxide capacitance, which is a directly proportional to the effective channel length:

$$(C_{I2} - C_{A2}) = C_o^a L_{effCV} = C_o^a (L_m - \Delta L_{effCV}) \tag{4.49}$$

where ΔL_{effCV} is the effective channel length reduction obtained using the C-V method. Then, the slope of the $(C_I - C_A)$ versus L_m plot gives $C_o^a = 2.18\text{x}10^{-15}$ F/μm², and the intercept of the straight line to the L_m axis gives $\Delta L_{effCV} = 0.55$ μm. The oxide capacitance obtained from this method (i.e., 2.18 fF/μm²) is notably larger than the simulated value (i.e., 1.37 fF/μm²). This error arises from the crude assumption used in the method that the capacitance between the inverted channel and the substrate is negligible [23].

The following approach is equivalent to Sheu-Ko's method. First, C_o^a is obtained by using (4.49) for two different values of L_m and taking the difference between them:

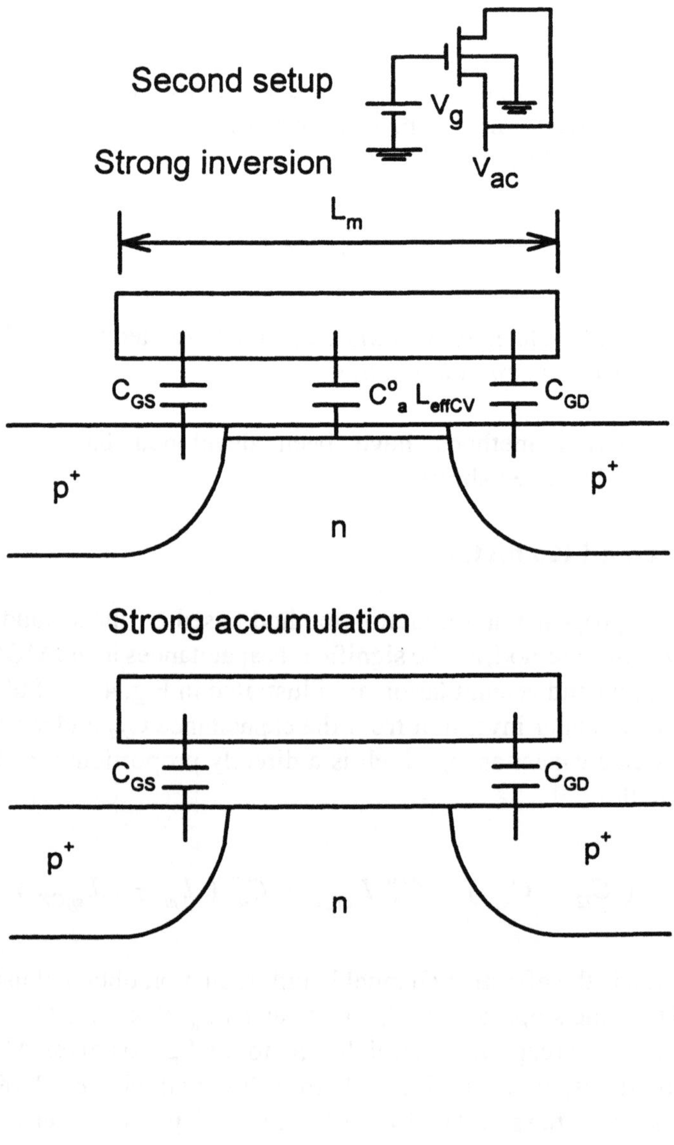

Figure 4.20 : Schematic showing the capacitances using setup 2 under inversion and accumulation considered by the Sheu-Ko's method.

$$C_o^a = \left[\frac{\Delta(C_{I2} - C_{A2})}{\Delta L_m}\right] = \left[\frac{\Delta C_I}{\Delta L_m}\right] \quad . \tag{4.50}$$

Then, L_{effCV} is evaluated from (4.50) as

$$L_{effCV} = \frac{(C_{I2} - C_{A2})}{C_o^a} \quad . \tag{4.51}$$

4.3.3 Vitanov et al. method

In the work of Vitanov and coworkers [16, 19], only one device (i.e., one L_m) in the accumulation region but two different setups (i.e., setup 1 and 2) were used to obtain L_{effCV}. First, let us focus on setup 1 in the accumulation region, (i.e., $V_g > 2$ V). According to this method, C_{A1} (in F/μm) is given by

$$C_{A1} = C_{GB} + C_{GS} + C_{GD} \tag{4.52}$$

where C_{GB}, C_{GS} and C_{GD} are the capacitances per unit device width (in F/μm) for the gate-body, gate-source, and gate-drain regions. These capacitances are shown schematically in Fig. 4.21. On the other hand, for setup 2 in the accumulation region, only C_{GS} and C_{GD} are present (see Fig. 4.21). Thus

$$C_{A2} = C_{GS} + C_{GD} \quad . \tag{4.53}$$

Next, C_{GB} is assumed to be equal to the oxide capacitance, which is calculated by subtracting C_{A2} from C_{A1}:

$$C_{GB} = C_o^a L_{effCV} = C_{A1} - C_{A2} \quad . \tag{4.54}$$

Using the simulated data given in Figs. 4.17 and 4.18, we find that this method yields a negative capacitance because $C_{A2} > C_{A1}$. This failure arises from the fact that the capacitances for the source-body and drain-body junction regions were neglected in (4.53).

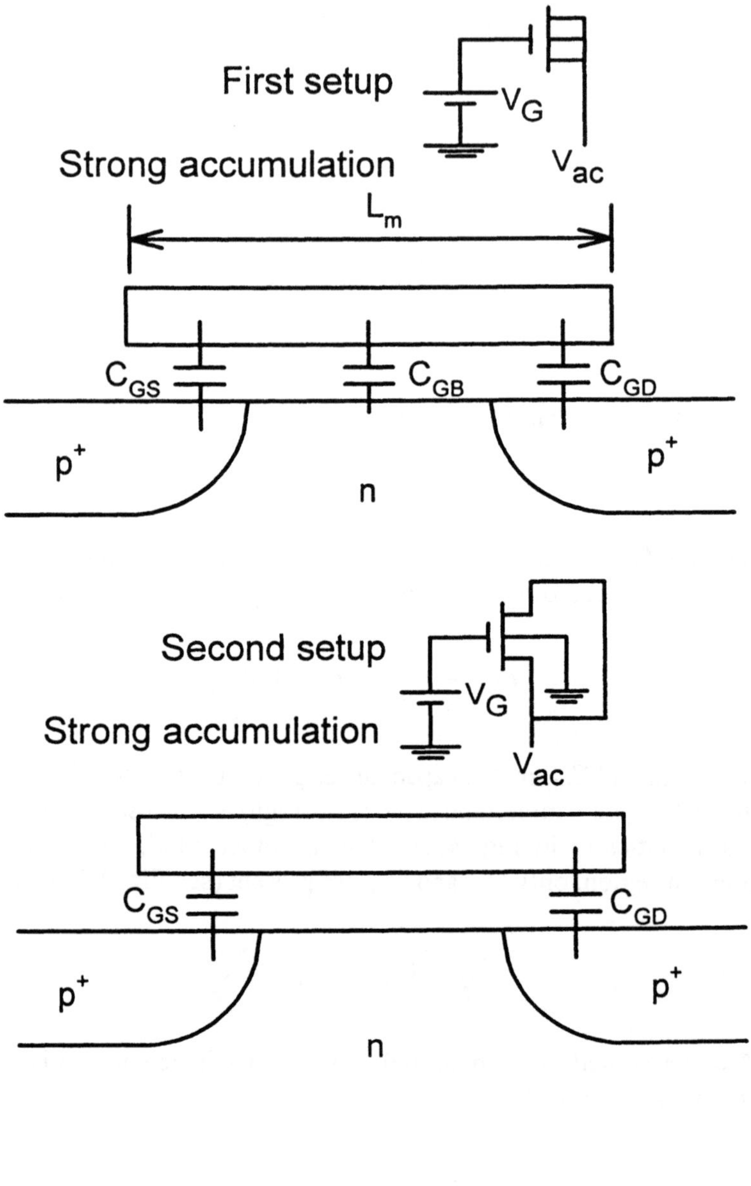

Figure 4.21 : Schematic showing the capacitances using setup 1 and setup 2 under accumulation considered by the Vitanov et al. method.

4.3.4 Lee method

The method developed by Lee [20] uses various devices (i.e., various L_m) and a single setup (i.e., setup 2) to obtain L_{effCV}. According to this method, L_{effCV} is obtained from

$$L_{effCV} = \frac{C_{I2} - C_{I2on}}{C_o^a} \tag{4.55}$$

where C_{I2on} is the capacitance at which the C-V curves for different L_m start to deviate from each other. The precision of this method is thus sensitive to the somewhat subjective determination of such a point. For a MOSFET with L_m = 0.75 μm, (4.55), in combination with the results presented in Fig. 4.18, yields L_{effCV} = 0.13 μm, or ΔL_{effCV} = 0.62 μm. This value is notably shorter than L_{met} (i.e., 0.25 μm) is therefore questionable.

4.3.5 Guo et al. method

The method by Guo et al. [21], like the Lee's method [20], uses various devices (i.e., various values of L_m) and setup 2 to obtain L_{effCV}. First, a large device (i.e., L_m= 10 μm) is measured and the maximum capacitance in the inversion region, called C_{I2L} (subscript I2L denotes inversion for setup 2 and a large device), is determined. Second, C_{gcs}^a is obtained by

$$C_{gcs}^a = \frac{C_{I2L}}{W L_m} \quad . \tag{4.56}$$

where W is the channel width. Third, the minimum value of the capacitance in the accumulation region for the small devices (see Fig. 4.18), called C_{A2m} (subscript A2m denotes accumulation for setup 2 and the minimum value), is also selected. This capacitance is considered as the outer fringe capacitance contributed by the finite thickness of the gate. Fourth, the mask channel length can be expressed by

$$L_m = \frac{C_{I2m} - C_{A2m}}{C_{gcs}^a} \quad . \tag{4.57}$$

where C_{I2m} is the minimum value of the capacitance in the inversion region. Finally, ΔL_{effCV} is evaluated by

$$\Delta L_{effCV} = \frac{C_{I2on} - 2\,C_{A2m}}{2\,W\,C_{gcs}^{a}} \tag{4.58}$$

where C_{I2on} is the capacitance at which the capacitance versus the gate voltage curves for different mask channel lengths start to deviate from each other.

4.3.6 Latif et al. method

The method proposed by Latif et al. [23] also uses various devices (i.e., various values of L_m) and setup 2 to extract L_{effCV}. This method accounts for capacitances that the Sheu-Ko's method neglected, as illustrated in Fig. 4.22. It suggests that subtracting C_{I2} and C_{A2} in inversion gives $C_{gcs}^{a} = (C_o^{a} + C_s^{a})$ (see Fig. 4.22), and the effective channel length can be obtained from

$$L_{effCV} = \frac{C_{I2m} - C_{A2m}}{C_{gcs}^{a}} \tag{4.59}$$

where the value of C_{gcs}^{a} is determined from (4.45). The main difference between (4.51) and (4.59) is that the former neglects C_o^{a} and the latter includes such a capacitance.

The simulation results in Fig. 4.18 and (4.59) give L_{effCV} = 0.29 μm, 0.54 μm, and 0.79 μm, for L_m = 0.75 μm, 1 μm and 1.25 μm, respectively, or $\Delta L_{effCV} = L_m - L_{effCV}$ = 0.46 μm for the three L_m considered.

4.4 Simulation-based method

The effective channel length of MOSFET can also be extracted from the physical insight, such as free-carrier concentrations, electric field distribution, and energy band diagrams, obtainable from device simulations.

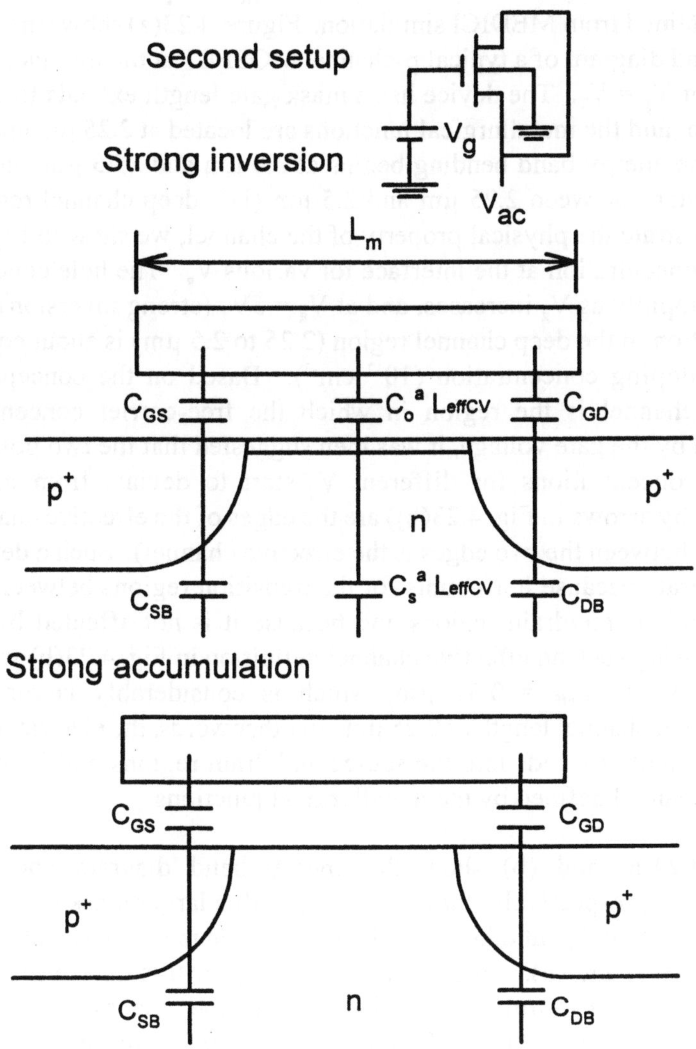

Figure 4.22 : Schematic showing the capacitances using setup 2 under inversion and accumulation considered by the Latif et al. method.

4.4.1 Narayanan et al. method

Narayanan et al. [4-5] estimated the value of ΔL_{eff} through the means of physical insight obtained from MEDICI simulation. Figure 4.23(a) shows the simulated energy band diagram of a typical p-channel MOSFET at the interface along the channel for $V_g = V_T$. The device has a mask gate length extends from 2.0 μm to 2.75 μm, and the metallurgical junctions are located at 2.25 μm and 2.5 μm. Clearly, the energy band bending begins at 2.0 μm and 2.75 μm and remains fairly constant between 2.25 μm and 2.5 μm (i.e., deep channel region). To further illustrate the physical property of the channel, we show in Fig. 4.23(b) the hole concentration at the interface for various V_g. The hole concentration increases rapidly as V_g increases, and at $V_g = 5V_T$ (strong inversion), the hole concentration in the deep channel region (2.25 to 2.5 μm) is about equal to the substrate doping concentration (10^{17} cm^{-3}). Based on the concept that the effective channel is the region in which the free-carrier concentration is controlled by the gate voltage, it was then suggested that the two points where the hole concentrations for different V_g start to deviate from each other (indicated by arrows in Fig. 4.23(b)) are the edges of the effective channel (i.e., the region between the two edges is the effective channel). Such a definition is more accurate because it accounts for the transition regions between the deep channel and source/drain regions and because it is not affected by the gate voltage. Using such an effective channel definition in Fig. 4.23(b) yields ΔL_{eff} = 0.24 μm and L_{eff} = 0.51 μm, which is considerably larger than the metallurgical channel length of 0.25 μm. In other words, the electrical effective channel actually extends into the source and drain regions and is beyond the physical channel defined by the metallurgical junctions.

Figures 4.24(a) and (b) show the energy band diagram and the hole concentration, respectively, for a MOSFET with a larger mask channel length (i.e., L_m = 1.25 μm), but otherwise identical device makeup as that considered in Figs. 4.23(a) and (b). To be consistent, ΔL_{eff} for this device should be the same as the L_m = 0.75 μm device. This is indeed the case, as evidenced by the results shown in Fig. 4.24(b), which gives ΔL_{eff} = 0.24 μm (or L_{eff} = 1.01 μm).

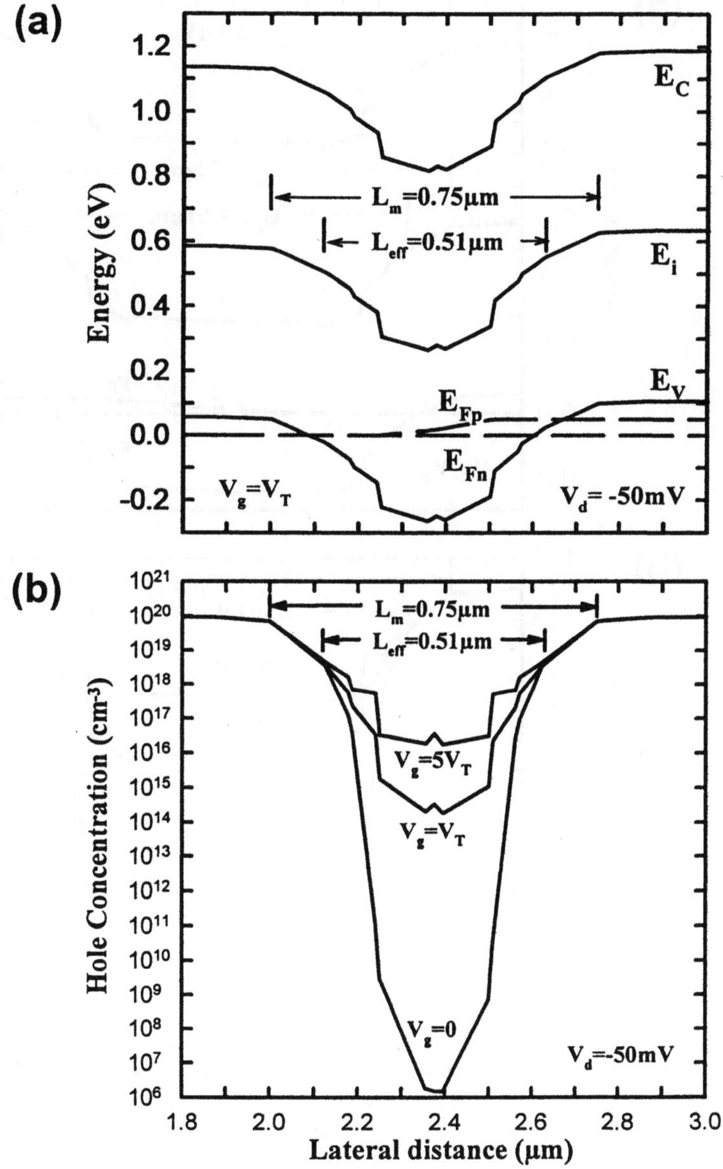

Figure 4.23 : (a) Energy band diagram, and (b) hole concentration at the interface of the MOSFET with L_m = 0.75 µm.

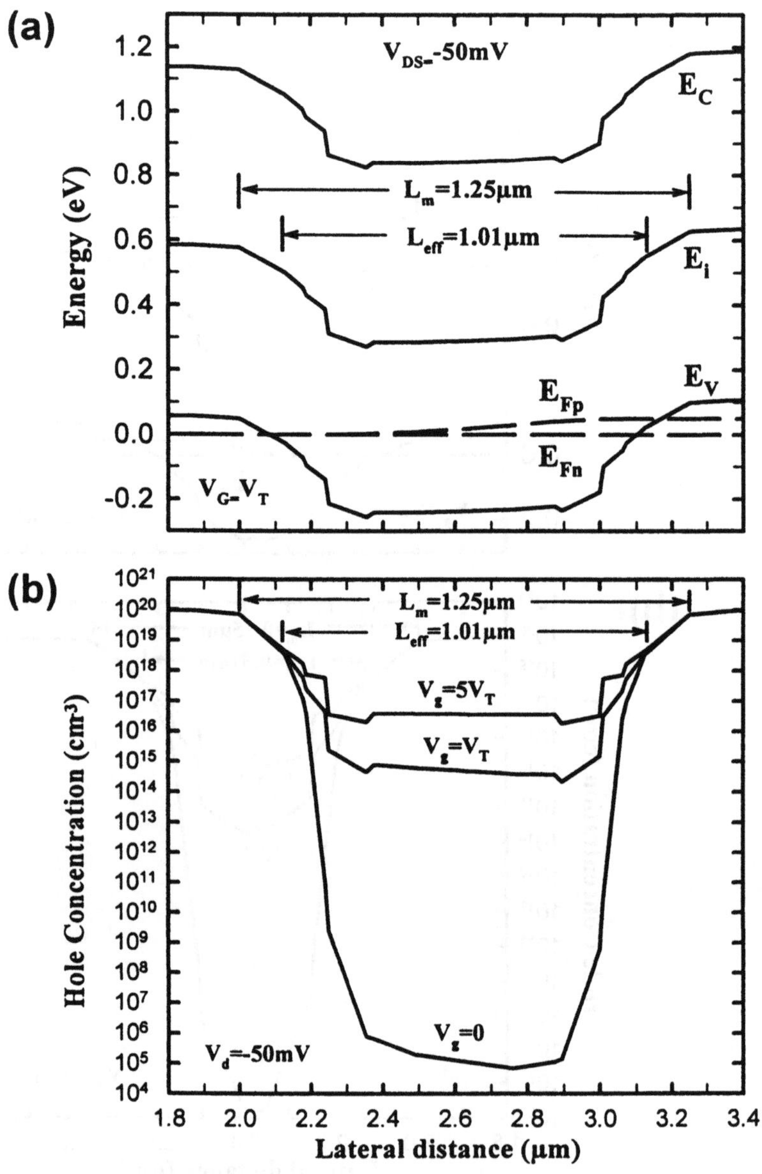

Figure 4.24 : (a) Energy band diagram, and (b) hole concentration at the interface of the MOSFET with L_m = 1.25 μm.

4.4.2 Niu et al. method

Niu et al. [24] also proposed a method to determine L_{eff} through the means of physical insight obtained from simulations. While Niu et al. agreed with the physical reasoning of Narayanan's method [4-5], they felt that it is somewhat objective and arbitrary to determine the effective channel based on the two points where the free-carrier concentrations for different V_g start to deviate from each other.

Niu's method is based on the assumption that the diffusion current is negligible for a MOSFET biased in strong-inversion. Therefore, the following behavior should be found along the effective channel: 1) the inversion carrier concentration is nearly constant; 2) the lateral electric field (i.e., $-d\psi/dx$) is also nearly constant to keep a constant drift current; 3) the electrostatic potential ψ varies linearly with respect to the lateral distance x; and 4) the second derivative of the electrostatic potential with respect to x should be zero (i.e., $d^2\psi/dx^2 = 0$). Then, Niu proposed that the edges of the effective channel should be defined at the points where $d^2\psi/dx^2$ are maximum.

To test this method, we have simulated p-channel LDD MOSFETs using a device simulator called MICROTEC run on a PC platform [53]. The devices had a substrate doping concentration of 10^{17} cm^{-3}, and the same source and drain make-up as follows: a heavily doped p^+-type Gaussian profile with a peak doping concentration of 10^{20} cm^{-3}, and a lightly doped p^--type Gaussian profile with a peak doping concentration of 10^{18} cm^{-3}. The mask gate length extends from 2.0 to 3.3 µm, and the metallurgical junctions are located at 2.18 µm and 3.11 µm (i.e., $L_{met} = 0.93$ µm).

Figure 4.25(a) shows the doping profile along the channel at the interface, and the inversion free-carrier density simulated for different gate voltages are illustrated in Fig. 4.25(b). Based on the Narayanan's method, L_{eff} is found to be about 1.2 µm, and the determination of the boundaries of the effective channel is somewhat subjective because the precise points where the curves start to deviate from each other are not very clear.

Figures 4.26(a)-(c) show ψ, $d\psi/dx$ and $d^2\psi/dx^2$, respectively, at the interface along the channel for $V_g = -3$ V and $V_d = -0.05$ V. We see in Fig. 4.26(a) that ψ varies approximately linearly with respect to x along the effective channel, but there are two different slopes because of the presence of the LDD regions. Four positive peaks and two negative peaks for $d^2\psi/dx^2$ are shown in Fig. 4.26(c).

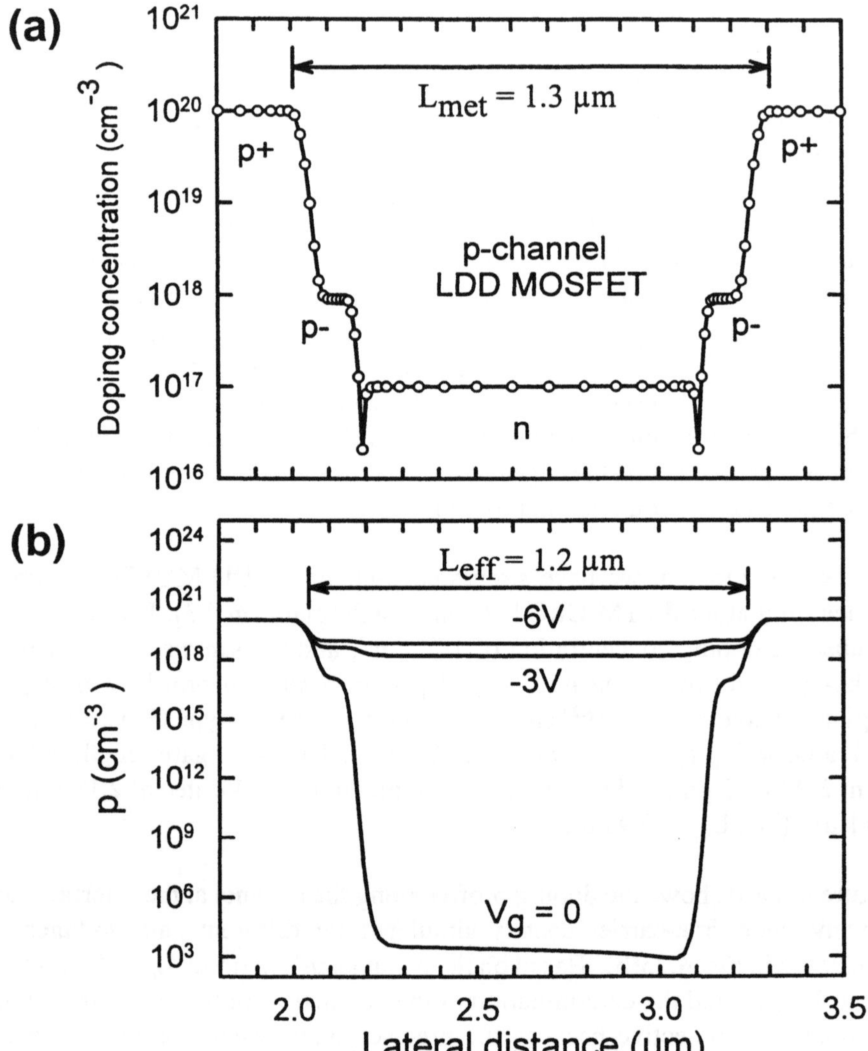

Figure 4.25 : (a) Impurity doping concentration at the Si-SiO_2 interface along the channel of the simulated LDD MOSFET with $L_m = 1.3$ μm. (b) Hole concentration at the interface of the MOSFET for various gate biases.

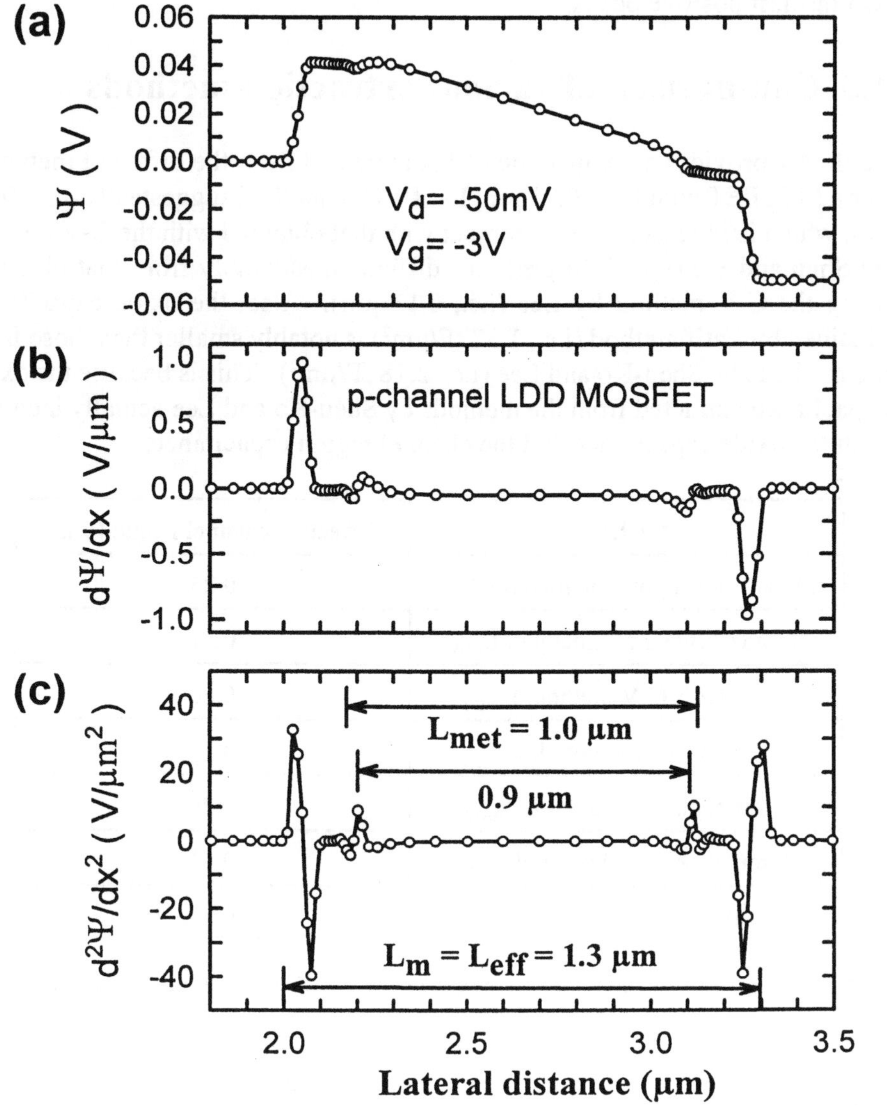

Figure 4.26 : (a) Electrostatic potential ψ, (b) first derivative of the electrostatic potential with respect to x (i.e., $d\psi/dx$), and (c) second derivative of the electrostatic potential with respect to x (i.e., $d^2\psi/dx^2$) at the interface of the MOSFET with L_m = 1.3 μm.

Using the two closest positive peaks to define the effective channel, one will obtain a value of 0.9 μm. This value is incorrect because it is smaller than L_{met} = 0.93 μm. A more reasonable value of L_{eff} = 1.3 μm is obtained by using the two farthest positive peaks.

4.5 Comparison of various extraction methods

Table 4.1 provides a comparison of L_{eff} extracted from the different methods. For a MOSFET with L_m = 0.75 μm, Latif's C-V method suggested L_{effCV} = 0.29 μm, which is in reasonable agreement with that obtained with the C-V method by Sheu and Ko (i.e., 0.20 μm), but differs considerably from that obtained using the C-V method by Lee (i.e., 0.13 μm). Also, the oxide capacitance obtained by Latif method (i.e., 1.37 fF/μm²) is notably smaller than those from the methods by Sheu-Ko and Lee (i.e., 2.18 fF/μm²). This is because the oxide capacitance extracted from the methods by Sheu-Ko and Lee actually includes both the oxide capacitance and the channel region capacitance.

Method	Effective channel length (μm)
Metallurgical junction method, L_{met}	0.25
Terada-Muta's I-V method, L_{effIV}	0.51
Sheu-Ko's C-V method, L_{effCV}	0.20
Lee's C-V method, L_{effCV}	0.13
Latif et al. C-V method, L_{effCV}	0.29
Simulation-based method, L_{effd}	0.51
Shift & ratio method, $L_{effS\&R}$	0.51

Table 4.1 : Effective channel length of a 0.75-μm MOSFET extracted from the various methods

The Terada-Muta's I-V method, on the other hand, yielded a much larger effective channel length (i.e., L_{effIV} = 0.51 μm) than its C-V counterpart, as shown in Table 4.1. This value is consistent with L_{effd} determined from the simulation-based method [4-5] based on the physics that L_{eff} is the length of a channel region in which the inversion free-carrier density is controlled by the gate voltage. The same effective channel length has also been extracted from the S&R I-V method (i.e., $L_{effS\&R}$ = 0.51 μm). Notice that all L_{effd}, L_{effIV}, and $L_{effS\&R}$ are larger than the length L_{met} defined by the source and drain metallurgical junctions, resulting from the fact that the source and drain regions adjacent to the channel are actually part of the effective channel governing the MOSFET electrical behavior. Conversely, the C-V methods yield an effective channel length close to L_{met}, which is inconsistent from the physics point of view.

The problem of the C-V method lies in the inconsistencies imbedded in (4.45) and (4.59). Let us first focus on (4.45). The capacitance per unit area can be modeled by the change of capacitances divided by the change of L_m, the method used in (4.45), only if all capacitances involved have the same length L_m. This is not true in (4.45), however, as the length associated with C_o^a is L_m, whereas the length associated with C_s^a is L_{eff}. Another inconsistency of the C-V method occurs in $C_{gcs} = C_{I2} - C_{A2}$ in (4.59). Because the drain and source are highly doped, having accumulation in these regions is extremely difficult. Thus, the effective channel for the accumulation region is approximately confined by L_{met}. In other words, the effective channel length associated with C_{A2} is L_{met}. On the other hand, the effective channel length for the inversion region is L_{eff}, which is larger than L_{met}. The two different effective channel lengths thus invalidate the direct substraction of C_{I2} and C_{A2}. These two inconsistencies in (4.45) and (4.59) lead to the incorrect L_{effCV}.

The preceding discussions clearly indicate that the flaw of the C-V method results from the different lengths associated with the oxide capacitance (i.e., L_m), the channel region capacitance in accumulation region (i.e., L_{met}), and the channel region capacitance in inversion region (i.e., L_{eff}). Since the difference of the three lengths decreases with increasing L_m, one should expect that the C-V method would become more accurate as L_m is increased. This is indeed the case. Our calculations show that $L_{effIV}/L_{effCV} = 1.8$, 1.4, and 1.3 for 0.75, 1.0 and 1.25 μm MOSFETs, respectively.

It is apparent from the preceding analysis that every existing extraction method has its shortcomings, and much work is still needed to develop a more accurate method for the determination of the effective channel length of MOS devices. This is particularly important, yet challenging, for modern deep-submicron MOSFETs in which the quantum mechanical effects become prominent and need to be accounted for in developing the extraction method.

REFERENCES

[1] J. J. Liou, *Advanced Semiconductor Device Physics and Modeling*, Artech House, Inc., Boston, 1994.

[2] K. K. Ng and W. T Lynch, "Analysis of the gate-voltage-dependent series

resistance of MOSFET's," IEEE Trans. Electron Devices, vol. ED-33, pp. 965-972, July 1986.

[3] S. S. Bhattacharya, E. R. Worley and R. A. Williams, Proc. IEEE Int. Conf. Microelec. Test Struct., San Diego, California, March 1994, pp. 195-197.

[4] R. Narayanan, A. Ortiz-Conde, J. J. Liou, F. J. García Sánchez, and A. Parthasarathy, "Two-dimensional numerical analysis for extracting the effective channel length of short-channel MOSFET," Solid-St. Electron., vol. 38, pp. 1155-1159, June 1995.

[5] R. Narayanan, A. Ortiz-Conde, J. J. Liou, F. J. García Sánchez, and A. Parthasarathy, "Effective channel length of submicrom MOSFETs: numerical simulation, physical mechanisms, and extraction methods," Int. Conf. Solid St. and Int. Cir. Tech., Beijing, China, Oct. 1995, pp. 292-294.

[6] Y. Taur, Y.-J. Mii, R. Logan and H.-S. Wong, "On effective channel length in 0.1-μm MOSFET's" IEEE Electron Dev. Lett., vol. 16, pp.136-138, April 1995.

[7] Y. Taur, "CMOS technology evolution: from 1 μm to 0.1 μm," Proc. Int. Conf. on Solid State Integrated-Circuit Technology, Beijing, China, Oct. 1995, pp. 312-316.

[8] K. Terada,, and H. Muta, "A new method to determine effective MOSFET channel length," Jap. J. Appl. Phys., vol. 18, pp. 953-959, May 1979.

[9] J. G. J. Chern, P. Chang, R. F. Motta, and N. Godinho,"A new method to determine MOSFET channel length," IEEE Electron Dev. Lett., vol. EDL-1, pp. 170-173, Sept. 1980.

[10] K. K. Ng and J. R. Brews, "Measuring the affective channel length of MOSFETs," IEEE Circ. Dev., vol. 6, pp. 33-38, Nov. 1990.

[11] Y. Taur, D. S. Zicherman, D. R. Lombardi, P. J. Restle, C. H. Hsu, H. Y. Hanafi, M. R. Wordeman, B. Davari and G. G. Shahidi, "A new "shift and ratio" method for MOSFET channel-length extraction," IEEE Electron Device Letters, vol. EDL-13, pp. 267-269, May 1992.

[12] K. Terada, "Suppression of measurement errors in effective-MOSFET-channel-length extraction," Proc. IEEE Int. Conf. Microelec. Test Struct., March, 1992, pp. 208-212.

[13] F. J. García Sánchez, A. Ortiz-Conde, M. García Núñez, and R. L. Anderson, "Extraction of series resistance and effective gate length for short-channel PMOS devices at liquid nitrogen temperature," Proc. IEEE Int. Conf. Microelec. Test Struct., San Diego, California, March 1994, pp. 190-194.

[14] F. J. García Sánchez, A. Ortiz-Conde, M. García Núñez, and R. L. Anderson, "Extracting the series resistance and effective channel length of short-channel MOSFETs at liquid nitrogen temperature," Solid-St. Electron., vol. 37, pp. 1943-1948, Dec. 1994.

[15] Y.-S. Jean and C.-Y. Wu, "A new extraction algorithm for the metallurgical channel length of conventional and LDD MOSFET`s," IEEE Trans. Electron Devices, vol. ED-43, pp. 946-953, June 1996.

[16] P. Vitanov, V. Schwabe, and I. Eisele, "Electrical characterization of feature sizes and parasitic capacitances using a single test structure," IEEE Trans. Electron Devices, vol. ED-31, pp. 96-100, Jan. 1984.

[17] B. J. Sheu and P. K. Ko, "A capacitance method to determine channel lengths for

conventional and LDD MOSFET's," IEEE Electron Dev. Lett., vol. EDL-5, pp. 491-493, Nov. 1984.

[18] J. Y. -C. Sun, M. R. Wordeman, and S. E. Laux, "On the accuracy of channel-length characterization of LDD MOSFET's," IEEE Trans. Electron Devices, vol. ED-33, pp. 1556-1562, Oct. 1986.

[19] P. Vitanov, T. Dimitrova, R. Kamburova, and K. Filljov, "Capacitance method for determination of LDD MOSFET geometrical parameters," Solid-St. Electron., vol. 35, pp. 985-991, July 1992.

[20] S.-W. Lee, "A capacitance-based method for experimental determination of metallurgical channel length of submicron LDD MOSFET's," IEEE Trans. Electron Devices, vol. ED-41, pp. 403-412, March 1994.

[21] J. -C. Guo, S. S. -S. Chung, and C. C. -H. Hsu, "A new approach to determine the effective channel length and drain-and-source series resistance of miniaturized MOSFET's," IEEE Trans. Electron Devices, vol. ED-41, pp. 1811-1818, Oct. 1994.

[22] M.-S. Jo, J.-H. Kim, S.-K. Kim, H.-S. Yoon and D.-H. Lee, "A capacitance method to determine the metallurgical gate-to-source/drain overlap length of submicrom LDD MOSFET's," Proc. IEEE Int. Conf. Microelec. Test Struct., Japan, 1990, pp. 151-155.

[23] Z. Latif, A. Ortiz-Conde, J. J. Liou, and F. J. García Sánchez, "A study of the validity of capacitance-based method for extracting the effective channel length of MOSFETs," IEEE Trans. Electron Devices, vol. 44, pp. 340-343, Feb. 1997.

[24] G. F. Niu, R. M. Chen, and G. Ruan, "Extraction of effective channel length (L_{eff}) of deep submicrom MOSFET's from numerically simulated surface potential," Solid-St. Electron., vol. 41, pp. 1377-1382, Aug. 1997.

[25] D. K. Schroeder, *Semiconductor Material and Device Characterization*, Wiley, New York, 1990.

[26] L. Chang and J. Berg, "A derivative method to determine a MOSFET's effective channel length and width electrically," IEEE Electron Dev. Lett., vol. EDL-7, pp. 229-231, April 1986.

[27] J. Whitfield, "A modification on 'An improved method to determine MOSFET channel length," IEEE Electron Device Letters, vol. EDL-6, pp. 109-110, March 1985.

[28] P. I. Suciu and R. L. Johnston, "Experimental derivation of the source and drain resistance of MOS transistors," IEEE Trans. Electron Devices, vol. ED-27, pp. 1846-1848, Sept. 1980.

[29] F. H. De La Moneda, H. N. Kotecha and M. Shatzkes, "Measurements of MOSFET constants," IEEE Electron Dev. Lett., vol. EDL-3, pp. 10-12, Jan. 1982.

[30] S. T. Hsu, "A simple method to determine series resistance and k factor of an MOS field effect transistor," RCA Rev., vol 44, pp. 424-429, Sept. 1983.

[31] K. -L. Peng, S. -Y. Oh, M. A. Afromowitz and J. L. Moll, "Basic parameter measurements and channel broadening effect in the submicrometer MOSFET," IEEE Electron Dev. Lett., vol. EDL-5, pp. 473-475, Nov. 1984.

[32] B. J. Sheu, C. Hu, P. K. Ko, and F. -C. Hsu, "Source-and-drain series resistance of LDD MOSFET's," IEEE Electron Dev. Lett., vol. EDL-5, pp. 365-367, Sept.

1984.

[33] G. J. Hu, Chang and Y. -T. Chia, "Gate-voltage-dependent effective-channel-length and series resistance of LDD MOSFET's," IEEE Trans. Electron Devices, vol ED-34, pp. 2469-2475, Dec. 1987.

[34] C. C. McAndrew and P. A. Layman, "MOSFET effective channel length, threshold voltage, and series resistance determination by robust optimization," IEEE Trans. Electron Devices, vol. ED-39, pp. 2298-2311, Oct. 1992.

[35] W. Fikry, G. Ghibaudo, H. Haddara, S. Cristoloveanu, and M. Dutoit, "Method for extracting deep submicrometer MOSFET parameters," Electron. Lett., vol. 31, pp. 762-764, April 1995.

[36] S. Hong and K. Lee, "Extraction of metallurgical effective channel length in LDD MOSFET's," IEEE Trans. Electron Devices, vol. 42, pp. 1461-1466, Aug. 1995.

[37] H. -H. Li and C. -Y. Wu, "A novel extraction technique for the effective channel length of MOSFET devices," IEEE Trans. Electron Devices, vol. ED-42, pp. 856-863, May 1995.

[38] P. R. Karlsson and K. O. Jeppson, "An efficient method for determining threshold voltage, series resistance and effective geometry of MOS transistors", IEEE Trans. Semiconductor Manufacturing, vol. 9, pp.215-222, May 1996.

[39] K. Takeuchi, N. Kasai, T. Kunio and K. Terada, "An effective channel length determination method for LDD MOSFET," IEEE Trans. Electron Devices, vol. 43, pp. 580-587, April 1996.

[40] MEDICI Manual, Technology Modeling Associates, Inc., Palo Alto, CA, 1993.

[41] M. García Núñez, A. Ortiz-Conde, F. J. García Sánchez, and R. L. Anderson, "On MOSFET parameter extraction at liquid nitrogen temperature," VIII Congresso da Socidade Brasileira de Microelectronica, vol. 8, pp. VI.7-12, (Campinas, Brasil), Sept. 1993.

[42] A. Ortiz-Conde, J. J. Liou, F. J. García Sánchez, M. García Núñez, and R. L. Anderson, "Series resistance and effective channel length extraction of n-channel MOSFET at 77 K," Electron. Lett., vol. 30, pp. 670-672, April 1994.

[43] Z. Latif, J. J. Liou, A. Ortiz-Conde, F. J. García Sánchez, W. Wong, and Y.G. Chen, "Analysis of the validity of methods used for extracting the effective channel of short-channel LDD MOSFETs," Solid-St. Electron., vol. 39, pp. 1093-1094, July 1996.

[44] Z. Latif, A. Ortiz-Conde, J. J. Liou, F. J. García Sánchez and W. Wong, "Failure of effective-channel length extraction methods due to the effect of the relative doping level of source and drain in short-channel LDD MOSFETs," Proc. of IEEE HKEDM, Hong Kong, June 1996, pp. 91-93.

[45] R. S. Muller and T. I. Kamins, *Devices Electronics for Integrated Circuits*, Wiley, New York, 1986.

[46] G. Pellegrini and R. L. Anderson, "Metal-oxide-semiconductor field-effect transistor characteristics as influenced by carrier mobility variation along the channel" J. Appl. Phys., vol. 72, pp. 3606-3609, Oct. 1992.

[47] W. F. Clark, B. El-Kareh, R. G. Pires, S. L. Titcomb and R. L. Anderson, "Low temperature CMOS - a brief review," IEEE Trans. Comp. Hybrid Man. Tech., vol. 15, pp. 397-404, June 1992.

[48] C. -L Huang and G. S. Gildenblat, "An accurate engineering model of an n-channel MOSFET for 60-300K temperature range", Solid State Electron., vol. 33, pp. 1309-1318, 1990.

[49] M. Shur, T. A. Fjeldly, T. Ytterda, and K. Lee, "Unified MOSFET model," Solid-State Electron., vol. 35, pp. 1795-1802, 1992.

[50] K. Lee, M. Shur, T. A. Fjeldly, and T. Ytterdal, *Semiconductor Device Modeling for VLSI*, Prentice-Hall, Englewood Cliffs, 1993.

[51] C. S. Wen, M. Guldahl, L. P. Sadwick, R. Kent, and H. Gaffur, "Measurement and parameter extraction of sub-microm VLSI MOSFET test structure," Proc. IEEE Int. Conf. Microelec. Test Struct., March 1992, pp. 196-201.

[52] S. Jain, " Measurement of threshold voltage and channel length of submicron MOSFET`s," IEE Proc. Pt. I, vol. 135, pp. 162-164, Dec. 1988.

[53] MICROTEC Manual, Siborg Systems, Inc., Waterloo, Ontario, 1996.

Chapter 5

Extraction of the source and drain series resistances of MOSFETs

5.1 Introduction

A MOSFET can be considered, from the modeling point of view, as an intrinsic device in series with the drain resistance R_D and the source resistance R_S, as shown in Fig. 5.1. These resistances influence the device operating characteristics and complicate the extraction of the device intrinsic model parameters, which ideally should be independent of these parasitic resistances.

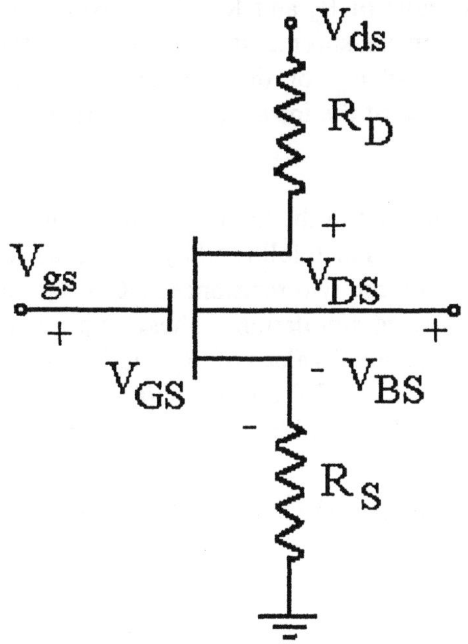

Figure 5.1 : MOSFET equivalent circuit with drain and source series resistances

Assuming the current that flows through the device channel is the same as that flows through the drain and source series resistances, that is, assuming there are no gate and substrate currents, then the MOSFET's intrinsic gate-source voltage V_{GS}, drain-source voltage V_{DS}, and body-source voltage V_{BS} can be defined in terms of the drain current I_d and the extrinsic (or external) voltage counterparts V_{gs}, V_{ds}, and V_{bs} as

$$V_{GS} = V_{gs} - I_d R_S \tag{5.1}$$

$$V_{DS} = V_{ds} - I_d \left(R_S + R_D \right) \tag{5.2}$$

$$V_{BS} = V_{bs} - I_d R_S \tag{5.3}$$

It is apparent from the above equations that the correct calculations of the MOSFET intrinsic voltages, and thus the MOSFET characteristics, depend heavily on the knowledge of R_S and R_D. Moreover, extracting the intrinsic device's parameters from measurements of the extrinsic variables requires either the knowledge of R_S and R_D, or the availability of a method capable of performing the extraction of the intrinsic model parameters independent of R_S and R_D [1-3].

In this chapter, the extraction of the total drain and source resistance ($R_D + R_S$) will first be covered. This is then followed by the discussions on the extraction of the difference between the two resistances ($R_D - R_S$) based on the physical insight provided by device simulation. These two quantities will allow the determination of the individual values of R_D and R_S. The device focused here will be the conventional MOSFET, and extraction of R_D and R_S for the lightly-doped drain (LDD) MOSFET will be treated in Chapter 6.

5.2 Extraction of total drain and source series resistance

Obtaining the individual values of the source and drain series resistances requires either the knowledge of their sum ($R_D + R_S$) and difference ($R_D - R_S$), or the ability to extract the two parameters separately. The main focus of this section is the extraction of ($R_D + R_S$). The extraction of ($R_D - R_S$) will be treated in the next section.

5.2.1 Extraction methods

The widely used Terada-Muta method to extract the effective channel length (discussed in details in Chapter 4) can also be used to extract $(R_D + R_S)$ [4-5]. This method is briefly reviewed below. First, the total drain-to-source resistance R_m of several devices with different mask channel lengths are measured or simulated [6-10] in strong inversion condition and with the devices biased in the linear region using a small applied drain voltage [11]. Under these conditions, and assuming $V_{gs} = V_{GS}$, the total resistance can be expressed as

$$R_m = \frac{V_{ds}}{I_d} \approx \left(R_D + R_S \right) + \frac{\left(L_m - \Delta L_{eff} \right)}{W \mu C_o \left(V_{gs} - V_T \right)} \tag{5.4}$$

where C_o is the oxide capacitance per unit area, μ is the effective channel mobility, L_m is the mask channel length, $(L_m - \Delta L_{eff})$ is the effective channel length, W is the channel width, and V_T is the threshold voltage. The second term on the right-hand side of (5.4) is the resistance associated with the channel region of MOSFET. According to (5.4), plotting the measured or simulated R_m versus L_m for different values of $(V_{gs} - V_T)$, having previously extracted V_T, should produce a family of straight lines, all intersecting at one point of which the abscissa yields ΔL_{eff} and the ordinate yields $(R_D + R_S)$ [6]. This plot, together with the intersection of the straight lines, is illustrated in Fig. 5.2. Note that the Terada-Muta method requires a set of several MOS devices having different L_m but otherwise identical device make-up.

Several other techniques have been proposed to extract the total drain and source resistance, using either a single device or a set of devices with different channel lengths [12-17]. For example, a method [18] has been proposed to use the nonlinear optimization, together with an iterative linear regression procedure, to extract the threshold voltage, the effective geometry, and the total parasitic series resistance. The method uses one set of data obtained in the linear region of several MOSFETs having different geometries.

Methods that extract $(R_D + R_S)$ from a single device are always preferable when the aim is to use this sum in conjunction with the extracted $(R_D - R_S)$ to obtain the individual source and drain resistances. A procedure has been developed based on the conventional MOSFET theory and using a single device [19-20]. It determines the source or drain series resistance either from the device dc characteristics at V_{ds} approaching zero, or from the device frequency response subject to an ac signal with small magnitude and low frequency [20]. However, such a procedure assumes symmetrical drain and source configurations, and therefore becomes questionable when $R_S \neq R_D$.

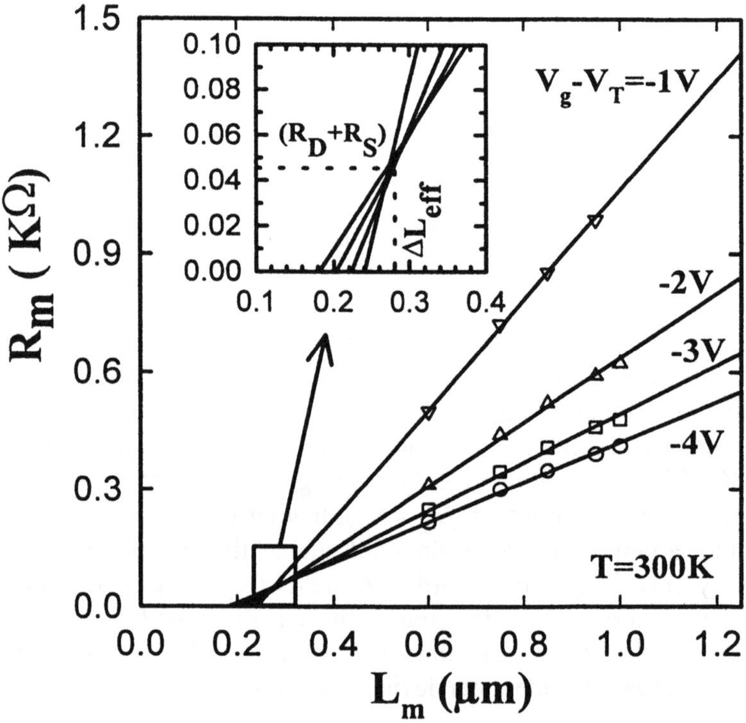

Figure 5.2 : Total resistance versus mask channel plot, where the intersection of the straight lines to the y-axis yields the total drain and source resistance

Measurements of the extrinsic drain conductance $g_d = \partial I_d/\partial V_{ds}$, gate transconductance $g_m = \partial I_d/\partial V_{gs}$, and body (substrate) transconductance $g_b = \partial I_d/\partial V_{bs}$ are also useful in obtaining the drain and source resistances. These parameters can be related to their intrinsic counterparts (i.e., g_{m0} , g_{d0}, g_{b0}) using a simple circuit theory. Assuming the drain current I_d passing through R_S, R_D, and the intrinsic MOSFET channel is the same (i.e., zero current passing through the gate and body terminals), and considering R_S and R_D are gate-voltage dependent and R_D is also drain-voltage dependent, yield [21]:

$$\frac{g_m}{g_{m0}} = \frac{1 - \frac{\left(g_{m0} + g_{b0}\right)}{g_{m0}} I_d \frac{\partial R_S}{\partial V_{gs}} - \frac{g_{d0}}{g_{m0}} I_d \frac{\partial \left(R_S + R_D\right)}{\partial V_{gs}}}{1 + \left(g_{m0} + g_{b0}\right) R_S + g_{d0}\left(R_S + R_D\right)}, \tag{5.5}$$

$$\frac{g_d}{g_{d0}} = \frac{1 - I_d \frac{\partial R_D}{\partial V_{ds}}}{1 + \left(g_{m0} + g_{b0}\right) R_S + g_{d0}\left(R_S + R_D\right)} \tag{5.6}$$

and

$$\frac{g_b}{g_{b0}} = \frac{1 - \frac{\left(g_{m0} + g_{b0}\right)}{g_{b0}} I_d \frac{\partial R_S}{\partial V_{bs}} - \frac{g_{d0}}{g_{b0}} I_d \frac{\partial \left(R_S + R_D\right)}{\partial V_{bs}}}{1 + \left(g_{m0} + g_{b0}\right) R_S + g_{d0}\left(R_S + R_D\right)}. \tag{5.7}$$

For conditions where the voltage dependencies of R_S and R_D may be neglected, (5.5), (5.6) and (5.7) are reduced to the following single expression:

$$\frac{g_m}{g_{m0}} \approx \frac{g_d}{g_{d0}} \approx \frac{g_b}{g_{b0}} \approx \frac{1}{1 + \left(g_{m0} + g_{b0}\right) R_S + g_{d0}\left(R_S + R_D\right)} \tag{5.8}$$

For long channel MOS transistors, g_{m0} can be neglected when operating in the linear region at a very small drain-to-source voltage, and g_{d0} can be neglected in the saturation region. For short channel devices, however, these terms are important and cannot be omitted.

From (5.8), and using a simple device model without considering R_S and R_D being bias dependent, the ratio of the drain conductance to the gate transconductance is [22]:

$$\frac{g_d}{g_m} \approx \frac{g_{d0}}{g_{m0}} \approx \frac{V_{GS} - V_T}{V_{DS}} \tag{5.9}$$

where V_T is the threshold voltage. Replacing the intrinsic voltages by their extrinsic counterparts, together with (5.1) and (5.2), yields

$$V_{gs} - \frac{g_d}{g_m}\left[V_{ds} - I_d\left(R_D - R_S\right)\right] = I_d\left(1 - \frac{2\,g_d}{g_m}\right)R_S + V_T \quad . \tag{5.10}$$

The left-hand side of this expression, when plotted at various bias points versus the multiplicand of R_S on the right-hand side of (5.10), should produce a straight line with a slope equals to R_S. In addition, the intercept of the line at the voltage axis gives the value of V_T. The value of R_D can then be found, provided the value of $(R_D - R_S)$ is known.

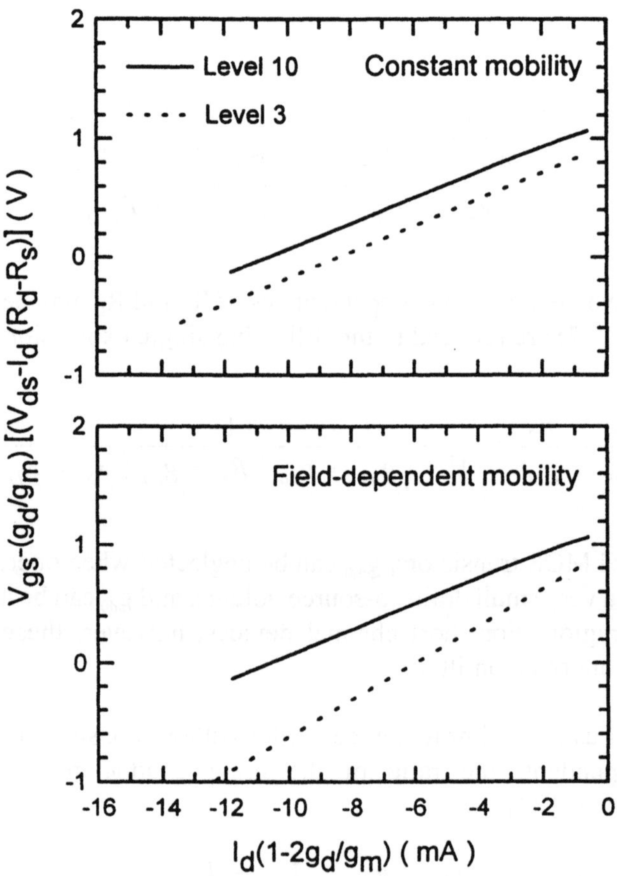

Figure 5.3 : Results showing the characteristics of equation (5.10) simulated using AIM-SPICE with different MOSFET models (i.e., level-10 and level-3 models) and different free-carrier mobility models.

To illustrate this method, we have simulated the results of the left-hand side of (5.10) versus the first term on the right-hand side of (5.10) using AIM-SPICE simulator with level-10 and level-3 MOSFET models, as shown in Fig. 5.3. In addition to the two different models, both the constant and field-dependent free-carrier mobilities were considered. As mentioned earlier, the slope of the line gives R_S and the intercept of the line at the x-axis gives V_T. It is apparent that the level-10 results are independent of the type of mobility used and thus are more reliable than the level-3 counterpart. Figure 5.4 shows the simulated I_d, g_m, and g_d versus V_g characteristics. Again, a notable discrepancy is found between the results simulated using level-10 and level-3 MOSFET models.

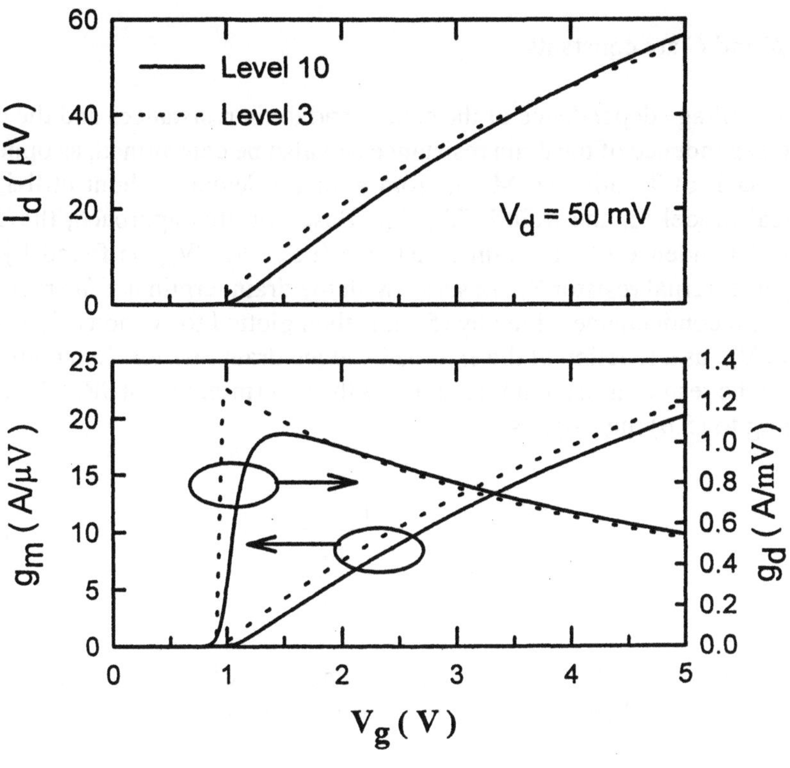

Figure 5.4 : Characteristics of I_d, g_m, and g_d versus V_g simulated using AIM-SPICE and level-10 and level-3 models.

5.2.2 Bias dependencies of the total drain and source resistance

In general, both the source and drain series resistances R_S and R_D are gate-voltage dependent, and the drain resistance R_D is drain-voltage dependent as well. Two-dimensional device simulations indicated that the drain resistance increases with decreasing gate voltage and/or increasing drain voltage [24]. Conventional methods such as that of Terada and Muta [6] have been used to study the dependence of the total source and drain resistance ($R_D + R_S$) on gate voltage. Using this technique, Hu et al. [23] found that the gate-bias dependent total series resistance of MOSFETs can be expressed in the form

$$\left(R_S + R_D \right) = K - C \log \left(V_{gs} - V_T \right) \tag{5.11}$$

where K and C are constants.

The gate-voltage dependence of the source and drain resistances and the drain-voltage dependence of the drain resistance can also be determined, as oppose to the approach of Terada and Muta, from a single device without utilizing an analytical model for the MOSFET [21]. Based on this approach, the drain-voltage dependence of the drain resistance (i.e., $\partial R_D/\partial V_{ds}$) is found by first adding an external resistor R_{ext} in series with the drain terminal. The reciprocal of the drain conductance, given by (5.6), is then plotted for various V_{ds} and at a constant V_{gs}, as a function of the externally added drain resistor. Calculating the slopes of the resulting straight lines permits the determination of $\partial R_D/\partial V_{ds}$, since according to (5.6) the slope is

$$\frac{\partial g_d^{-1}}{\partial R_{ext}} = \frac{1}{1 - I_d \dfrac{\partial R_D}{\partial V_{ds}}} \; . \tag{5.12}$$

On the other hand, the gate-voltage dependence of the drain and source resistances (i.e., $\partial(R_D + R_S)/\partial V_{gs}$) can be determined from:

$$\frac{\partial\left(R_S + R_D \right)}{\partial V_{gs}} = \left[\frac{g_{m0}}{g_{d0}} - \frac{g_m}{g_d}\left(1 - \frac{\partial R_D}{\partial V_{ds}} \right) \right] \frac{I_d^{-1}}{2 + \dfrac{g_{m0} + g_{b0}}{g_{d0}}} \; . \tag{5.13}$$

The ratio $(g_{mo} + g_{bo})/g_{do}$ in (5.13) can be found by adding the external resistor R_{ext} in series with the source terminal, and then calculating the slope of the reciprocal drain conductance versus R_{ext}. The mathematical expression is given by

$$\frac{g_{m0} + g_{b0}}{g_{d0}} = \frac{\partial g_d^{-1}}{\partial R_{ext}} \left(1 - \frac{\partial R_D}{\partial V_{ds}} \right) . \tag{5.14}$$

The remaining unknown term in (5.13), the ratio g_{mo}/g_{do}, can be obtained at a low drain bias from

$$\frac{g_{d0}}{g_{m0}} = \frac{1 + k}{\dfrac{g_{m0} + g_{b0}}{g_{d0}}} , \tag{5.15}$$

where k is a factor that accounts for the body effect:

$$k = \frac{\partial V_T}{\partial V_{sb}} . \tag{5.16}$$

Alternatively, the gate-voltage dependence of $(R_D + R_S)$ can be modeled semi-empirically by an expression of the type [21]:

$$\left(R_S + R_D \right) = a_0 + \frac{a_1}{a_2 + \left(V_{gs} + V_T \right)} , \tag{5.17}$$

where the coefficients a_1 and a_2 can be determined by fitting $\partial(R_D + R_S)/\partial V_{gs}$, obtained according to (5.13) and plotted as a function of the gate voltage, to the second term on the right-hand side of (5.17). The coefficient a_0 represents the total series resistance $(R_D + R_S)$ that is gate-voltage independent. Based on this concept, Guo et al. [15] developed a model for the gate-bias dependent $(R_D + R_S)$:

$$\left(R_S + R_D\right) = R_0 + 0.5 \left[\alpha \left(V_{gs} - V_T\right)^{-\beta} \right] + 0.5 \left[\alpha \left(V_{gs} - V_T - V_{ds}\right)^{-\beta} \right] , \tag{5.18}$$

where α and β are the channel doping concentration dependent parameters, R_0 is a residual resistance at very high gate voltages. Note that the second term on the right-hand side of (5.18) represents the gate-voltage dependent R_S, and the third term on the right-hand side of (5.18) represents the gate- and drain-voltage dependent R_D, which is analogous to the second term when the MOSFET is operating in the linear region (i.e., V_{ds} is small).

5.3 Difference in drain and source series resistances

In extracting the drain and source series resistances, it is a common practice to assumed that the parasitic resistances associated with the drain and source regions of MOSFETs are approximately equal to each other, and the extraction of the intrinsic parameters of the device normally proceeds, with the knowledge of $(R_D + R_S)$, using $R_S \approx R_D \approx (R_S + R_D)/2$. However, this assumption becomes invalid when the drain and source regions of the device are not totally symmetrical. Such an asymmetry results in a difference in the drain and source resistances $(R_D - R_S)$ and can affect considerably the current-voltage characteristics of MOSFETs in both the linear and saturation regions of operation.

This difference in the drain and source resistances arises mainly from processing, layout, and/or electrical stressing, and it becomes more prominent in the case of deep-submicron devices. This is because the relative importance of the parasitic resistances over the intrinsic components is increased as the geometry of the device shrinks.

Obviously, the techniques used to extract $(R_D + R_S)$, presented in Sec. 5.2, alone are not capable of extracting the individual values of R_D and R_S. In this section, the methods for extracting $(R_D - R_S)$ of MOSFETs will be presented and discussed. The individual values of R_D and R_S can then be obtained from the knowledge of $(R_D + R_S)$ and $(R_D - R_S)$.

A commonly used technique for extracting $(R_D - R_S)$ is based on performing measurements of an MOS device, first connected in the "normal configuration" in which the source and body are grounded, and then measuring again in the "inverted configuration" in which the source and drain terminals are interchanged, as shown in Figs. 5.5(a) and (b), respectively. Two extraction methods, namely the reciprocal transconductance method and gate-voltage shift method, have been developed based on this approach and are presented below.

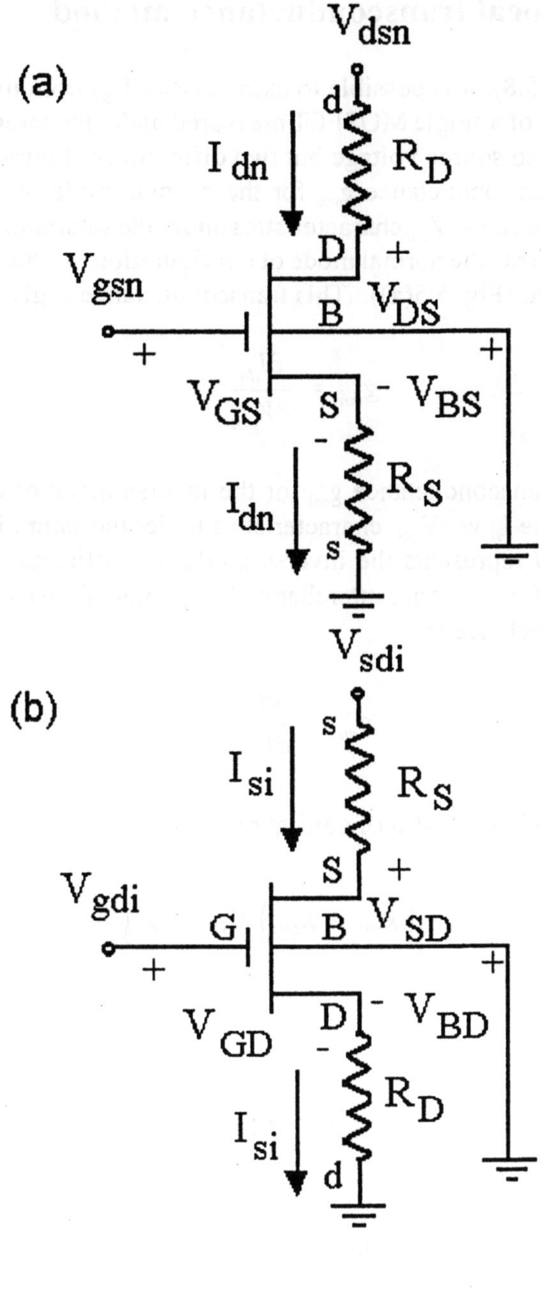

Figure 5.5 : (a) Normal configuration with the source and body grounded, and (b) inverse configuration with the drain and body grounded.

5.3.1 Reciprocal transconductance method

As suggested in (5.8), it is possible to extract (R_D - R_S) from the extrinsic gate transconductance of a single MOSFET measured under the saturation operation at the same drain to source voltage but two different configurations. First, the extrinsic gate transconductance g_{mn} for the normal mode of configuration is measured from the I_{dn} vs. V_{gsn} characteristics under the saturation region (i.e., the subscript *n* represents the normal mode of configuration in which the source and body are grounded (Fig. 5.5(a)). This transconductance is given by

$$g_{mn} = \frac{\partial I_{dn}}{\partial V_{gsn}} \quad . \tag{5.19}$$

The other gate transconductance g_{mi} for the inverse mode of configuration is measured from the I_{si} vs. V_{gdi} characteristics under the saturation region, (i.e., where subscript *i* represents the inverse mode of configuration in which the source and drain functions are interchanged (Fig. 5.5(b)). Analogous to (5.19), such a transconductance is

$$g_{mi} = \frac{\partial I_{si}}{\partial V_{gdi}} \quad . \tag{5.20}$$

Applying (5.8) to both modes of configuration yields

$$\frac{g_{m0}}{g_{mn}} = 1 + \left(g_{m0} + g_{b0} \right) R_S + g_{d0} \left(R_S + R_D \right) \tag{5.21}$$

and

$$\frac{g_{m0}}{g_{mi}} = 1 + \left(g_{m0} + g_{b0} \right) R_D + g_{d0} \left(R_S + R_D \right) \quad . \tag{5.22}$$

It should be noted that the intrinsic variables are the same for both modes of configuration, and only R_S and R_D asymmetry is present in the device. Neglecting the body effect and intrinsic body transconductance in (5.21) and (5.22) and subtracting one equation from the other reveal that the difference between the drain and source resistances is equal to the difference between the inverse and normal reciprocal extrinsic gate transconductances [25]:

$$\left(R_D - R_S \right) = \frac{1}{g_{mi}} - \frac{1}{g_{mn}} \quad . \tag{5.23}$$

However, this expression would only provide a rough estimate of the drain and source resistance asymmetry, since neglecting the intrinsic body transconductance in the saturation region is not generally justifiable and may result in a large error [26-27].

If the body effect is included by retaining the intrinsic body transconductance, the difference between the drain and source resistances is

$$\left(R_D - R_S \right) = \frac{\dfrac{1}{g_{mi}} - \dfrac{1}{g_{mn}}}{1 + \dfrac{g_{b0}}{g_{m0}}} \quad . \tag{5.24}$$

In is clear from (5.24) that, in addition to measuring the normal and inverse extrinsic gate transconductances in saturation, it is necessary to know the ratio of the intrinsic body transconductance to the intrinsic gate transconductance (i.e., g_{b0}/g_{m0} term in the denominator of (5.24)) before (R_D - R_S) can be determined. Three different procedures to calculate this term have been developed and are presented below.

5.3.1.1 VER procedure

This procedure involves connecting a variable external resistor (VER) alternatively in series with the source (denoted by R_{xS}) and with the drain (denoted by R_{xD}) terminals (i.e., $R_{xS} = R_{xD} \equiv R_{xV}$). The gate transconductance of the device, in the normal mode of configuration (source and body grounded), is then measured under these two connections, at the same drain current but several different values of R_{xV}. Using (5.8), the reciprocals of the two transconductances $(1/g_{mn})_{RxS}$ and $(1/g_{mn})_{RxD}$ can be expressed by [22,28-29]

$$\left(\frac{1}{g_{mn}} \right)_{R_{xS}} = \frac{1}{g_{m0}} + \left(R_S + R_{xS} \right) \left(1 + \frac{g_{b0}}{g_{m0}} \right) + \frac{g_{d0}}{g_{m0}} \left(R_D + R_S + R_{xS} \right), \tag{5.25}$$

for the connection where the VER is connected to the source, and

$$\left(\frac{1}{g_{mi}}\right)_{R_{xD}} = \frac{1}{g_{m0}} + R_S\left(1 + \frac{g_{b0}}{g_{m0}}\right) + \frac{g_{d0}}{g_{m0}}\left(R_D + R_{xD} + R_S\right) \quad , \tag{5.26}$$

for the connection where the VER is connected to the drain.

These two normal-mode gate transconductances would be linear functions of R_{xV} if g_{m0}, g_{d0} and g_{b0} can be assumed to be roughly constant within the range of the measurement. In that case, when plotted versus R_{xV}, these conductances can be fitted to straight lines. It then follows from (5.25) and (5.26) that subtracting the slopes of the two resulting straight lines produces the term $(1 + g_{b0}/g_{m0})$ needed in (5.24) to calculate the drain and source resistance asymmetry:

$$1 + \frac{g_{b0}}{g_{m0}} = slope\ of\left(\frac{1}{g_{mn}}\right)_{R_{xS}} - slope\ of\left(\frac{1}{g_{mn}}\right)_{R_{xD}} \quad . \tag{5.27}$$

5.3.1.2 CER procedure

A simpler alternative procedure to obtain the term $(1 + g_{b0}/g_{m0})$ in (5.24) is to measure the two extrinsic gate transconductances, as before, but with a single constant external resistor (CER) (i.e., $R_{xS} = R_{xD} \equiv R_{xC}$), instead of a variable one. Taking the difference between the two reciprocal transconductances, given in (5.25) and (5.26), and dividing this difference by the value of the external resistance, we have

$$1 + \frac{g_{b0}}{g_{m0}} = \frac{\left(\frac{1}{g_{mn}}\right)_{R_{xS}} - \left(\frac{1}{g_{mn}}\right)_{R_{xD}}}{R_x} \quad . \tag{5.28}$$

The advantages of the CER procedure are that it only uses a fixed value external resistor, it does not need the straight-line approximation used in the VER procedure and, therefore, it does not require the assumption of constant g_{m0}, g_{d0} and g_{b0}.

5.3.1.3 EBT procedure

It is also possible to obtain the term $(1 + g_{b0}/g_{m0})$ in (5.24) without measurements involving any, variable or fixed, external resistors connected to the source or to the drain terminals. The procedure consists of performing a direct measurement of the extrinsic body transconductance (EBT) g_{bn} in the normal mode configuration, as defined by

$$g_{bn} = \frac{\partial I_{dn}}{\partial V_{bsn}} \quad , \tag{5.29}$$

or the extrinsic body transconductance in the inverse mode configuration, g_{bi}, as defined by

$$g_{bi} = \frac{\partial I_{si}}{\partial V_{bdi}} \quad . \tag{5.30}$$

Applying (5.8) to the normal and inverse mode configurations, the ratio of the intrinsic body transconductance to the intrinsic gate transconductance can be expressed as:

$$\frac{g_{b0}}{g_{m0}} \approx \frac{g_{bn}}{g_{mn}} \approx \frac{g_{bi}}{g_{mi}} \quad . \tag{5.31}$$

The EBT procedure is simple in the sense that it does not make use of any external resistor, nor does it rely on the straight-line fitting scheme. However, as will be shown later, such a procedure is sensitive to the current level and processes the largest error among the three procedures.

5.3.1.4 Comparison of the three procedures for reciprocal transconductance method

Here we compare the results of the reciprocal transconductance method obtained from the VER, CER, and EBT procedures. Results of (R_D - R_S) for an *n*-channel MOSFET were extracted using the AIM-SPICE level-10 simulation and the three above mentioned procedures.

The transistor's parameters used in simulation were: 331.5 cm^2V^{-1} surface mobility, 10^{17} cm^{-3} substrate doping density, 0.43 μm metallurgical junction depth, 1.38 x 10^{-7} F·cm^{-2} oxide capacitance, 1 V threshold voltage, 331.5 $V^{-1}cm^{-2}s^{-1}$ bulk mobility, and asymmetrical drain and source resistances of $R_S = 100\ \Omega$ and $R_D = 200\ \Omega$ (i.e., the correct extracted value for (R_D - R_S) should be 100 Ω). The extraction was performed in the saturation operation at a drain voltage of 5 V, and different values of drain current up to 1 mA. Figure 5.6 shows the calculated R_D and R_S asymmetries obtained by applying the three procedures.

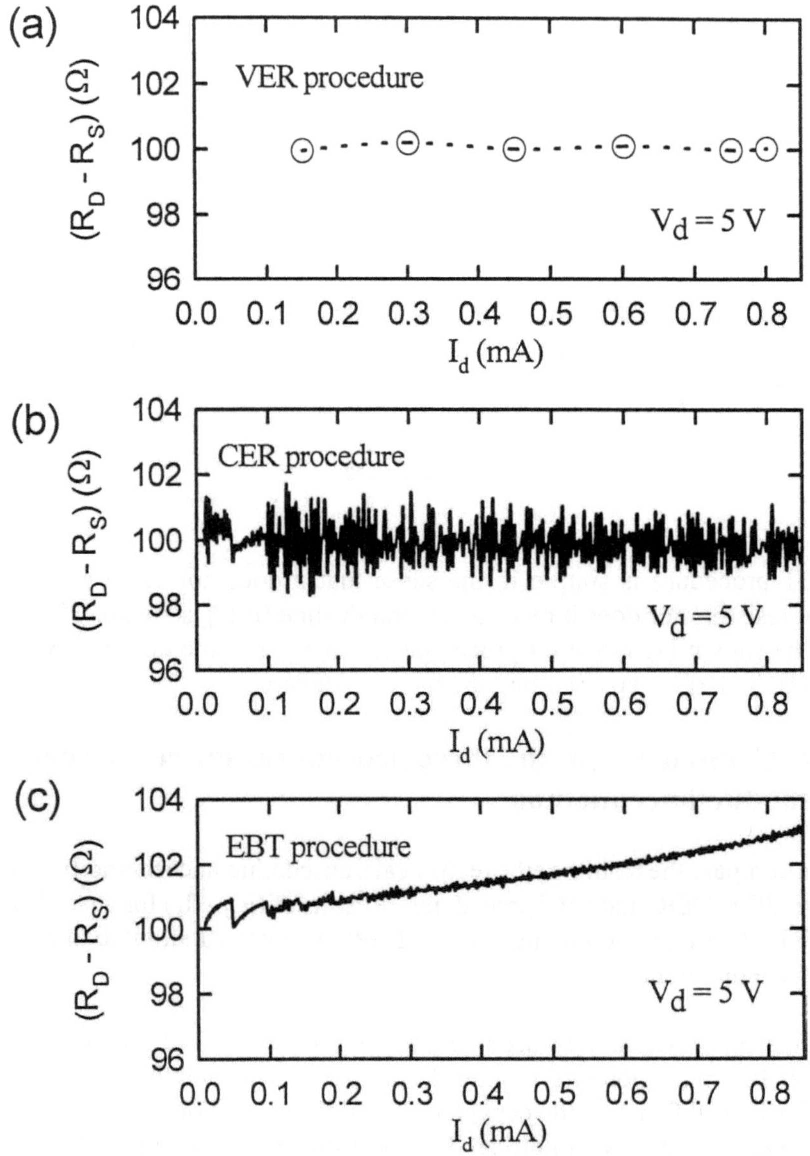

Figure 5.6 : Drain and source resistance difference extracted using the reciprocal transconductance method with three different procedures: a) VER, b) CER, and c) EBT.

For the VER procedure, eight different values of the variable resistor R_{xV} ranging from 5 to 160 Ω were used to calculate the slopes of the straight-line fitted reciprocal transconductance versus R_x. The data points shown in Fig. 5.6(a) correspond to the resulting (R_D - R_S), extracted in this way at only six representative values (circled in Fig. 5.6(a)) of drain current because of the lengthy fitting and slope calculation involved. For the CER procedure, a single value of R_{xC} = 10 Ω was used. Because no straight line fitting is necessary, the resistance asymmetry was calculated at every point of the operating drain current, about 500 values in this case, as shown in Fig. 5.6(b). Similarly, for the EBT procedure, the body transconductance was obtained at every point of operating drain current under the normal configuration, using a body voltage variation of δV_{bsn}, and (R_D - R_S) was calculated and shown in Fig. 5.6(c).

Comparing against the correct value of (R_D - R_S) = 100 Ω, the VER procedure produces the smallest maximum extraction error, of around 0.1 %, because it includes an averaging step inherent to the straight-line fitting scheme. In the case of the CER procedure, the results exhibit a random maximum error of about 2 %, which is within a typical range of error for measurements. If a data smoothing step were included in the CER procedure, as that inherented in the VER procedure, the extraction accuracy would be similar to that of the VER procedure. On the other hand, the EBT procedure, which requires measuring the body transconductance directly, presents a progressive error that increases with increasing drain current level, up to around 3 % at 1 mA in this case. This can be attributed to the fact that a higher current results in a larger voltage drop in the drain and source resistances, and that the EBT procedure is sensitive to such a voltage drop.

5.3.2 Gate-voltage shift method

The gate-voltage shift method relies, as does the reciprocal transconductance method discussed above, on measuring a single transistor when it is connected alternatively in the normal and inverse configurations. However, instead of measuring the difference between normal and inverse reciprocal gate transconductances, it is based on measuring the shift of the gate voltage needed to maintain the same magnitude of drain current when the device is connected in the inverse and normal configurations [30-31]. Consider a MOSFET in the normal configuration, with the source and body grounded, and also in the inverse configuration, with the drain and source interchanged. The drain current in I_{dn} in the normal configuration can be expressed as a general function of the intrinsic voltages as:

$$I_{dn} = f\left[\left(V_{GS} - V_{Tn}\right), V_{DS}\right] \quad , \tag{5.32}$$

where f is a function defined by a particular MOSFET model, V_{Tn} is the threshold voltage in the normal configuration, and the body voltage dependence has been implicitly incorporated. The function f does not make any other *a priori* assumptions as to the model describing the relationship between drain current and applied voltages. The intrinsic gate-to-source and drain-to-source voltages can be expressed, from (5.1) and (5.2), in terms of their extrinsic counterparts as

$$V_{GS} = V_{gsn} - I_{dn} R_S \tag{5.33}$$

and

$$V_{DS} = V_{dn} - I_{dn} (R_S + R_D) \quad . \tag{5.34}$$

Here V_{gsn} and V_{dsn} represent the extrinsic gate-source and drain-source voltages, respectively, in the normal configuration. Analogously, the source current in the inverse configuration is given by

$$I_{si} = f\left[\left(V_{GD} - V_{Ti} \right), V_{SD} \right] \quad , \tag{5.35}$$

where V_{Ti} is the threshold voltage in the inverse configuration, and V_{GD} and V_{SD} are the intrinsic gate-drain and source-drain voltages, respectively. These voltages can be related to their extrinsic counterparts by

$$V_{GD} = V_{gdi} - I_{si} R_D \tag{5.36}$$

and

$$V_{SD} = V_{sdi} - I_{si} (R_S + R_D) \quad , \tag{5.37}$$

where V_{gdi} and V_{sdi} are the extrinsic gate-drain and source-drain voltages, respectively, in the inverse configuration. If the device in both configurations is biased with the same source-drain voltage (i.e., $V_{sdi} = V_{dsn}$) and V_{gdi} is adjusted until the source current in the inverse configuration is equal to that in the normal configuration (i.e., $I_{si} = I_{dn} \equiv I_d$), then the normal and inverse intrinsic gate voltage overdrive must be the same:

$$\left(V_{GS} - V_{Tn} \right) = \left(V_{GD} - V_{Ti} \right) \quad . \tag{5.38}$$

Substituting (5.33) and (5.36) into (5.38) yields

$$I_d \left(R_D - R_S \right) = \left(V_{gdi} - V_{gsn} \right) - \left(V_{Ti} - V_{Tn} \right) \quad . \tag{5.39}$$

If the device is biased in the linear region, the term (V_{Ti} - V_{Tn}) in the above equation is small because (V_{DB} - V_{SB}) is small. Therefore it can be approximated by the first term of its Taylor series expansion as:

$$\left(V_{Ti} - V_{Tn} \right) \approx I_d \left(R_D - R_S \right) \frac{dV_T}{dV_{SB}} \quad . \tag{5.40}$$

Substituting (5.40) into (5.39) gives

$$\left(R_D - R_S \right) = \frac{\left(\dfrac{V_{gdi} - V_{gsn}}{I_d} \right)}{1 + \dfrac{dV_T}{dV_{SB}}} \quad . \tag{5.41}$$

The term ($1 + dV_T/dV_{SB}$) in (5.41) takes into account the dependence of the threshold voltage on the source-to-body voltage V_{SB}. To obtain such a dependence, one can measure the dependence of the gate voltage on V_{SB} instead. This is because the drain current is proportional to V_{gs} and V_T as

$$I_d \propto V_{gs} - I_d R_S - V_T \quad . \tag{5.42}$$

Thus, at a constant drain current,

$$\frac{dV_T}{dV_{sb}} = \frac{dV_{gsn}}{dV_{sb}} \quad . \tag{5.43}$$

Substituting (5.43) into (5.41), we have

$$\left(R_D - R_S \right) = \frac{\left(\dfrac{V_{gdi} - V_{gsn}}{I_d} \right)}{1 + \dfrac{dV_{gsn}}{dV_{sb}}} \quad . \tag{5.44}$$

The denominator of (5.44) can be obtained by measuring the gate voltage change needed to respond to a small change in the body voltage in order to maintain the same drain current. This term can be easily determined using a circuit involving an operational amplifier shown in Fig. 5.7.

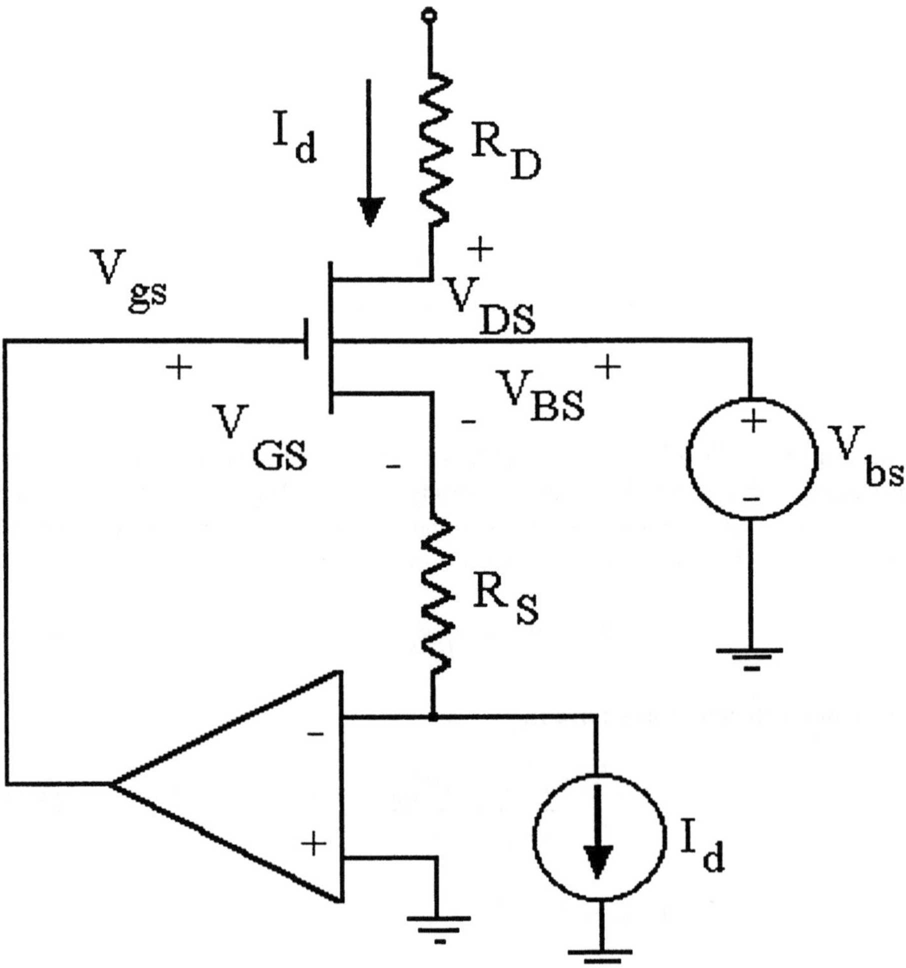

Figure 5.7 : Circuit for measuring the derivative of the threshold voltage with respect to the body-source voltage at a constant drain current.

Figure 5.8 shows the calculated drain and source resistance asymmetry obtained from the gate-voltage shift method and results simulated from AIM-SPICE level-10 model for the same *n*-channel MOSFET used in Fig. 5.6. The transistor's parameters are the same as those indicated in Sec. 5.3.1.4, including asymmetrical source and drain resistances of $R_S = 100\ \Omega$ and $R_D = 200\ \Omega$. The extraction was performed in the linear region of operation at a drain voltage of 100 mV, and for 360 values of drain current up to 70 μA. Clearly, the method is very accurate, as the simulated $(R_D - R_S)$ is identical to that specified in AIM-SPICE.

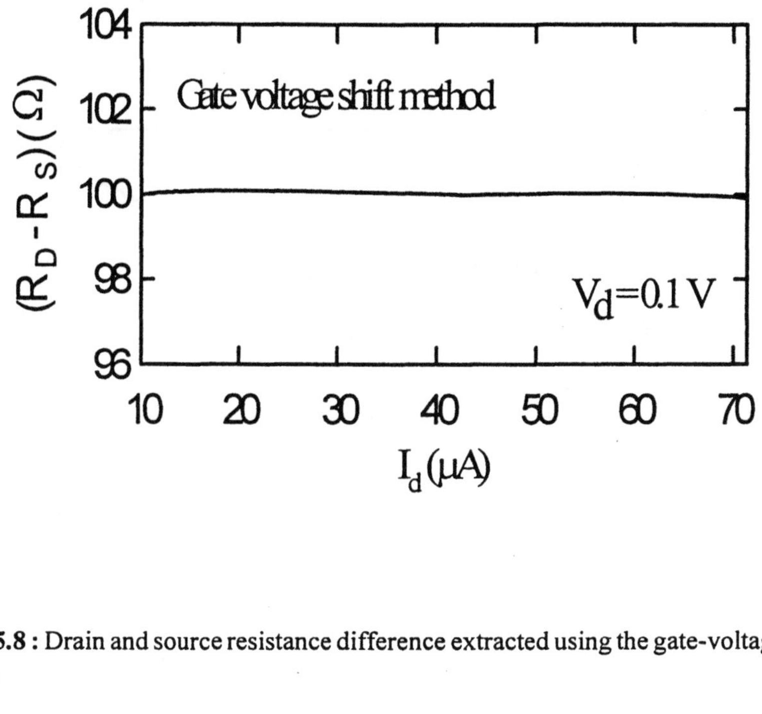

Figure 5.8 : Drain and source resistance difference extracted using the gate-voltage shift method.

In an attempt to examine the effects of the term dV_{gsn}/dV_{sb} in (5.44) on the accuracy of the gate-voltage shift method, we have simulated another *n*-channel MOSFET using AIM-SPICE level-1 and level-10 models. The MOS transistor's parameters used here were the same as before, except that the asymmetrical source and drain resistances in this case are $R_S = 10\ \Omega$ and $R_D = 20\ \Omega$ (i.e., the correct value of $(R_D - R_S)$ is $10\ \Omega$). The threshold voltage dependence on the source-to-body voltage, $dV_{Tn}/dV_{SB} = dV_{gsn}/dV_{sb}$, was first simulated as a function of drain current, from the circuit in Fig. 5.7, using AIM-SPICE level-1 and level-

10 models. A constant value of 0.73 is obtained using the level-1 model, corresponding to the same value that could be calculated using the conventional MOSFET theory: $dV_{Tn}/dV_{SB} = 2\epsilon q N_A/C_{ox}$ [32]. The more comprehensive level-10 model yields a higher dV_{Tn}/dV_{SB}, rising from 1 to 2.4 as the drain current increases from 0 to 15 μA.

Figure 5.9 presents (R_D - R_S) extracted using AIM-SPICE level-1 and level-10 models and the gate-voltage shift method with and without including the effects of dV_{Tn}/dV_{SB}. The results indicate that the gate-voltage shift method is erroneous and becomes MOSFET model dependent if the effects of dV_{Tn}/dV_{SB} are not accounted for. On the other hand, when the body-voltage dependence is included, the correct result of 10 Ω is obtained, and the extraction method becomes insensitive to the type of model selected in simulation.

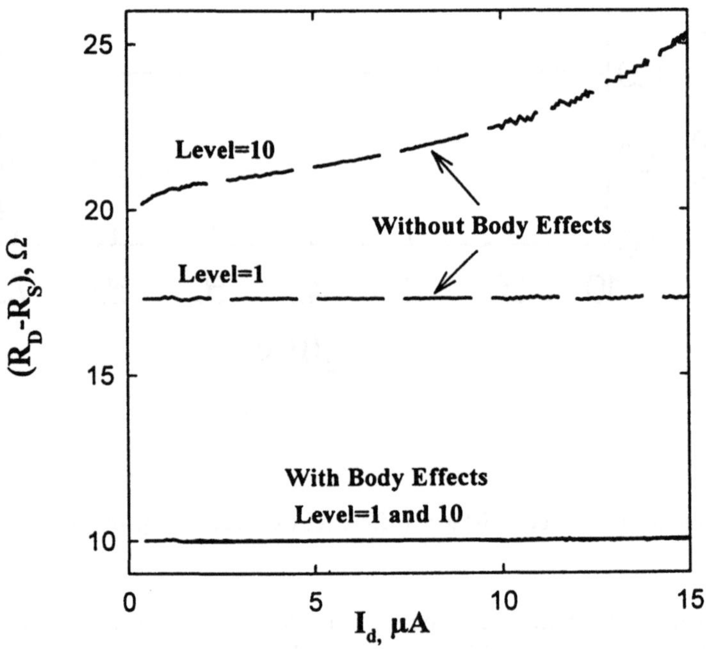

Figure 5.9 : Drain and source resistance difference extracted using the gate-voltage shift method with and without considering the body effect factor. The extraction was carried out in the circuit simulator AIM-SPICE with level-1 and level-10 MOSFET models and R_S =10 Ω and R_D = 20 Ω.

5.4 Physical mechanisms contributing to the drain and source asymmetry

Numerical simulations can be used to investigate the main physical mechanisms contributing to the asymmetry in source and drain resistances in short-channel MOSFETs [33]. To this end, the drain and source resistance difference (R_D - R_S) has been extracted from *p*-channel MOSFETs using the gate-voltage shift method and the results simulated from the two-dimensional device simulator MEDICI [34]. The following three cases are considered: (i) devices with different source and drain contact resistances; (ii) devices with different source and drain doping densities; and (iii) devices with a gate misalignment. The devices used in simulation, the schematic of which is shown in Fig. 5.10, have mask channel lengths L_m of 1.25 and 10 μm, source and drain make-ups consisting of *p*-type Gaussian profiles with peak doping concentrations N_{AS} for the source and N_{AD} for the drain, junction depth of 0.34 μm, and a lateral extent of 0.75 of the vertical extent. The distances separating the source/drain contacts and the gate are represented by L_{GS} and L_{GD}, respectively. The gate oxide thickness is 25 nm (i.e., the oxide capacitance is $1.38\text{x}10^{-7}$ F/cm^2) and the substrate doping concentration is 10^{17} cm^{-3}. The distributed metal-semiconductor contact resistances for the source and drain are denoted by R_{CS} and R_{CD}, respectively.

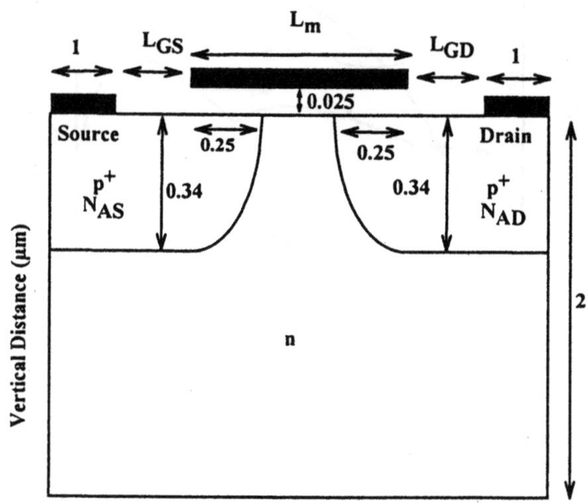

Figure 5.10: Device structure of the p-channel MOSFET used in simulating the physical mechanisms contributing to the drain and source asymmetry.

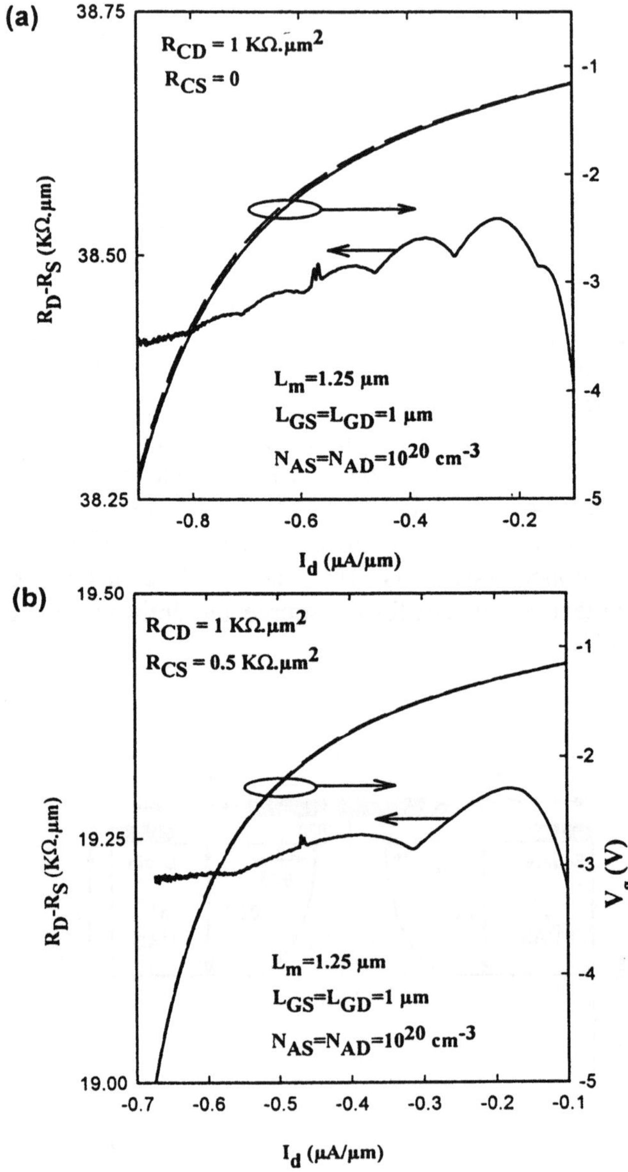

Figure 5.11: Drain and source resistance difference (R_D - R_S) extracted for a 1.25-μm MOSFET with (a) R_{CS} = 0 and R_{CD} = 1 KΩ.μm and (b) R_{CS} = 0.5 KΩ.μm and R_{CD} = 1 KΩ.μm and otherwise symmetrical drain and source make-up.

The effective channel length L_{eff} is smaller than L_m, due to the presence of the lateral penetration of the drain and source regions underneath the gate, but is somewhat larger than the physical channel length L_{met} defined by the source and drain metallurgical junctions, since the inverted free carriers are extended into the source and drain regions [35]. For example, for the MOS device with L_m = 1.25 μm, L_{met} = 0.75 μm, and $L_{eff} \approx$ 0.85 μm.

Figures 5.11(a) and (b) present (R_D - R_S), V_{gsn}, and V_{gdi} as functions of I_d for devices with a drain contact resistance R_{CD} = 1 kΩ.μm and two different source contact resistances (a) R_{CS} = 0 and (b) R_{CS} = 500 Ω.μm. The devices have the same source and drain doping densities, $N_{AS} = N_{AD} = 10^{20}$ cm^{-3}, and there is no gate misalignment (i.e., $L_{GD} = L_{GS}$ = 1 μm). The channel length is 1.25 μm and other parameters are as noted before. A small drain voltage of 50 mV is used to prevent the channel from pinch-off. Note that slightly different values of V_{gsn} and V_{gdi} are needed to achieve the same I_d in the normal and inverse configurations. The figure shows that (R_D - R_S) is reduced approximately in half, from a maximum value of about 38 kΩ.μm to about 19 kΩ.μm, when (R_{CD} - R_{CS}) is reduced from 1 kΩ.μm to 500 Ω.μm. It is apparent that the extracted (R_D - R_S) is much larger than the contact resistance difference. This is because of the distributed nature of the series resistance in the drain and source regions, which in effect amplifies the difference of R_D and R_S in the two regions.

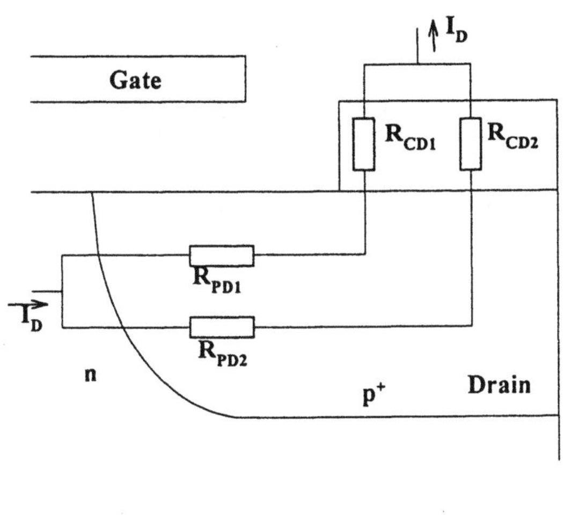

Figure 5.12 : Schematic showing the distributed nature of the resistance in the drain region by using two parallel branches, each with a contact resistance and a series resistance associated with the diffusion region.

To illustrate this, let us focus on the drain region, and the distributed nature of the drain resistance is represented by two parallel branches, each with a contact resistance R_{CD} and a resistance R_{PD} associated with the p-type diffusion region, as shown in Fig. 5.12. The total drain resistance is

$$R_D = \frac{1}{\frac{1}{R_1} + \frac{1}{R_2}} \tag{5.45}$$

where

$$R_1 = R_{RPD1} + R_{RCD1} \; ; \; and \;\; R_2 = R_{RPD2} + R_{RCD2} \tag{5.46}$$

Note that R_{CD1} can be assumed the same as R_{CD2}, but R_{PD1} is smaller than R_{PD2} due to the following two factors: 1) the current path for R_{PD1} is shorter than that for R_{PD2}; and 2) the doping density in the upper diffusion region represented by R_{PD1} is higher than that in the lower diffusion region represented by R_{PD2}. Applying the same concept for the source region, and after some algebraic manipulation, we have

$$\left(R_D - R_S \right) = R_{CDS} + \frac{X_1}{X_2 \left(R_{PD1} + R_{PD2} \right) \left(R_{CD1} + R_{CD2} \right)} \tag{5.47}$$

where

$$R_{CDS} = R_{CD} - R_{CS} \tag{5.48}$$

$$X_1 = \left(R_{PD1} \, R_{CD2} - R_{CD1} \, R_{PD2} \right)^2 \tag{5.49}$$

and

$$X_2 = R_{PD1} + R_{PD2} + R_{CD1} + R_{CD2} \tag{5.50}$$

Since X_1 and X_2 are not zero and are positive, $(R_D - R_S)$ is larger than $(R_{CD} - R_{CS})$.

It is important to point out that for the operation region with very low drain currents (i.e., moderate and weak inversion), the extracted values of $(R_D - R_S)$ are questionable because of the invalidity of the linear extrapolation method, used in obtaining (5.40), in such operations. Also note that the drain current has a unit of μA/μm and the contact resistances and extracted $(R_D - R_S)$ have a unit of kΩ.μm, where μm is the third dimension of the MOSFET (i.e., MOSFET width) not accounted for in the two-dimensional device simulator MEDICI. In other words, the current in the unit of μA and the resistances in the unit of kΩ can be obtained by multiplying and dividing, respectively, their values shown in the figures by the width (in μm) of the MOSFET under consideration.

To investigate the effects of the channel length on (R_D - R_S), the results for an analogous long-channel device with L_m = 10 μm are extracted and shown in Fig. 5.13. Here, a trend almost identical to that in Fig. 5.11(a) is found, and the maximum (R_D - R_S) (i.e., 38.5 kΩ.μm) is about the same as that shown in Fig. 5.11(a). This suggests that the method is channel-length independent, an observation consistent with the definition that R_D and R_S are parasitic resistances associated with the source and drain regions but not with the channel.

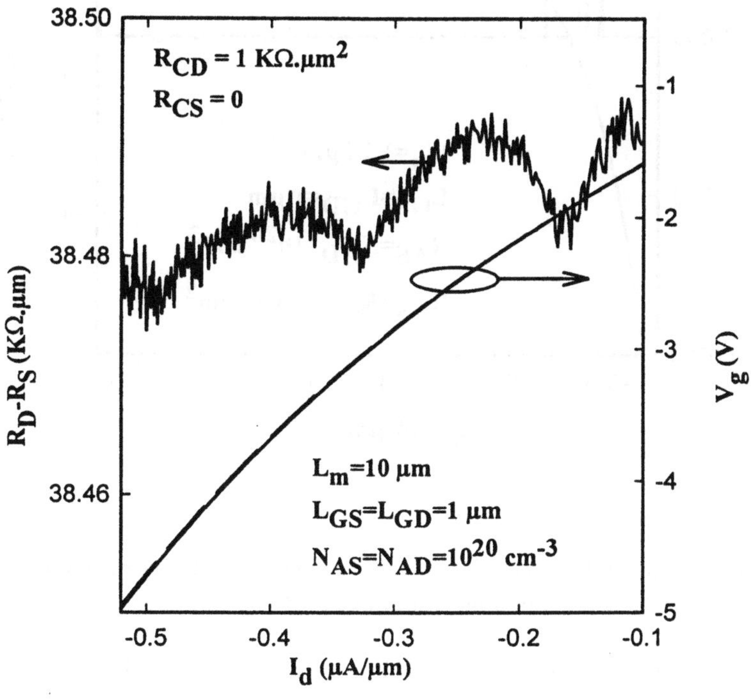

Figure 5.13 : Drain and source resistance difference (R_D - R_S) extracted for a long-channel device having the same device make-up as the device in Fig. 5.9(a) but a longer channel length of L_m = 10 μm.

If a device with completely symmetrical drain and source regions is considered (i.e., same contact resistances, same doping densities, and no gate misalignment), then (R_D - R_S) should approach zero. This is indeed the case, as evidenced by the results in Fig. 5.14, which show that (R_D - R_S) is zero except for the presence of some small impulses caused by numerical errors.

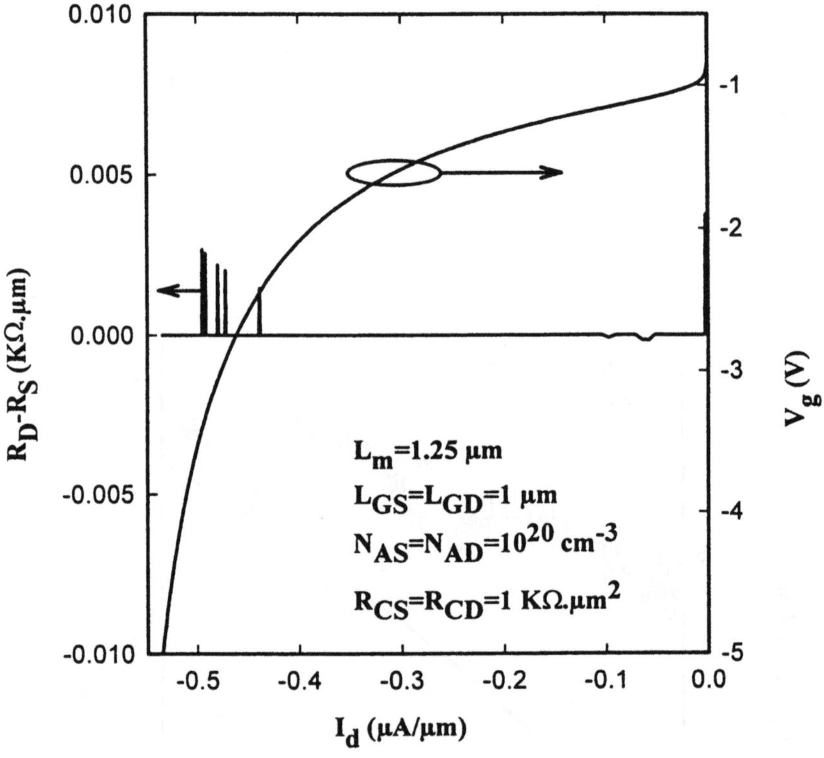

Figure 5.14 : Drain and source resistance difference (R_D - R_S) extracted for a device with a completely symmetrical drain and source make-up.

To help analyzing the effects of the difference in doping densities in the source and drain regions on (R_D - R_S), Figs. 5.15(a) and (b) present (R_D - R_S), V_{gsn}, and V_{gdi} as functions of I_d for two devices, both with asymmetrical doping of N_{AS} = 10^{20} cm^{-3} and N_{AD} = 10^{19} cm^{-3}, L_{GD} = L_{GS} = 1 μm, and L_m =1.25 μm, but one with R_{CS} = R_{CD} = 0 and the other with R_{CS} = R_{CD} = 1 kΩ.μm.

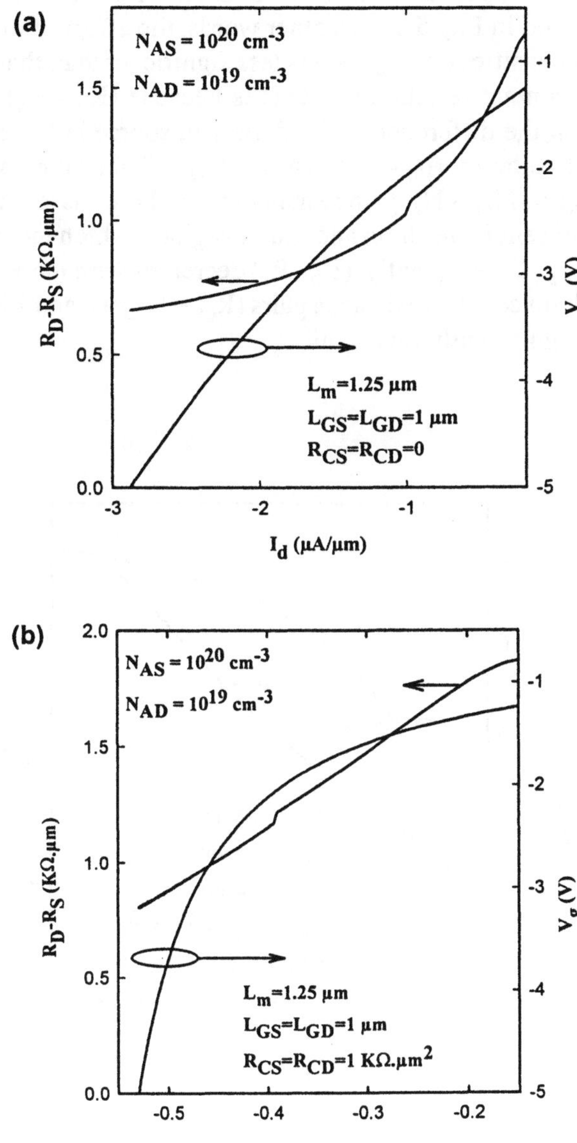

Figure 5.15 : Drain and source resistance difference (R_D - R_S) extracted for two devices, both with asymmetrical drain and source doping densities but otherwise symmetrical source and drain make-up, with (a) $R_{CD} = R_{CS} = 0$ and (b) $R_{CD} = R_{CS} = 1k\Omega.\mu m$.

The figures show that (R_D - R_S) in both cases decreases as the magnitude of I_d increases (or the magnitude of the gate voltage increases) and reaches a minimum value around 0.6 kΩ.μm, which is about 25 times smaller than the minimum value found in Fig. 5.11. In other words, the effect of different drain and source doping densities on (R_D - R_S) is less significant than that of different drain and source contact resistances. This is because (R_D - R_S) in this case depends entirely on the difference in the drain and source hole densities (p_D - p_S). At small gate voltages, (p_D - p_S) ≈ (N_{AD} - N_{AS}). But as the gate voltage is increased, (p_D - p_S) < (N_{AD} - N_{AS}) since many excess holes associated with the drain current are present in the drain and source regions, which increase p_D more significantly than p_S. Consequently, (R_D - R_S) decreases with increasing I_d. The different source/drain contact resistance pairs (R_{CS} = R_{CD} = 0 and 1 kΩ.μm) does not seem to alter significantly this result.

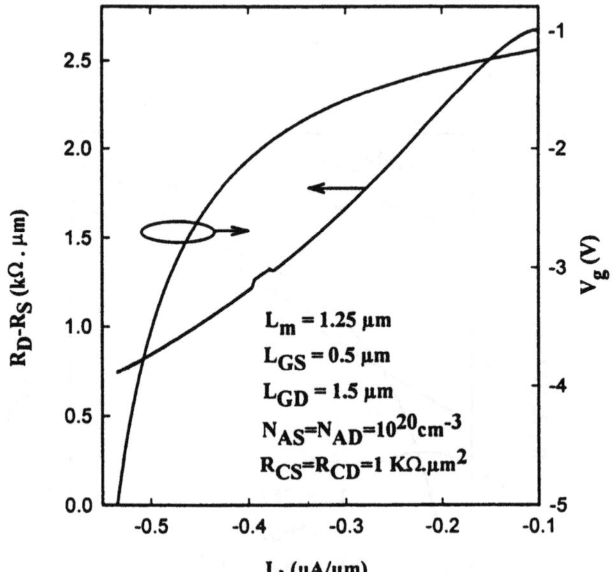

Figure 5.16 : Drain and source resistance difference (R_D - R_S) extracted for a device with gate misalignment of 0.5 μm toward the source (i.e., L_{GS} = 0.5 μm and L_{GD} = 1.5 μm).

The effect of gate misalignment can be studied using a device with L_{GS} = 0.5 μm and L_{GD} = 1.5 μm (i.e., a gate misalignment of 0.5 μm toward the source), and otherwise symmetrical drain and source regions. The resistance asymmetry (R_D

- R_S), V_{gsn}, and V_{gdi} simulated from such a device are shown in Fig. 5.16. Like the trend seen in the previous figures, (R_D - R_S) decreases as I_d is increased. Again, the value of (R_D - R_S) is much smaller than those caused by the different drain and source contact resistances. These results suggest that, in the absence of electrical stressing effects, drain and source resistance asymmetry is originated mainly from the difference in the drain and source contact resistances, and not from the gate misalignment, nor from the difference in source and drain doping densities.

REFERENCES

[1] F. J. García Sánchez, A. Ortiz-Conde, G. De Mercato, J. J. Liou, and L. Rech, "Eliminating parasitic resistances in parameter extraction of semiconductor device-models," Proc. First IEEE Intl. Caracas Conf. Devices, Circuits and Syst., Caracas, Venezuela, Dec. 1995, p. 298.

[2] F. J. García Sánchez, A. Ortiz-Conde, and J. J. Liou, "A parasitic series resistance-independent method for device-model parameter extraction," IEE Proc.-Circuits Devices Syst., vol. 143, No. 1, pp. 68-70, Feb. 1996

[3] P. R. Karlsson and K. O. Jeppson, "Extraction of series resistance-independent of MOS transistor model parameters," IEEE Trans. Electron Devices, vol. ED-13, pp. 581-583, Nov. 1992.

[4] K. K. Ng and J. R. Brews, Jr, "Measuring the effective channel length of MOSFETs," IEEE Cir. & Dev. Magazine, vol. 6, pp. 33-38, 1990.

[5] D. K. Schroeder, *Semiconductor Material and Device Characterization*, Wiley, New York, 1990.

[6] K. Terada and H. Muta, "A new method to determine effective MOSFET channel length," Japn. J. Appl. Phys., vol. 18, pp. 953-959, 1979.

[7] J. Chern, P. Chang, R. Motta, and N. Godinho, "A new method to determine MOSFET channel length," IEEE Electron. Dev. Lett., vol. 1, pp. 170-173, 1980.

[8] P. I. Suciu and R. L. Johnston, "Experimental derivation of source and drain resistance of MOS transistors," IEEE Trans. Electron Device, vol. ED-27, pp. 1846-1848, Sept 1980.

[9] J. Whitfield, "A modification on an improved method to determine MOSFET channel length," IEEE Electron. Devices Letters, vol. EDL-6, pp. 109-110, March 1985.

[10] Md. Rofiqul Hassan, J. J. Liou, A. Ortiz-Conde, F. J. García Sánchez, and E. Gouveia Fernandes, "Drain and source resistances of short-channel LDD MOSFETs," Solid-St. Electron., vol. 41, pp. 778-780, 1997.

[11] Y. -S. Jean and C. -Y. Wu, "A new extraction algorithm for the metallurgical channel length of conventional and LDD MOSFET`s," IEEE Trans. Electron Device, vol. ED-43, pp. 946-953, June 1996.

[12] C. C. McAndrew and P. A. Layman, "MOSFET effective channel length, threshold voltage, and series resistance determination by robust optimization," IEEE Trans. Electron. Dev., vol. ED-39, pp. 2298-2311, Oct. 1992.

[13] Y. Taur, D. S. Zicherman, D. R. Lombardi, P. J. Restle, C. H. Hsu, H. Y. Hanafi, M. R. Wordeman, B. Davari and G. G. Shahidi, "A new "shift and ratio" method

for MOSFET channel-length extraction," IEEE Electron. Devices Letters, vol. EDL-13, pp. 267-269, May 1992.

[14] F. J. García Sánchez, A. Ortiz-Conde, M. García Núñez, and R. L. Anderson, "Extracting the series resistance and effective channel length of short-channel MOSFETs at liquid nitrogen temperature," Solid-St. Electron., vol. 37, pp. 1943-1948, Dec. 1994.

[15] J. -C. Guo, S. S. -S. Chung and C. C. -H. Hsu, "A new approach to determine the effective channel length and drain-and-source series resistance of miniaturized MOSFET's," IEEE Trans. Electron. Dev., vol. ED-41, pp. 1811-1818, Oct. 1994.

[16] P. R. Karlsson and K. O. Jeppson, "A direct method to extract effective geometries and series resistances of MOS transistors," Proc. IEEE Int. Conf. Microelec. Test Struct., vol. 7, pp. 184-189, March 1994.

[17] K. O. Jeppson, A. W. Bogren, and P. R. Karlsson, "A new method of determining the effective channel width and its dependence on the gate voltage," Proc. IEEE Int. Conf. Microelec. Test Struct., vol. 9, pp. 151-159, March 1996.

[18] P. R. Karlsson and K. O. Jeppson, "An efficient method for determining threshold voltage, series resistance and effective geometry of MOS transistors," IEEE Trans. Semiconductor Manufacturing, vol. 9, pp. 215-222, May 1996.

[19] L. Selmi, E. Sangiorgi, and B. Ricco, "Parameter extraction from I-V characteristics of single MOSFET's," IEEE Trans. Electron Devices, vol. ED-36, pp. 1094-1101, June 1989.

[20] L. Selmi and B. Ricco, " Frequency-resolved measurements for the characterization of MOSFET parameters at low longitudinal field," IEEE Trans. Electron Devices, vol. ED-42, pp. 315-320, Feb. 1995.

[21] J. A. M. Otten and F. M. Klaassen, "A novel technique to determine the gate and drain bias dependent series resistance in drain engineered MOSFETs using one single device," IEEE Trans. Electron Devices, vol. ED-43, pp. 1478-1488, Sept. 1996.

[22] A. Raychaudhuri, M. J. Deen, M. I. H. King, and J. Kolk, "Finding the asymmetric parasitic source and drain resistances from the ac conductances of a single MOSFET," Solid-State Electron., vol. 39, No. 6, pp. 909-913, 1996.

[23] G. J Hu, C. Chang, R. F. Motta, and N. Godinho, "Gate-voltage-dependent effective channel length and series resistance of LDD MOSFETs," IEEE Trans. Electron Devices, vol. ED-34, pp. 2469-2475, 1987.

[24] G. S. Samudra, B. P. Seah, and C. H. Ling, "Determination of LDD MOSFET drain resistance from device simulation," Solid-St. Electron., vol. 39, pp. 753-758, 1996.

[25] K. Lee, M. Shur, T. A. Fjeldly, and T. Ytterdal, *Semiconductor Device Modeling for VLSI*, Prentice Hall, Englewood, NJ, 1993.

[26] S. Y. Chou and D. A. Antoniadis, "Relationships between measured and intrinsic transconductances of FETs," IEEE Trans. Electron Devices, vol. ED-34, pp. 448-450, Feb. 1987.

[27] S. Cserveny, "Relationships between measured and intrinsic conductances of MOSFETs," IEEE Trans. Electron Devices, vol. ED-37, pp. 2413-2414, Nov. 1990.

[28] A. Raychaudhuri, J. Kolk, M. J. Deen, and M. I. H. King, "A simple method to extract the asymmetry in parasitic source and drain resistances from measurements on a MOS transistor," IEEE Trans. Electron Devices, vol. ED-42, pp. 1388-1390, July 1995.

[29] A. Raychaudhuri, M. J. Deen, M. I. H. King, and W. S. Jwan, "A simple method to qualify the LDD structure against the early mode of hot-carrier degradation," IEEE Trans. Electron Devices, vol. ED-43, pp. 110-115, Jan. 1996.

[30] A. Ortiz-Conde, J. J. Liou, and F. J. García Sánchez, "Simple method for extracting the difference between the drain and source series resistances in MOSFETs," Electron. Lett., vol. 30, pp. 1013-1015, June 1994.

[31] A. Ortiz-Conde, F. J. García Sánchez, and J. J. Liou, "An improved method for extracting the difference between the drain and source resistances in MOSFETs," Solid-State Electron., vol. 39, pp. 419-421, 1996.

[32] J. J. Liou, *Advanced Semiconductor Device Physics and Modeling*, Artech House, Boston, MA, 1994.

[33] A. Ortiz-Conde, J. J. Liou, R. Narayanan, and F. J. García Sánchez, "Determination of the physical mechanisms contributing to the difference between drain and source in short-channel MOSFETs," Solid-St. Electron., vol. 39, pp. 211-215, 1996.

[34] MEDICI Manual, Technology Modeling Associates, Inc., Palo Alto, CA, 1993.

[35] R. Narayanan, A. Ortiz-Conde, J. J. Liou, F. J. García Sánchez, and A. Parthasarathy "Two-dimensional numerical analysis for extracting the effective channel length of short-channel MOSFETs," Solid-St. Electron., vol. 38, pp. 1155-1159, 1995.

Chapter 6

Parameter extraction of lightly-doped drain (LDD) MOSFETs

Modern MOSFETs often incorporate a lightly-doped drain (LDD) region. Due to the presence of the LDD region, these so called LDD MOSFETs have a smaller electric field near the drain region and therefore a reduced hot-carrier effect over the conventional MOSFET [1-2]. This, however, comes with the expenses of an increase in the drain/source series resistances and therefore a reduced drain current level. Figures. 6.1(a)-(c) give the schematic of the cross section of conventional MOSFET, LDD MOSFET, and fully overlapped LDD (FOLD) MOSFET, respectively. It can be seen that the LDD and FOLD MOSFETs differ mainly in the gate structure; the LDD MOSFET has a typical polysilicon gate surrounded by the oxide sidewall, whereas the FOLD MOSFET has a larger gate consisting of a polysilicon gate and two spacers. Let us focus on the LDD MOSFET. The lightly and heavily doped drain and source regions are fabricated as follows. First, the lightly-doped n^- drain and source regions are formed by ion implantation defined by the edges of the polysilicon gate. The heavily-doped n^+ drain and source regions are then formed by a second ion implantation defined by the edges of the oxide sidewalls.

Because the free-carrier density in the portion of the n^- drain and source regions underneath the gate can be easily modulated by the gate bias, the drain and source series resistances and the effective channel length of the LDD device become gate-voltage dependent. This, when using the Terada-Muta method [3] developed intended for the conventional MOSFET to extract the LDD parameters, may result in a situation where no unique intersection can be found in the total resistance versus mask channel lenth plot, as shown in Fig. 6.2.

In this chapter, we will first investigate the validity of the Terada-Muta method for extracting the effective channel length L_{eff} of the LDD MOSFET. A measurement algorithm to extract the bias-dependent effective channel length and drain and source series resistance of LDD MOSFET is then discussed. Another L_{eff} extraction method which proposed a different concept that the effective channel length of the LDD MOSFET should be bias independent is also presented.

Extraction of other parameters of the LDD MOSFET, such as the drain and source resistances, metallurgical channel length, and threshold voltage, will also be addressed. Both measurement data and device simulation results will be used to facilitate and demonstrate the extraction process.

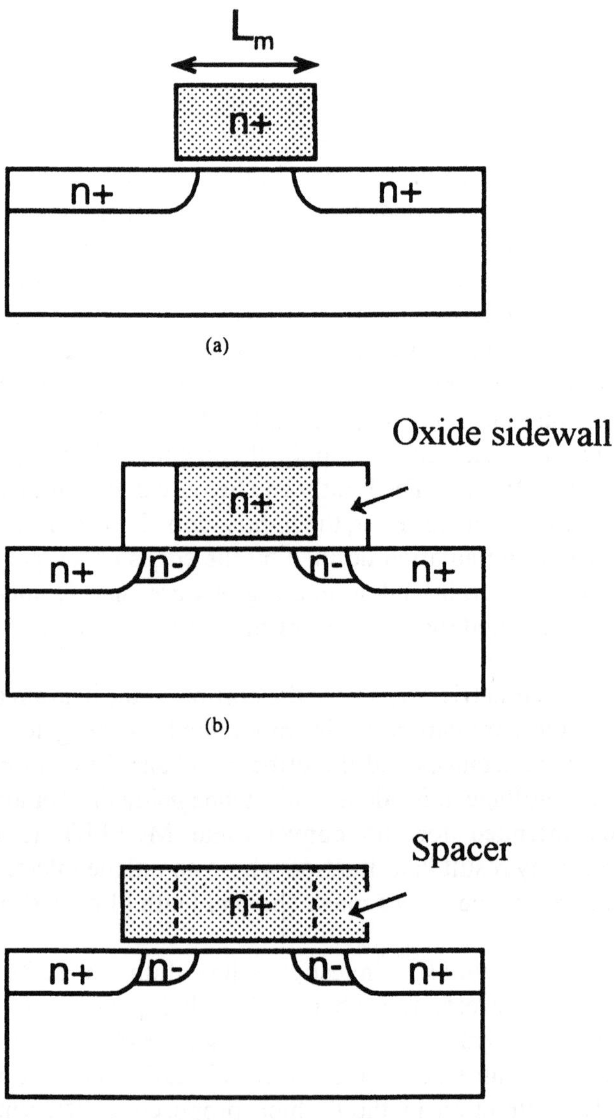

Figure 6.1 : Schematic of the (a) conventional MOSFET, (b) LDD MOSFET, and (c) fully overlapped LDD MOSFET (after Takeuchi et al. [10]).

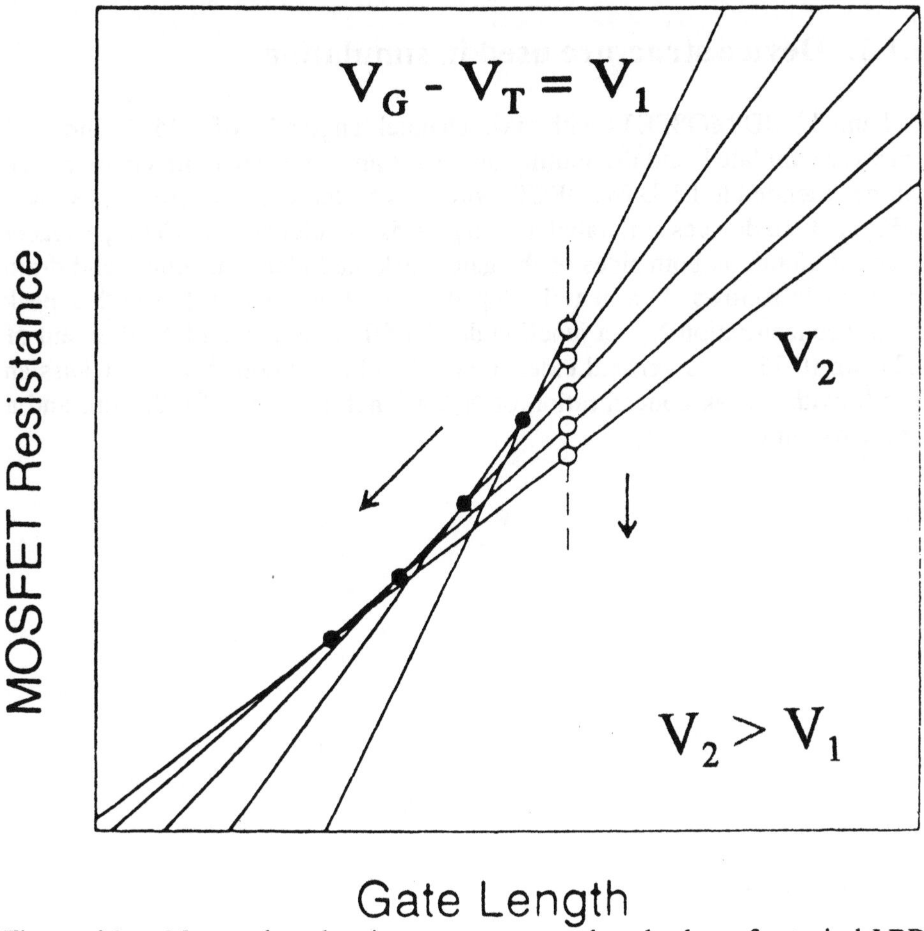

Figure 6.2 : Measured total resistance versus gate length plots of a typical LDD MOSFET (after Takeuchi et al. [10]).

6.1 Validity of the I-V extraction method for LDD MOSFETs

In this section, simulation results will be presented to study the effective channel length L_{eff} of typical LDD MOSFETs and to verify the validity of the conventional current-voltage method for extracting L_{eff}. The simulation is carried out using a two-dimensional device simulator MEDICI [4], which was discussed in details in Chapter 2. The effective channel length of LDD MOSFETs will be extracted using the Terada-Muta method [3] based on the simulated electrical characteristics. In addition, physical insight in the LDD MOSFETs, such as the free carrier density in the channel, will be provided to aid the correct definition of L_{eff}.

6.1.1. Device structure used in simulation

P-channel LDD MOSFETs with mask channel lengths L_m of 0.75, 1, and 1.25 μm were simulated, and the simulation results and device parameters extracted can be extended to LDD MOSFETs with L_m smaller than 0.75 μm. As shown in Fig. 6.3, all devices simulated have a gate oxide thickness of 10 nm, a spacer width of 60 nm on both sides of the gate mask, and identical source and drain make-up as follows: 1) a heavily doped p-type Gaussian profile with a peak doping concentration N_A^+, a junction depth of 0.12 μm, and a lateral extent of 0.09 μm (0.75 of its vertical extent); and 2) a lightly doped p-type Gaussian profile with a peak concentration of N_A^-, a junction depth of 0.08 μm, and a lateral extent of 0.06 μm.

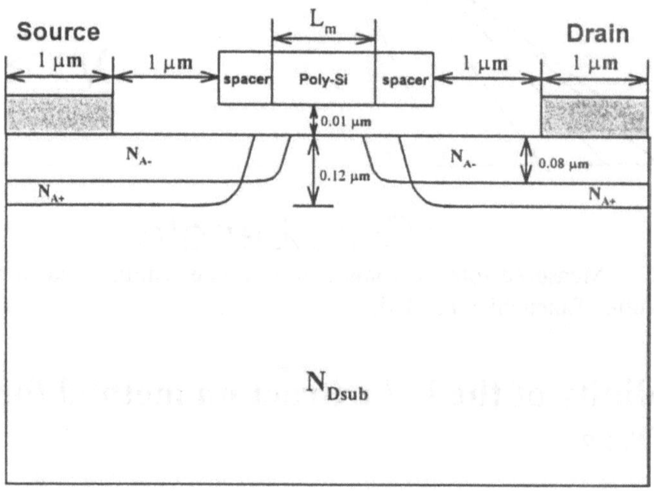

Figure 6.3 : Device structure of a p-channel LDD MOSFET used in simulation.

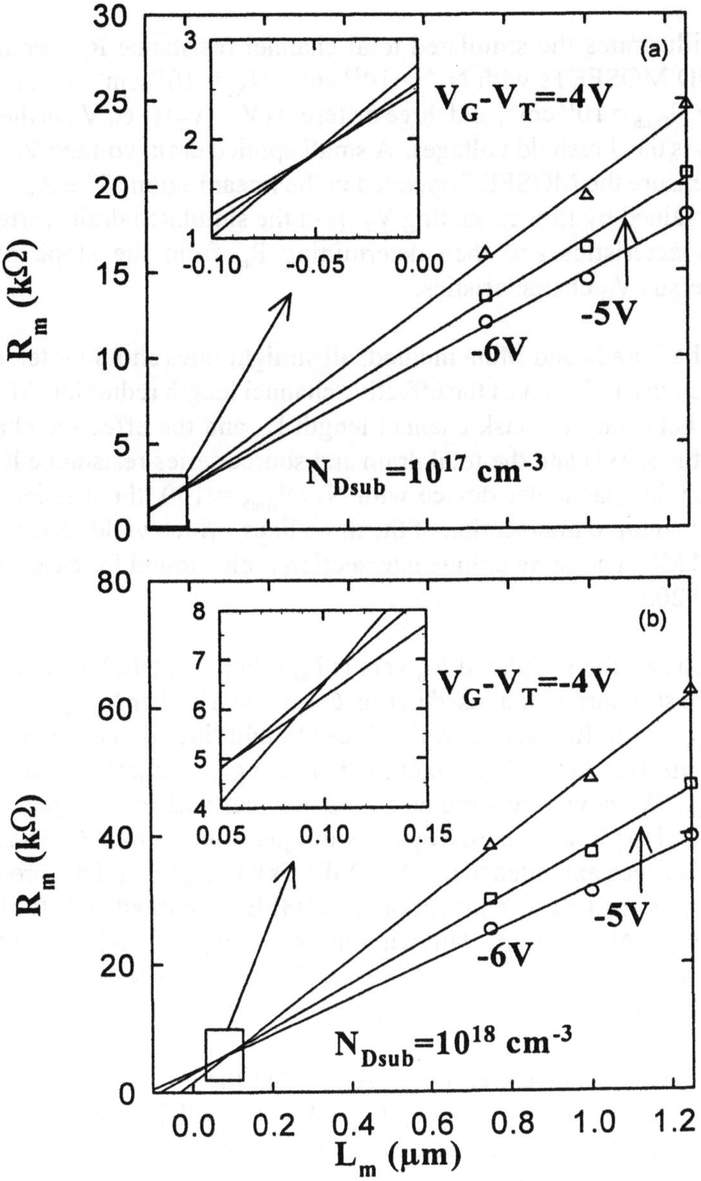

Figure 6.4 : Total channel resistance versus mask channel length for LDD MOSFETs with (a) $N_A^- = 10^{19}$ cm^{-3} and $N_{Dsub} = 10^{17}$ cm^{-3}; and (b) $N_A^- = 10^{19}$ cm^{-3} and $N_{Dsub} = 10^{18}$ cm^{-3}.

6.1.2 Simulation results and discussions

Figure 6.4(a) illustrates the simulated total channel resistance R_m versus L_m relation of LDD MOSFETs with $N_A^+ = 10^{20}$ cm^{-3}, $N_A^- = 10^{19}$ cm^{-3}, a substrate doping density $N_{Dsub} = 10^{17}$ cm^{-3}, and three different $(V_G - V_T)$ (i.e., V_G is the gate voltage and V_T is the threshold voltage). A small applied drain voltage V_D = -50 mV is used to ensure the MOSFET operated in the linear region. The R_m versus L_m plots are obtained by first extracting V_T from the simulated drain current I_D versus V_G characteristics and then determining R_m from the slope of the simulated I_D versus V_D characteristics.

According to the Terada and Muta method, all straight lines should intersect at one point, and such a point gives the effective channel length reduction ΔL (i.e., the difference between the mask channel length L_m and the effective channel length L_{eff}) on the x-axis and the total drain and source series resistance R_{ext} on the y-axis. For this particular device with N_A^-/N_{Dsub} = 100, there is indeed a macroscopically unique intersection of the three lines, which yields ΔL = -0.04 μm and R_{ext} = 2 kΩ. The same unique intersection is also found for cases of N_A^-/N_{Dsub} = 50 and 20.

Figure 6.4(b) shows the simulated R_m versus L_m relation for LDD MOSFETs having the same structure as that used in Fig. 6.4(a) but a higher $N_{Dsub} = 10^{18}$ cm^{-3} (i.e., N_A^-/N_{Dsub} = 10). Interestingly, the three straight lines do not intersect at one point, and the Terada and Muta method failed to give a unique solution for ΔL and thus L_{eff}. We have also simulated other devices with even higher N_{Dsub}, all of which failed to yield a macroscopically unique solution for ΔL. The same approach has also been extended to LDD MOSFETs having several different N_A^- and N_{Dsub} pairs, as well as to conventional MOSFETs without lightly doped regions (i.e., with N_A^+ regions but without N_A^- regions). All provided a macroscopically unique solution for ΔL, provided the ratio N_A^-/N_{Dsub} in LDD MOSFETs or N_A^+/N_{Dsub} in MOSFETs is larger than 10.

The preceding analyses have clearly suggested that the conventional Terada-Muta method is applicable for extracting L_{eff} of LDD MOSFETs and MOSFETs if N_A^- and N_A^+, respectively, are at least an order of magnitude higher than N_{Dsub}, and is questionable if otherwise.

Narayanan et al. [5] have reported a different extraction method, which is based on the information of inversion free carrier profiles in the channel obtained from device simulation and is not subjected to the above mentioned difficulty and limitation.

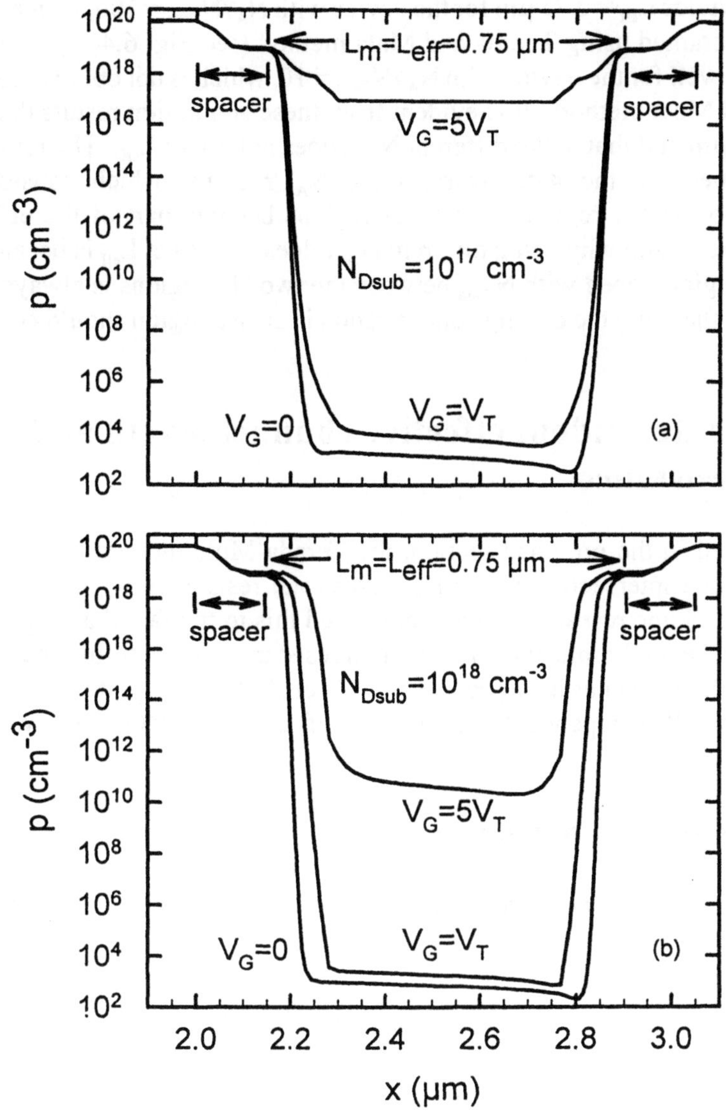

Figure 6.5 : Simulated hole concentrations at the SiO_2-Si interface of LDD MOSFETs with $L_m = 0.75$ µm and (a) $N_A^- = 10^{19}$ cm^{-3} and $N_{Dsub} = 10^{17}$ cm^{-3}; and (b) $N_A^- = 10^{19}$ cm^{-3} and $N_{Dsub} = 10^{18}$ cm^{-3}.

Figures 6.5(a) and (b) show the simulated hole concentration profiles at the SiO_2-Si interface of two LDD p-channel MOSFETs with $N_A^-/N_{Dsub} = 100$ and 10, respectively. Based on this method [5], L_{eff} is defined by a region in which the inversion density (i.e., hole density) is controlled by the gate voltages. Such a definition yields $L_{eff} = 0.75$ µm for the device with $N_A^-/N_{Dsub} = 100$, which agrees with that obtained using Terada and Muta method (see Fig. 6.4(a)), and $L_{eff} = 0.75$ µm as well for the device with $N_A^-/N_{Dsub} = 10$, which is not obtainable using Terada and Muta method. It is evident from these simulation results that $L_m = L_{eff} = 0.75$ µm and that using different N_{Dsub} does not affect L_{eff}. The reason $L_m = L_{eff}$ is because the gate overlaps the N_A^- regions, which causes hole accumulation in the regions, and these regions become part of the effective channel.The reason why using different N_{Dsub} does not affect L_{eff} is because the substrate region doped with N_{Dsub} between the two N_A^- regions is always a part of L_{eff}, and changing the doping concentration in such a region should not affect L_{eff}.

6.2 Bias-dependent effective channel length and series resistance

As discussed in the previous section, the Terada-Muta method often fails to yield a unique intersection in the L_m versus total resistance R_m plot for LDD MOSFETs. Some researchers have attributed this to the fact that L_{eff} and R_{ext} (i.e., R_{ext} is the total drain and source series resistance) are gate-bias dependent [6-7]. An algorithm has been proposed by Hu et al. [8] to alleviate the difficulty of extracting bias-dependent L_{eff} and R_{ext} of the LDD MOSFET, which is presented below.

6.2.1 Algorithm development

From the expression derived using a constant L_{eff}, R_m of an LDD MOSFET operated in the linear region can be generalized as

$$R_m(V_g) = R_{ch}(V_g) + R_{ext}(V_g) , \tag{6.1}$$

where V_g is the gate voltage and R_{ch} is the effective channel resistance, given by

$$R_{ch} = \frac{L_{eff}}{W\mu C_o(V_g - V_T - 0.5V_d)} , \tag{6.2}$$

where L_{eff} is the effective channel length, W is the channel width, µ is the free-carrier mobility, C_o is the oxide capacitance, V_T is the threshold voltage, and V_d is the drain voltage. Also,

$$L_{eff} = L_m - \Delta L . \tag{6.3}$$

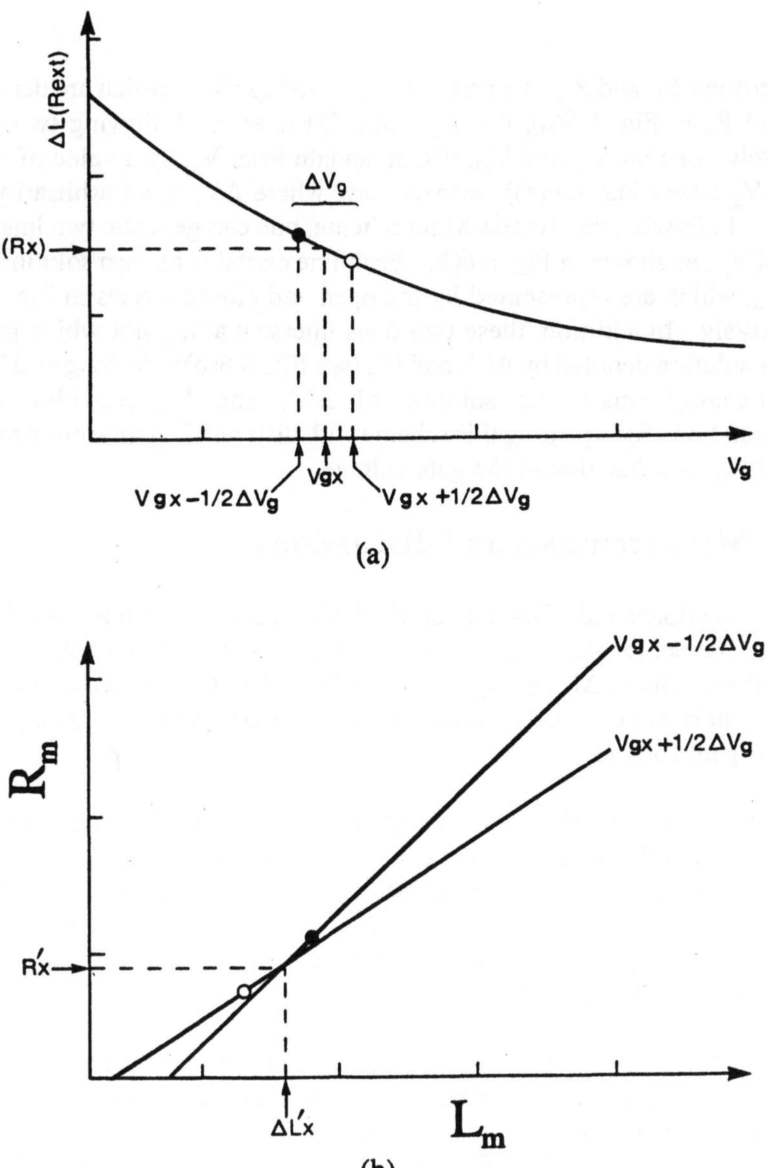

Figure 6.6 : Total resistance versus gate length plot for two closely separated gate biases (after Hu et al. [8]).

The above conventional approach is still valid for LDD devices provided modification to the extraction technique and a different interpretation of the result are properly made [8], which is discussed below. Simulation results [9] suggested a sublinear behavior of R_{ext} with respect to the gate voltage. It is further assumed that ΔL follows the same behavior, which is shown in Fig. 6.6(a).

To determine ΔL and R_{ext} at a particular gate voltage V_{gx}, which are denoted by ΔL_x and R_x in Fig. 6.6(a), the algorithm [8] uses the following two closely separately voltages V_{g1} and V_{g2}, which deviate from V_{gx} by a value of $+\Delta V_g/2$ and $-\Delta V_g/2$ (see Fig. 6.6(a)), respectively, where ΔV_g is an arbitrarily small voltage. Following the Terada-Muta scheme, one can generate two lines using V_{g1} and V_{g2}, as shown in Fig. 6.6(b). Each line contains its own solution of ΔL and R_{ext}, which are represented by the open and closed circles in Fig. 6.6(b), respectively. In addition, these two lines intersect at a point which provides another solution denoted by $\Delta L'_x$ and R'_x (see Fig. 6.6(b)). As long as ΔV_g used is sufficiently small, the solution of $\Delta L'_x$ and R'_x provides a good approximation. By repeating this scheme with different V_g pairs, one can obtain ΔL and R_{ext} as a function of the gate voltage.

6.2.2 Measurements and discussions

Both conventional and LDD n-channel MOSFETs with a channel width of 100 μm and channel lengths of 2, 2.5, 3, 3.5, and 4 μm are considered. Figure 6.7 shows the results of ΔL and R_{ext} extracted from the proposed algorithm [8] for the conventional MOSFETs. It is shown that both ΔL and R_{ext} vary only slightly with the gate voltage.

Two-dimensional simulation was also performed. Fig. 6.8 shows ΔL determined from the simulation results for the conventional MOSFETs, with ΔL_0 (i.e., $\Delta L_0 = L_m - L_{met}$, where L_{met} is the distance between the drain and source metallurgical junctions) used as a reference. Clearly, ΔL is nearly constant and is smaller than ΔL_0, indicating L_{eff} is larger than L_{met}. The physical insight of this difference has been discussed in Chapter 5.

For LDD MOSFETs, ΔL and R_{ext} extracted from the proposed algorithm [8] depend more strongly on the gate voltage, as illustrated in Fig. 6.9. Of equal importance to note is that ΔL shown here is smaller than those given in Fig. 6.7 for the conventional MOSFETs. This suggests that L_{eff} of the LDD MOFET is closer to L_m than that of the conventional MOSFET. Results extracted from the two-dimensional device simulation, given in Fig. 6.10, show a similar trend.

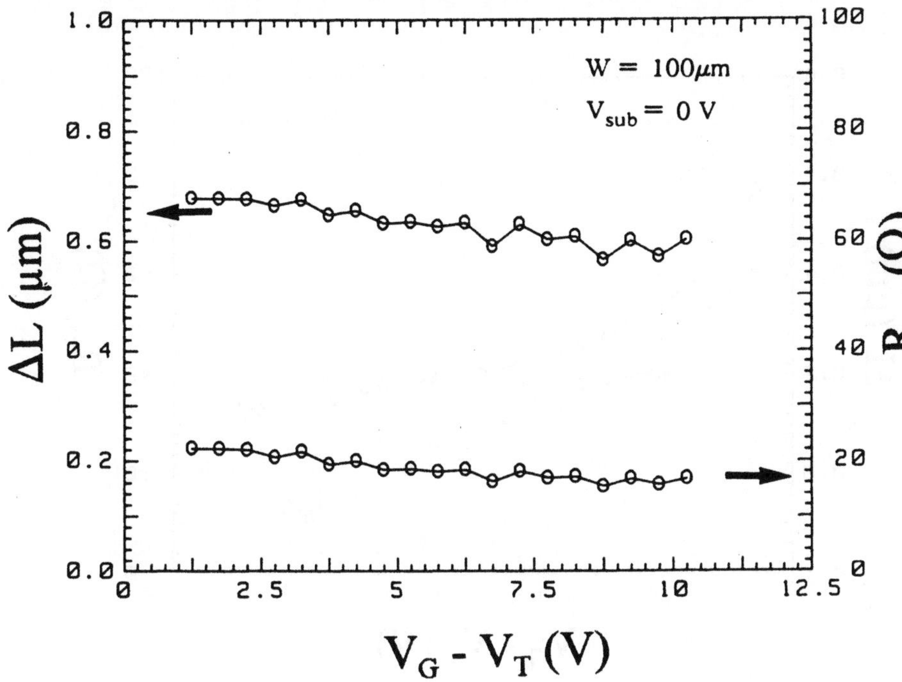

Figure 6.7 : ΔL and R_{ext} extracted from measurements of conventional MOSFET (after Hu et al. [8]).

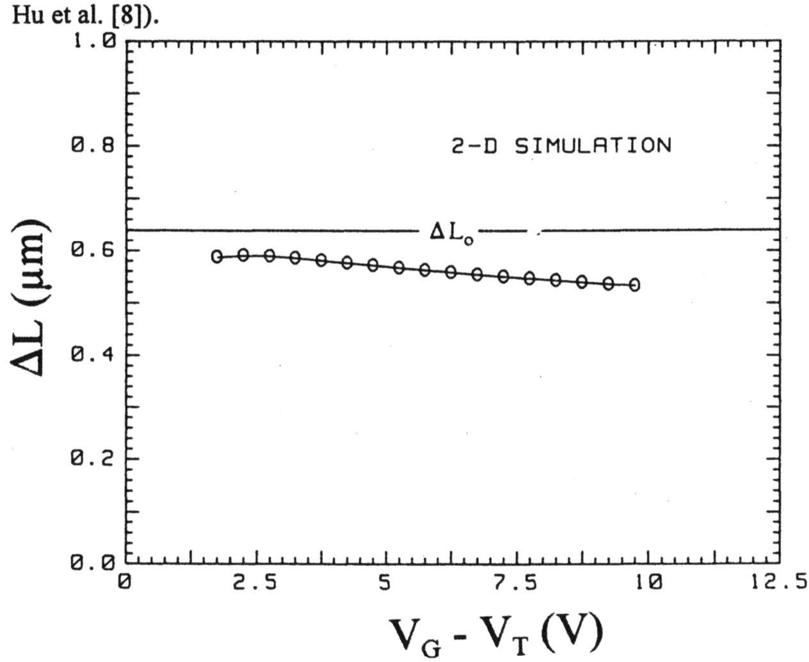

Figure 6.8 : ΔL and R_{ext} extracted from 2-D simulation of conventional MOSFET (after Hu et al. [8]).

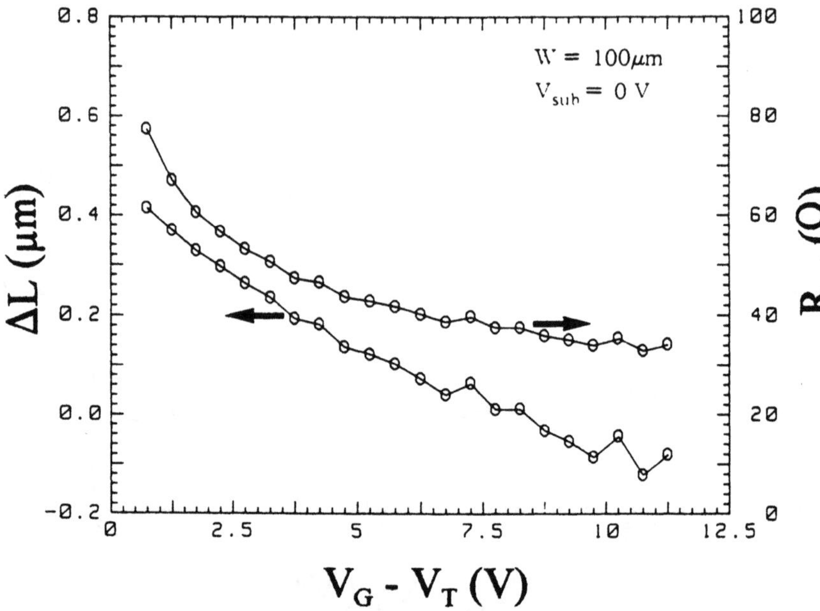

Figure 6.9 : ΔL and R_{ext} extracted from measurements of LDD MOSFET (after Hu et al. [8]).

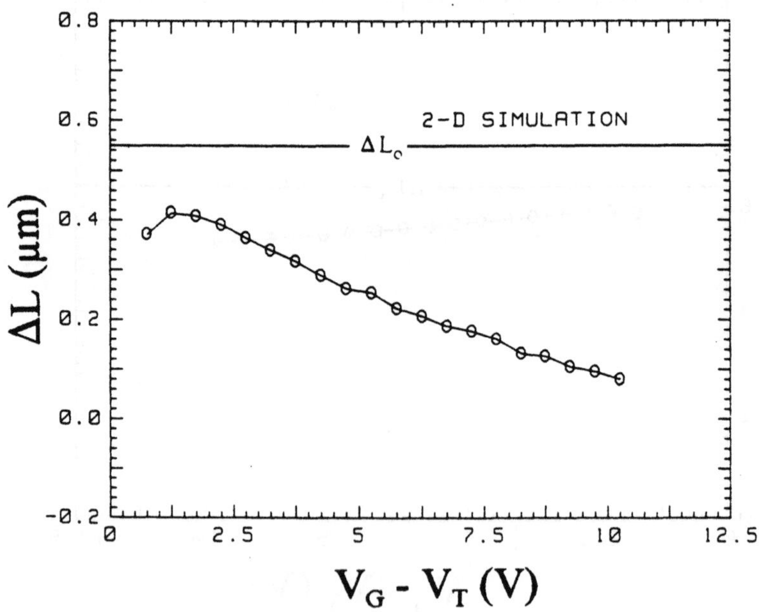

Figure 6.10 : ΔL and R_{ext} extracted from 2-D simulation of LDD MOSFET (after Hu et al. [8]).

6.3 Constant effective channel length determination method

In contrast to the extraction method discussed in Sec. 6.2, which yielded a bias-dependent L_{eff}, Takeuchi et al. [10] recently proposed a method from which L_{eff} and R_{ext} are determined as constant parameters for LDD MOSFETs. It was argued in [10] that, while L_{eff} and R_{ext} are bias-dependent in nature, in practice constant parameters are desirable for MOSFET modeling and characterization. It will shown below that, by using an appropriate definition for L_{eff}, such constant parameters are indeed available.

6.3.1 Method development

To meet the requirement of having a constant L_{eff} in LDD MOSFET, it was proposed that L_{eff} and R_{ext} are defined by imposing the following conditions [11]: 1) L_{eff} is constant, and 2) the bias dependence of R_{ext} is minimized. Variance of R_{ext} over a certain bias range was adopted as the measure of the bias dependence, since it is easy to handle mathematically. Considering that ΔL is constant, and R_{ext} is equal to R_m at $L_m = \Delta L$ according to the Terada-Muta method. Then ΔL can be determined as L_m at which R_m variance is minimized. In general, R_m is a linear function with respect to L_m and can be expressed in the form $y = ax + b$. Its variance is

$$Var(y) = Var(a)\{x + Cov(a,b)/Var(a)\}^2 \; Var(b) - Cov(a,b)^2 / Var(a) \quad , \tag{6.4}$$

where Var and Cov denote variance and covariance, respectively:

$$Var(u) = E(u^2) - E(u)^2 \tag{6.5}$$

$$Cov(u,v) = E(uv) - E(u)E(v) \quad . \tag{6.6}$$

Here E(u) is the expectation (i.e., average) of dummy variable u. Since Var(y) is a parabolic function of x, ΔL and R_{ext} are easily found from (6.4) as

$$\Delta L = -Cov(a,b)/Var(a) \tag{6.7}$$

$$R_{ext} = R_m(L_m - \Delta L) \quad . \tag{6.8}$$

6.3.2 Measurement procedure and results

The measurement procedure is as follows. First, R_m is measured for more than two samples with different L_m, over a certain range of gate voltage. Then, the parameters of a and b in $R_{ext} = y = ax + b$ expression are determined using the least square regression. Next, the a and b values are substituted into (6.7) and (6.8) to obtain a constant ΔL and a range of R_{ext}, as shown qualitatively in Fig. 6.11. Averaging R_{ext} values within the range yields a constant R_{ext}. Since L_{eff} and V_T are dependent on each other, a simple iterative procedure is needed to obtain self-consistent results.

Figure 6.12 shows (L_{eff} - L_{met}) versus the LDD implant dose of an LDD MOSFET and two FOLD (fully overlapped, see Fig. 6.1(c)) MOSFETs with different spacer widths. The process conditions and device structure of the LDD and FOLD devices are given in Table 6.1. It can be seen that L_{eff} and L_{met} are about the same only when the LDD dose is relatively high. All three devices have similar characteristics; for a relatively low LDD dose (i.e., the extreme case of LDD dose = 0 resembles the conventional MOSFET), L_{eff} is larger than L_{met}. This agrees with the finding presented in Chapter 4 that L_{eff} is larger than L_{met} for conventional MOSFETs.

Process conditions		
	Gate oxide thickness	10 nm
	Threshold voltage	0.4 V
	Substrate doping	$5x10^{16}$ cm^{-3} (uniform)
	n^+ implantation	As^+, 30 KeV, $3x10^{15}$ cm^{-2}
	n^- implantation	P^+, 30 KeV, 0.5-6.0$x10^{13}$ cm^{-2}
Source/Drain Structures		
	LDD	L_m = 0.6 μm, with 0.2 μm oxide sidewall
	FOLD1	L_m = 0.8 μm, with 0.1 μm gate-overlap (spacer)
	FOLD2	L_m = 1.0 μm, with 0.2 μm gate-overlap (spacer)

Table 6.1 : LDD MOSFET process conditions and structures (after Takeuchi et al. [10])

It is important to mention that the concept of constant L_{eff} proposed in [10] is consistent with the simulation-based method by Narayanan et al. [5], which defined physically the effective channel as a region in which the free-carrier concentration is influenced by the gate voltage. Based on this method, L_{eff} is independent of the gate voltage (see Sec. 6.1).

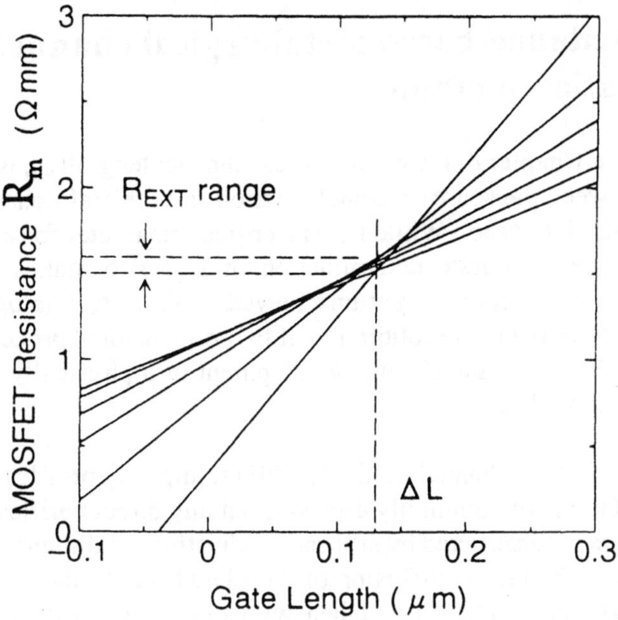

Figure 6.11 : Constant ΔL and R_{ext} range suggested by the method of Takeuchi et al. [10].

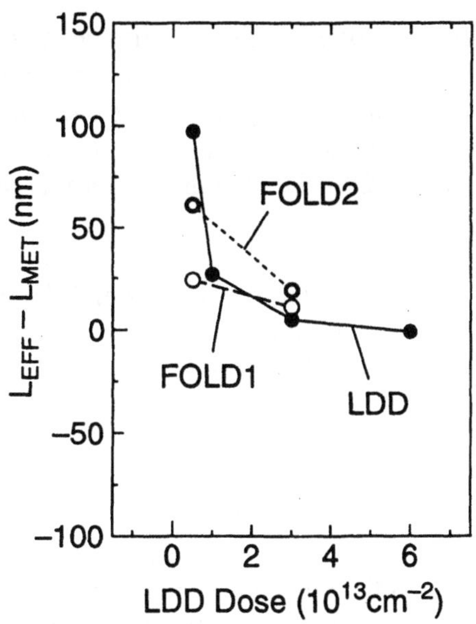

Figure 6.12 : Extracted L_{eff} - L_{met} for an LDD MOSFET and two FOLD MOSFETs (FOLD1 and FOLD2) with different spacer widths of 0.1 and 0.2 μm (after Takeuchi et al. [10]).

6.4 Capacitance-based metallurgical channel length determination method

As discussed in Chapter 4, the metallurgical channel length L_{met} is the physical separation between the source-channel junction and the drain-channel junction near the Si-SiO_2 interface. While L_{eff} is a critical parameter for characterizing the MOS devices, accurate determination of L_{met} also makes a significant contribution to two important yet unachieved goals in technology computer-aided design, namely (1) the calibration of two-dimensional process simulators such as SUPREM-IV, and (2) the development of a physically based device simulation methodology [12].

In simulation of short-channel LDD MOSFETs, impurity profiles generated by process simulators are commonly used without any direct verification, and the device structure is constructed by arbitrarily adjusting the dimension of the gate electrode and/or the lateral diffusion of the LDD tip until the simulated drain current at particular biases is matched with experimental data. Due to the complex dependencies of the current on device structure, this scheme may not result in correct MOS parameters, such as L_{met}, for device design and characterization.

A capacitance-based method for extracting L_{met} has been proposed [12]. Figure 6. 13(a) shows a schematic diagram of half of an n-channel LDD MOSFET, and Fig. 6.13(b) plots the capacitance (i.e., gate-source capacitance C_{gs} and source-substrate capacitance C_{sb}) versus voltage characteristics simulated using MEDICI for three LDD MOSFETs having different gate length L_{gate} but otherwise identical structure.

Simulation is performed by applying the dc bias and the ac signal to the gate with the source and substrate grounded and with the source and drain tied together. Also, only the source-side half of the LDD MOSFET is simulated for reduced computation.

The general trends are that C_{gs} is increased and C_{sb} is decreased with increasing gate-source voltage V_{gs} for n-channel MOS devices, and the opposite holds for p-channel MOSFETs. In Fig. 6.13(b), C_{gs} remains independent of L_{gate} for $V_{gs} < V_{gs}^{on}$, the voltage at which the channel inversion has just started. In general, V_{gs}^{on} is not exactly the same as the threshold voltage extracted from the linear region of drain current versus gate voltage curve.

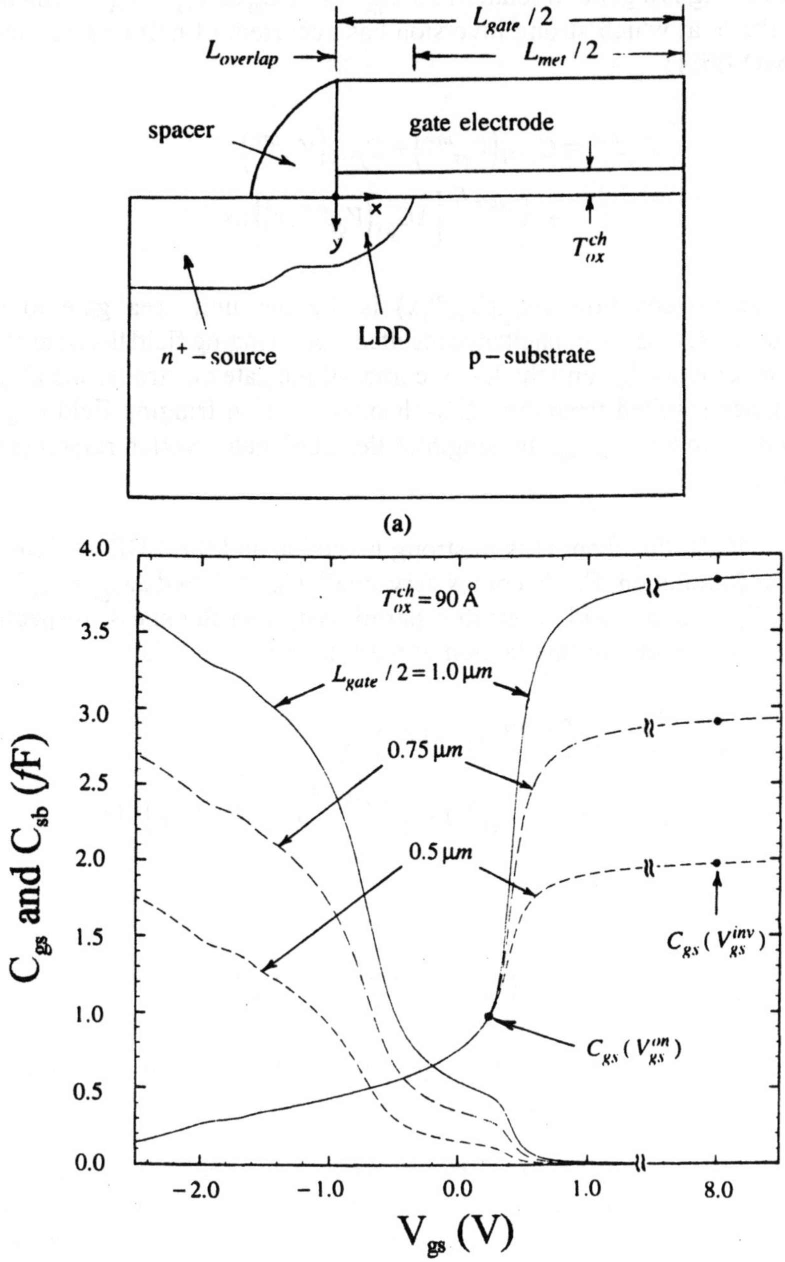

Figure 6.13 : (a) Schematic of an n-channel LDD MOSFET, and (b) simulated capacitance versus voltage characteristic of LDD MOSFETs with three different gate lengths (after Lee [12]).

The following is a general relation for C_{gs}^{inv}, i.e., C_{gs} at $V_{gs} = V_{gs}^{inv}$, which is the gate voltage at which strong inversion has occurred, of half of a symmetrical LDD MOSFET:

$$C_{gs}^{inv} = C_{fr\text{-}1}\left(V_{gs}^{inv}\right) + C_{fr\text{-}2}\left(V_{gs}^{inv}\right) + \int_0^{L_{gate}/2}\left[\Delta C_{gs}\left(V_{gs}^{inv}, \mathrm{x}\right)\right]\mathrm{dx} \quad (6.9)$$

In the above equation, $\Delta C_{gs}(V_{gs}^{inv},x)$ is the per unit area gate to source capacitance, $C_{fr\text{-}1}$ is the capacitance resulted from fringing field that is associated with the vertical edge and the lower corner of the gate electrode, and $C_{fr\text{-}2}$ is the capacitance resulted from the LDD-channel junction fringing field $E_{s\text{-}w}$ in the region beyond $x = L_{overlap}$, the length of the LDD-gate overlap region (see Fig. 6.13(a)).

At $V_{gs} = V_{gs}^{inv}$, the channel is in strong inversion and the LDD regions are in strong accumulation, $E_{s\text{-}w}$ becomes very small, $C_{fr\text{-}2} = 0$, and $\Delta C_{gs} = \epsilon_{ox}/T_{ox}$ (i.e., ϵ_{ox} and T_{ox} are the oxide dielectric permittivity and thickness, respectively). This, together with the satisfaction of the following condition

$$C_{fr\text{-}1}\left(V_{gs}^{inv}\right) + \int_0^{L_{overlap}}\left[\varepsilon_{ox}/T_{ox}\right]\mathrm{dx} = C_{fr\text{-}1}\left(V_{gs}^{on}\right) + C_{fr\text{-}2}\left(V_{gs}^{on}\right) + \int_0^{L_{overlap}}\left[\Delta C_{gs}\left(V_{gs}^{on}, \mathrm{x}\right)\right]\mathrm{dx} \quad (6.10)$$

yields

$$L_{met}/2 = \left(T_{ox}/\varepsilon_{ox}\right)\left(C_{gs}^{inv} - C_{gs}^{on}\right). \quad (6.11)$$

The assumptions $C_{fr\text{-}2}(V_{gs}^{inv}) = 0$, $\Delta C_{gs}(V_{gs}^{inv},x) = \epsilon_{ox}/T_{ox}$, and the condition stated in (6.11) have been verified and can be found in [12].

Figure 6.14 shows $L_{met}/2$ versus the oxide thickness extracted from this method (open circles) using results simulated from MEDICI for LDD MOSFETs with $L_{met} = 0.1695$ μm (indicated by the line). Clearly, the method is quite accurate for LDD MOSFETs having a thin oxide layer (i.e., less than about 70 Å), but underestimates L_{met} as the oxide layer thickness is increased.

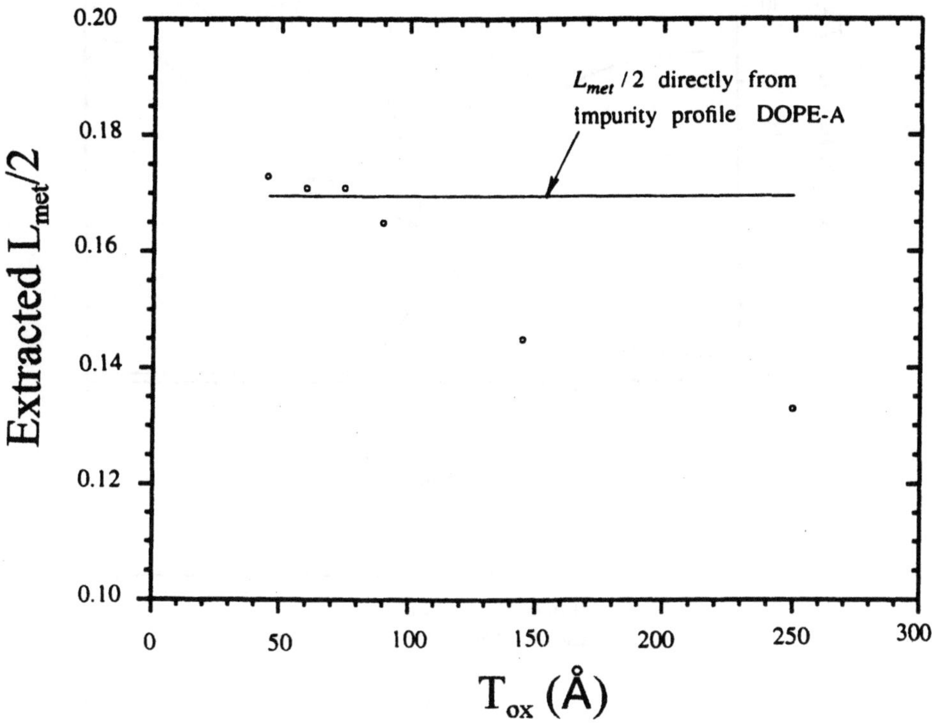

Figure 6.14 : Metallurgical channel length (open circles) extracted from the method based on results simulated from an LDD MOSFET with a metallurgical channel length of 0.1695 μm (indicated by the line) (after Lee [12]).

The capacitance versus voltage characteristics measured from an n-channel and p-channel LDD MOSFET are illustrated in Figs. 6.15(a) and (b), respectively. The extracted L_{met} for five different types of LDD MOSFETs are given in Fig. 6.16(a). Also included in the figure are the effective channel length L_{eff} of the devices extracted using the current-voltage method.

The main difference among the different types of LDD MOSFETs (i.e., DEV-A, DEV-B, and DEV-C) considered in Fig. 6.16(a) is the different impurity profiles (i.e., DOPE-A, DOPE-B, and DOPE-C), shown in Fig. 6.16(b), in the lightly-doped regions of these devices.

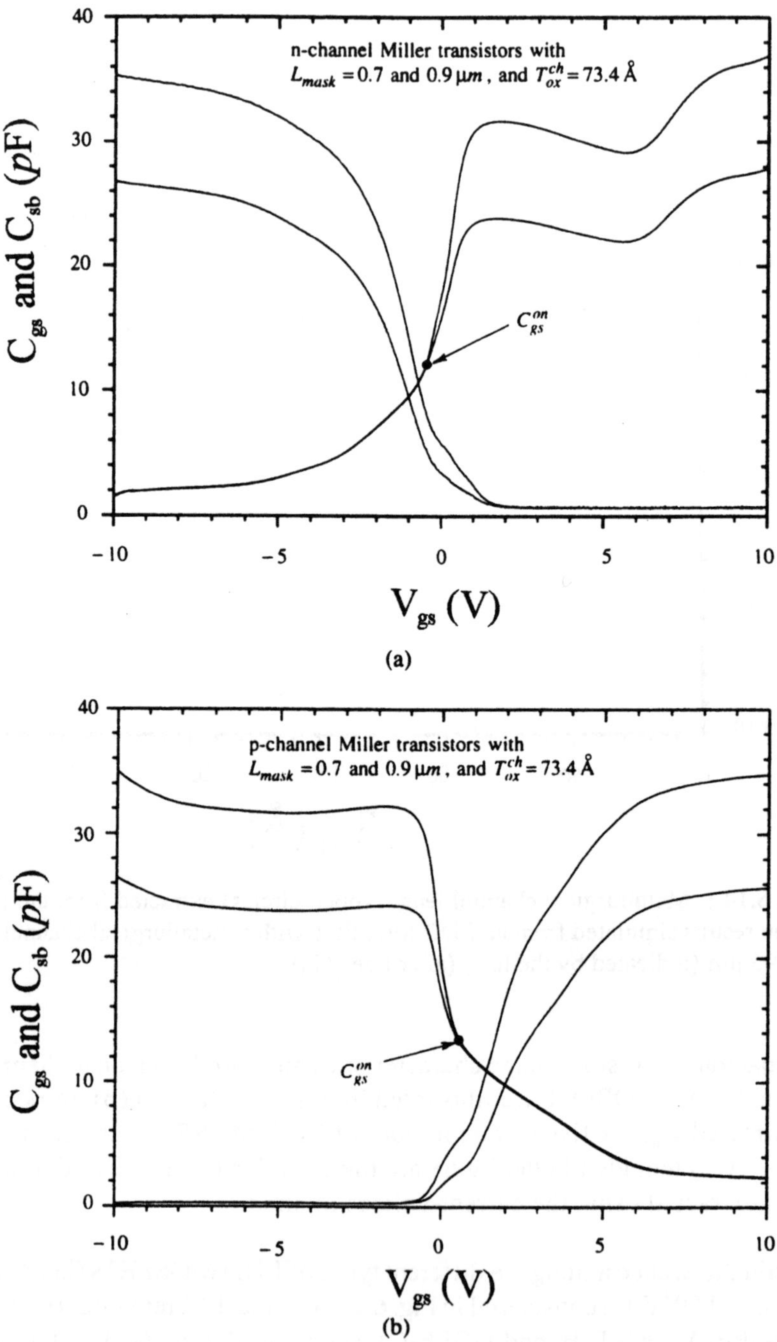

Figure 6.15 : Capacitance versus voltage characteristics measured from an (a) n-channel and (b) p-channel LDD MOSFET (after Lee [12]).

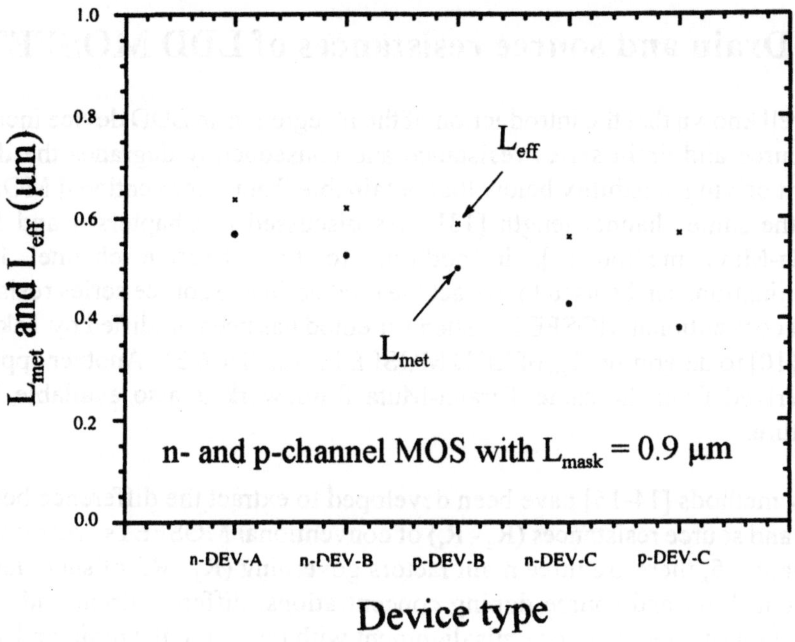

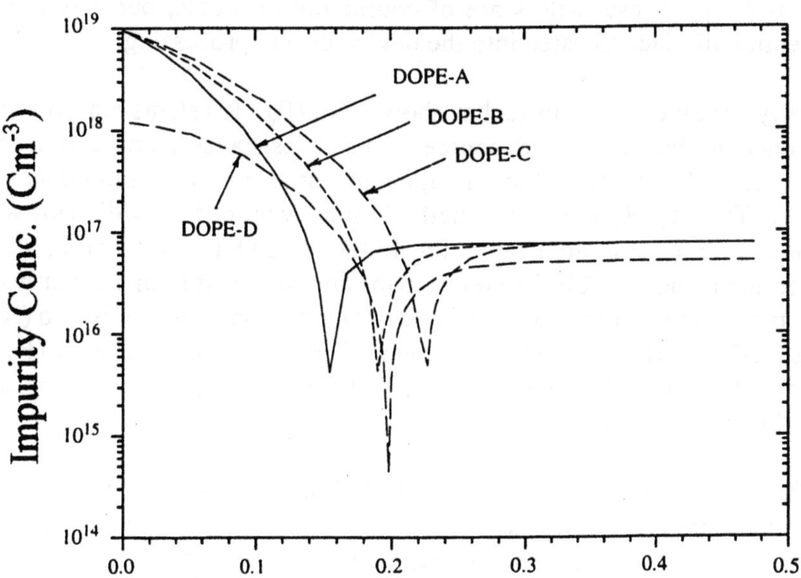

Figure 6.16 : (a) Comparison of metallurgical channel length and effective channel length extracted from measurement data for five different LDD MOSFETs, and (b) different doping profiles in the lightly-doped regions of the different devices used in (a) (after Lee [12]).

6.5 Drain and source resistances of LDD MOSFETs

It is well known that the introduction of the n^- region in an LDD device increases the source and drain series resistance and consequently degrades the device current driving capability below that obtainable from a conventional MOSFET with the same channel length [13]. As discussed in Chapters 4 and 5, the Terada-Muta method [3], in addition to the effective channel length determination, can be used to extract the total drain and source series resistance R_{ext} of conventional MOSFETs. Such a method has been modified by Takeuchi et al. [10] to determine R_{ext} of LDD MOSFETs (see Sec 6.3). Another approach [6] derived from the same Terada-Muta framework is also available in the literature.

A few methods [14-16] have been developed to extract the difference between drain and source resistances (R_d - R_s) of conventional MOSFETs. As discussed in Chapter 5, there are three main factors governing (R_d - R_s) of such devices: different drain and source doping concentrations, different drain and source contact resistances, and gate misalignment with respect to the drain and source contacts [17]. These factors are of course not desirable, but they are often unintentionally incorporated into the device during processing.

Recently, a numerical analysis has shown that (R_d - R_s) stems mainly from the difference in the drain and source contact resistances, and not from the difference in the source and drain doping densities nor the misalignment of the gate [4]. The (R_d - R_s) extraction method for conventional MOSFETs discussed in Chapter 5 is still applicable for LDD MOSFETs, but noting that the unintentional and possible difference in the doping densities in the lightly doped drain and source regions can contribute to an additional mechanism to (R_d - R_s) in the LDD MOSFET. The knowledge of (R_d - R_s), together with (R_d + R_s) extracted from the Takeuchi et al. method [10], allows one to determine the individual values of R_d and R_s of LDD devices.

Another method to extract (R_d - R_s) of LDD MOSFET has recently been reported in the literature [14]. It was developed from the measurements of the dynamic transconductances in the saturation region of operation of an MOSFET. For an MOS device with R_d and R_s, the transconductance g_m of the MOSFET is [14]

$$g_m = \frac{g_{m0}}{1+\left(g_{m0}+g_{b0}\right)R_s+g_{d0}\left(R_s+R_d\right)}\,, \tag{6.12}$$

where g_{m0}, g_{b0}, and g_{d0} are the magnitudes of the intrinsic conductances with respect to the gate, substrate, and drain biases, respectively.

Note that (6.12) neglects the variation of R_s and R_d with the gate voltage, but the channel pinch-off in the saturation region valids such an assumption. Any asymmetry in R_s and R_d will show up in the measured g_m, and hence the drain current versus gate voltage characteristics in saturation operation measured normally and with the drain and source interchanged. This is illustrated in results given in Fig. 6.17 measured normally (solid line) and measured with drain and source interchanged (dotted lines) for a 0.8 μm LDD MOSFET.

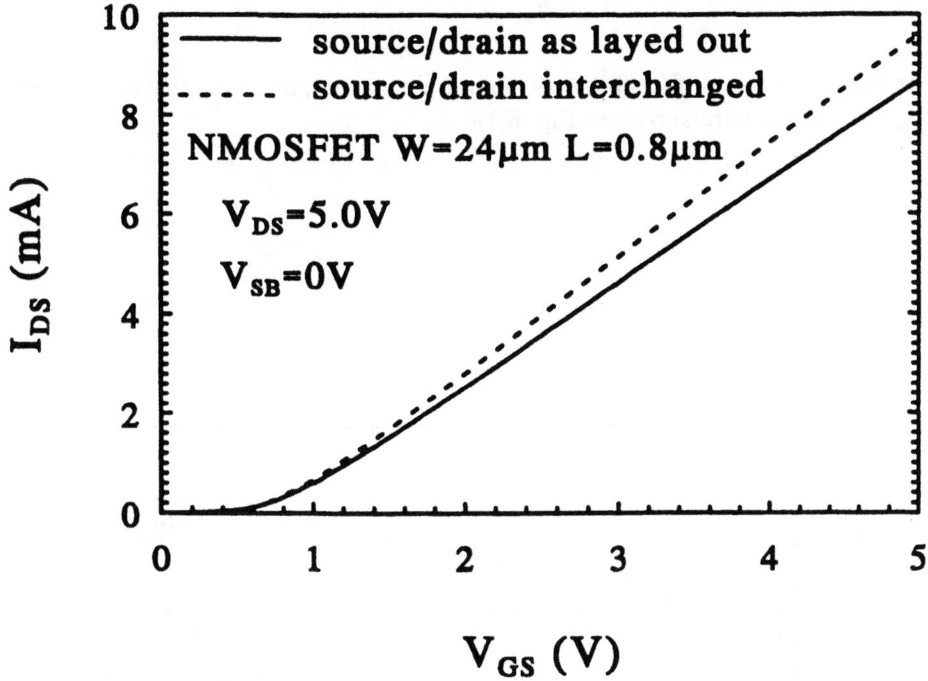

Figure 6.17 : Saturation transconductance characteristics of an LDD MOSFET with normal setup (solid line) and drain and source interchanged (dotted line) (Source: Raychaudhuri et al. [14]. Reprinted with permission).

Now, if an external resistance R_x is added intentionally to the source terminal, then the corresponding transconductance g_{ms} is [14]

$$\begin{aligned}\frac{1}{g_{ms}} = \frac{1}{g_{m0}} + R_s\left(1+\frac{g_{b0}}{g_{m0}}\right) + \left(\frac{g_{d0}}{g_{m0}}\right)(R_s + R_d) \\ + R_x\left(1+\frac{g_{b0}}{g_{m0}}+\frac{g_{d0}}{g_{m0}}\right)\end{aligned} \quad . \tag{6.13}$$

Conversely, if an external resistance R_x is added intentionally to the drain terminal, then the transconductance g_{md} is [14]

$$\frac{1}{g_{md}} = \frac{1}{g_{m0}} + R_s\left(1 + \frac{g_{b0}}{g_{m0}}\right) + \left(\frac{g_{d0}}{g_{m0}}\right)\left(R_s + R_d\right) + R_x\left(1 + \frac{g_{d0}}{g_{m0}}\right) \quad (6.14)$$

For different values of R_x and a constant drain current, one can measure g_{ms} and g_{md}, and the data are shown in Fig. 6.18.

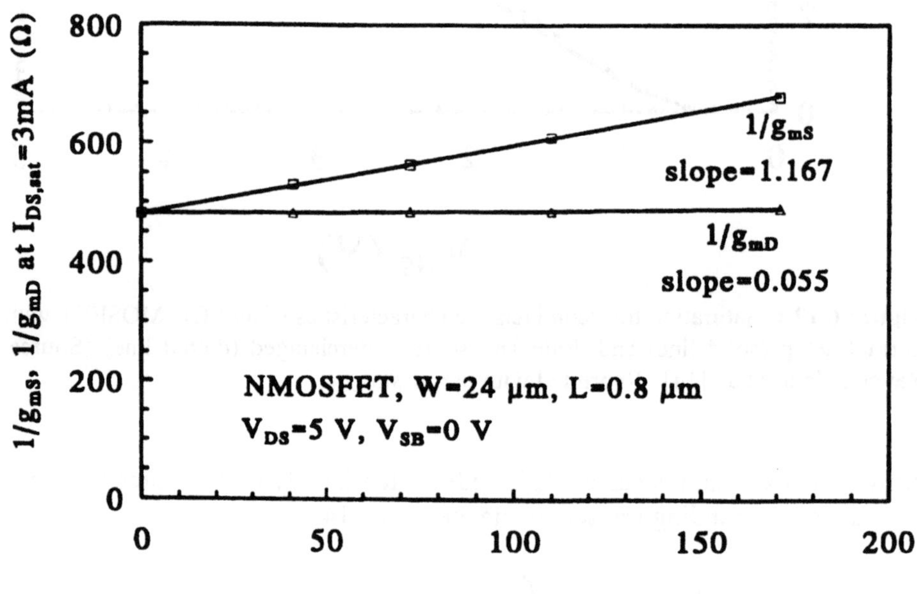

Figure 6.18 : Reciprocal of the transconductances versus the external resistance R_x measured from an LDD MOSFET (Source: Raychaudhuri et al. [14]. Reprinted with permission).

The slopes of the two least-square-fitted lines in Fig. 6.18 are

$$\left(1+\frac{g_{b0}}{g_{m0}}+\frac{g_{d0}}{g_{m0}}\right)=1.167 \quad \text{and} \quad \frac{g_{d0}}{g_{m0}}=0.055. \tag{6.15}$$

As a result, subtracting the two slopes, we get $(1 + g_{b0}/g_{m0}) = 1.112$.

If we now remove R_x and measure the same device again with the normal setup and with the drain and source interchanged, then

$$R_d - R_s = \frac{\dfrac{1}{g_{mr}}-\dfrac{1}{g_{mf}}}{1+\dfrac{g_{b0}}{g_{m0}}}, \tag{6.16}$$

where g_{mr} and g_{mf} are the transconductances under normal and exchanged setups, respectively. Substituting the value of $(1 + g_{b0}/g_{m0})$ into (6.16) yields $(R_d - R_s) = -39\ \Omega$ for this particular LDD MOSFET.

Next, we investigate the effects of the lightly doped regions on $(R_d - R_s)$ and determine the individual values of R_d and R_s of LDD MOSFETs. This is carried out by using the $(R_d - R_s)$ method based on the inverse and normal modes of configuration (i.e., discussed in Chapter 5) and the $(R_d + R_s)$ extraction method developed by Takeuchi et al. [10] (i.e., discussed in Sec. 6.3), as well as LDD MOSFET dc characteristics simulated from a two-dimensional device simulator MEDICI [4]. While using experimental data measured from LDD MOSFETs would be most desirable for extracting $(R_d - R_s)$ and $(R_d + R_s)$, device simulations do provide reasonable accuracy and great flexibility.

P-channel LDD MOSFETs with mask channel lengths L_m of 1, 0.75 and 0.5 μm and having the same device structure as that shown in Fig. 6.3 were considered. Based on the method developed in Chapter 5, $(R_d - R_s)$ can be calculated using the following two setups: (1) a MOSFET in the normal mode of configuration with the source and body grounded, and (2) the same MOSFET in the inverse mode of configuration with the drain and body grounded. Because of the different R_d and R_s, it will take two different gate voltages V_T in the normal and inverse modes to obtain the same drain current I_D in the two modes. Using such a concept, $(R_d - R_s)$ can be obtained from [16]:

$$R_d - R_s = \frac{V_{gi} - V_{gn}}{I_D\left(1 + \frac{dV_T}{dV_{SB}}\right)}, \tag{6.17}$$

where V_{gn} and V_{gi} are the extrinsic gate-source voltages in the normal and inverse modes, respectively, and the term dV_T/dV_{SB} accounts for the dependence of the threshold voltage with respect to the body voltage V_{SB}.

The value of $(R_d + R_s)$, on the other hand, can be determined using the Takeuchi et al. method (see Sec. 6.3), modified from the conventional Terada-Muta concept [3]. It was derived based on the current-voltage characteristics of an MOS transistor operating in the linear region. For a set of MOSFETs with different L_m, a straight line would be obtained by plotting the total measured resistance R_m versus L_m. The intersection of several such lines yields $(R_d + R_s)$.

The effects of different heavily-doped drain and source doping densities, different drain and source contact resistances, and gate misalignment on $(R_d - R_s)$ of conventional MOSFETs have been investigated in Chapter 5. It was found that the drain and source contact resistances difference is the main factor contributing to $(R_d - R_s)$ in conventional MOSFETs. Here we focus on the LDD MOSFET. In addition to the factor of different contact resistances, we will also examine the case of different lightly-doped drain and source doping densities, an asymmetry lie in the LDD MOSFET.

Figs. 6.19(a) and (b) illustrate $(R_d - R_s)$ as a function of I_D (in μA/μm, where μm is the dimension of LDD MOSFET width) for two LDD MOSFETs (LDDMOS-1 and LDDMOS-2); LDDMOS-1 and LDDMOS-2 have different drain/source contact resistances (R_{CD} and R_{CS}) and different p^- drain/source doping densities (N_{AD}^- and N_{AS}^-), respectively, but otherwise symmetrical drain and source make-ups. Also, three different mask channel lengths are considered in each case.

The results show that the contact resistances difference gives rise to a much larger $(R_d - R_s)$ than p^- doping densities difference and therefore is the dominant factor for $(R_d - R_s)$ in LDD MOSFETs.

Note also $(R_d - R_s)$ is larger than $(R_{CD} - R_{CS})$ (see Fig. 6.19(a)). This stems from the distributed nature of the contact and series resistances in the source and drain regions [17]. It should be pointed out that the extracted values of $(R_d - R_s)$ for the moderate and weak inversion regions (i.e., very low current levels) are questionable because the linear extrapolation method used is no longer valid in these regions.

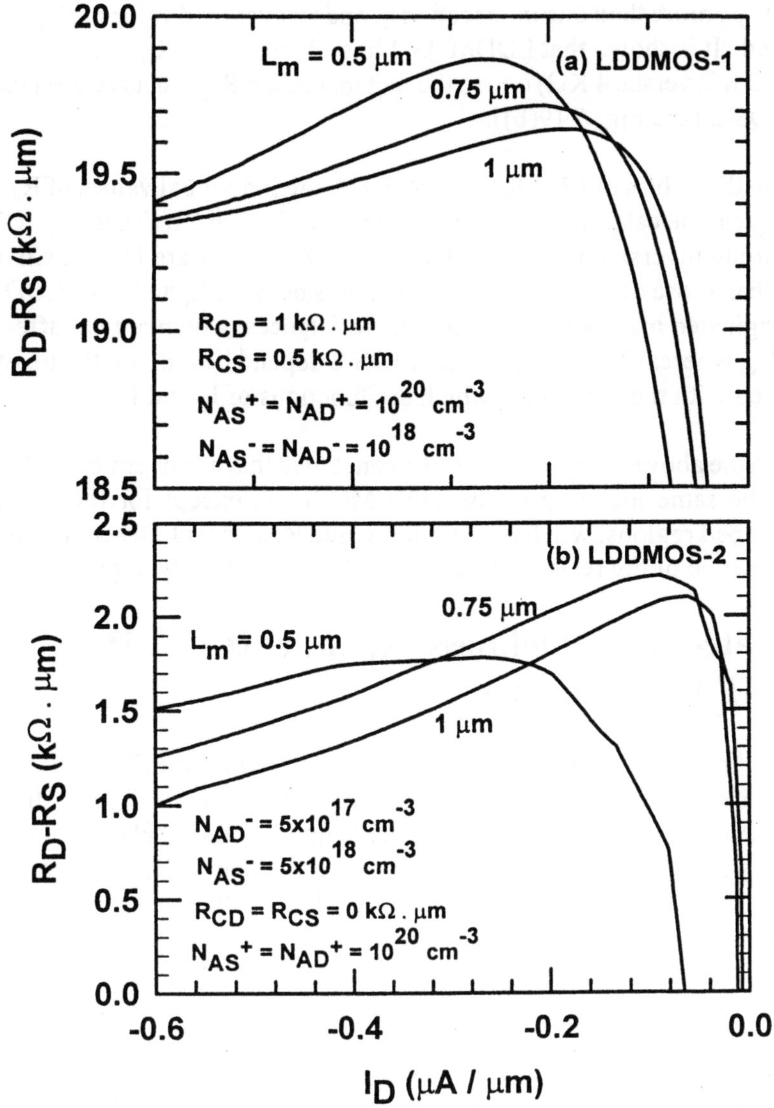

Figure 6.19: The difference between the drain and source resistances extracted from device simulation results for LDD MOSFETs with (a) asymmetrical drain/source contact resistances (LDDMOS-1), and (b) asymmetrical lightly doped drain/source densities (LDDMOS-2).

Figures 6.20(a) and (b) give the Terada-Muta plots of LDDMOS-1 and LDDMOS-2, respectively, in which the intersection of the lines to the x-axis yields ($R_d + R_s$). For LDD MOSFETs, the intersection of the lines may not be unique (i.e., more than one intersections), and an averaged value of ($R_d + R_s$) has been used. It is shown that LDDMOS-1 has a larger ($R_d + R_s$) than LDDMOS-2 (i.e., 62.5 KΩ versus 4 KΩ) due to the fact that $R_{CD} = R_{CS} = 0$ have been used for LDDMOS-2 (see Fig. 6.19(b)).

Based on the values of ($R_d - R_s$) and ($R_d + R_s$), the individual values of R_d and R_s versus I_D can be calculated, which are shown in Figs. 6.21(a) and (b). Clearly, in the strong inversion region, R_d and R_s of LDDMOS-1 are less sensitive to I_D and L_m than those of LDDMOS-2. This results because R_d and R_s of LDDMOS-1 are originated mainly from the contact resistances, which are not affected by I_D and L_m, whereas R_d and R_s of LDDMOS-2 depend heavily on the resistances associated with the p^- regions, which are functions of I_D and L_m.

Based on the above study, it can also be concluded that a conventional MOSFET having the same make-up as the LDD MOSFET, except for the absence of lightly doped regions, will have smaller R_d and R_s then its LDD counterpart due to the absence of the resistances associated with the lightly doped regions.

6.6 Gate-oxide thickness dependence of LDD MOSFET parameters

It is well known that the threshold voltage V_T of the MOSFET is a function of the gate oxide-layer thickness T_{ox} [18], but the effects of T_{ox} on other vital parameters, such as the effective channel length L_{eff} and total drain and source series resistance, are less clear. The oxide thickness dependence of these parameters is even more ambiguous for LDD MOSFETs. This section investigates the effects of oxide thickness variation on the device parameters of LDD MOSFETs based on the results simulated from a two-dimensional device simulator MEDICI. Relevant physical insight will also be discussed.

The LDD MOSFETs considered here have the same structure as that shown in Fig. 6.3, except that the oxide layer thickness is now a variable. Figure 6.22 illustrates the extracted V_T versus T_{ox} for LDD MOSFETs having two different L_m. The results are obtained from the linear extrapolation at the point of maximum slope on the drain current versus gate voltage V_G curve simulated from MEDICI (i.e., the linear-extrapolation method discussed in Chapter 3). A small applied drain voltage V_D = -50 mV is used to ensure that the MOSFETs are operated in the linear region.

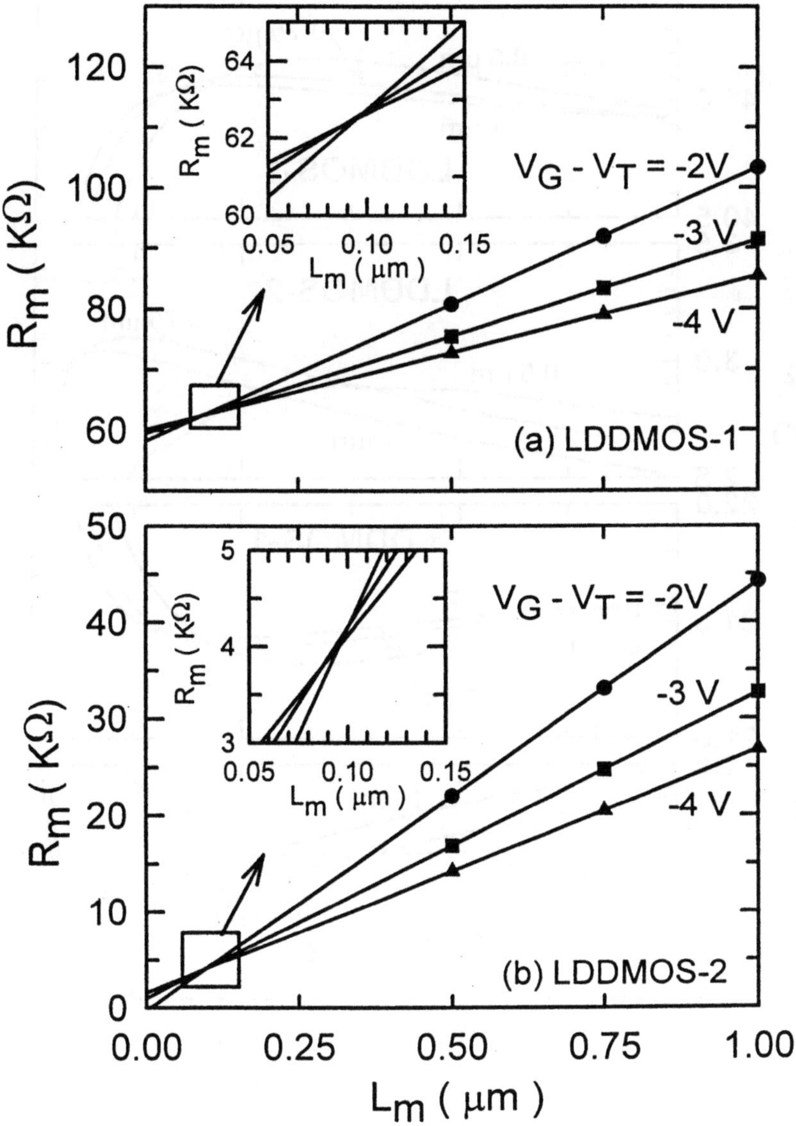

Figure 6.20: Plots obtained from Terada-Muta method for (a) LDDMOS-1 and (b) LDDMOS-2.

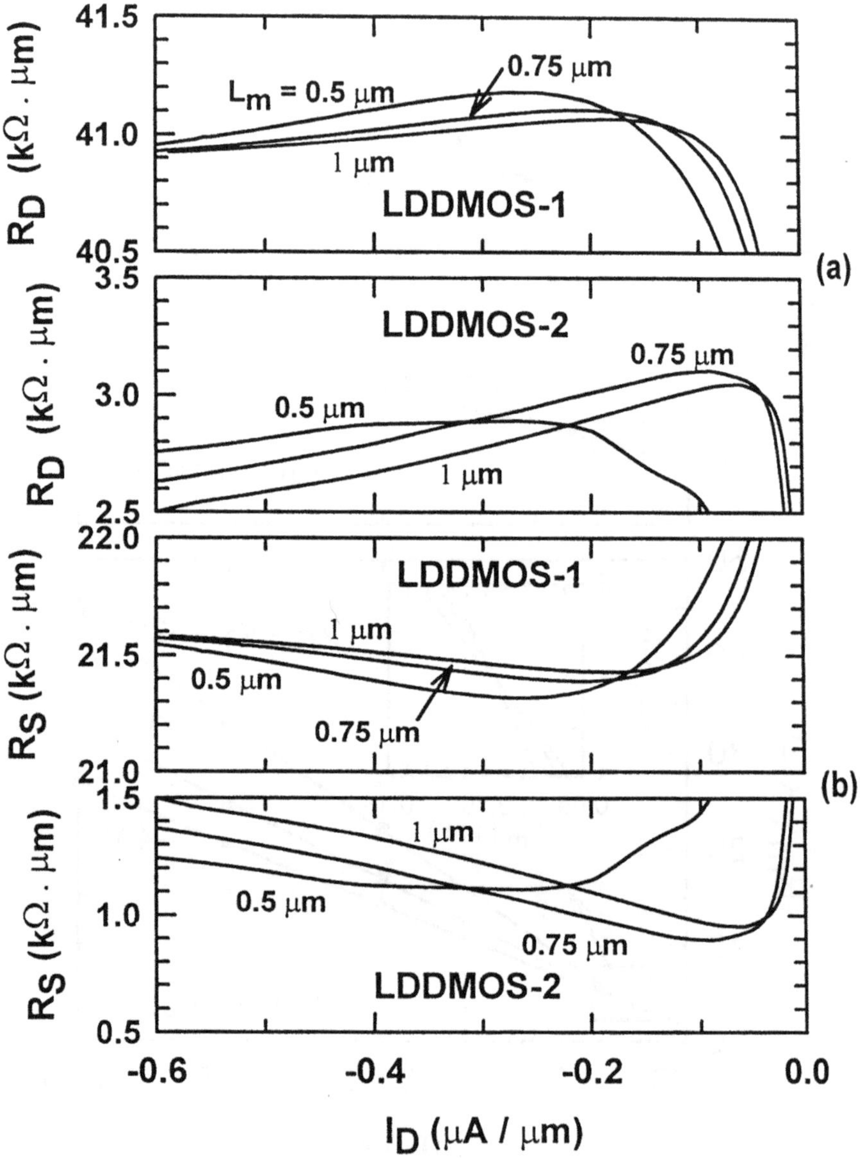

Figure 6.21: Values of (a) R_d and (b) R_s calculated based on the information of (R_d - R_s) and (R_d + R_s) for LDDMOS-1 and LDDMOS-2

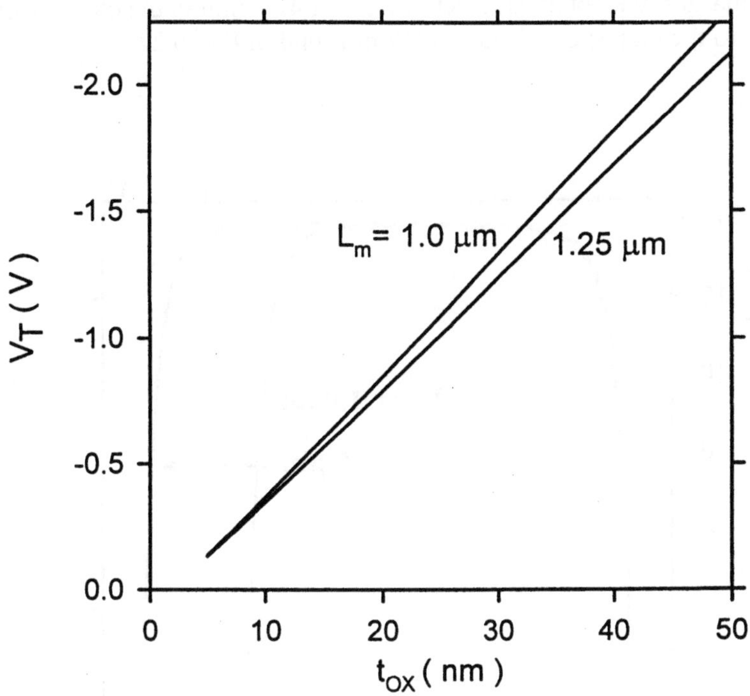

Figure 6.22 : Extracted threshold voltage versus gate-oxide thickness for LDD MOSFETs having two different mask channel lengths.

It can be seen in Fig. 6.22 that V_T depends linearly on T_{ox} and is a weak function of L_m. Note that the small decrease in V_T with increasing L_m is not normally seen in real devices and should be considered as errors associated with the linear- extrapolation method used to obtain V_T.

The physics underlying the linear dependance of V_T with respect to T_{ox} can be explained as follows. For a fixed V_G, the vertical electric field at the surface of the semiconductor along the channel is constant, as evidenced by the results shown in Fig. 6.23. When T_{ox} is increased, the vertical electric field along the channel reduces, which necessitates an increase in V_G to cause strong inversion and thus an increase in V_T. Since the amount of field reduced is related to the increased voltage drops in the oxide governing by the Gauss law, the slope of V_T versus T_{ox} should be approximately equal to the electric field in the oxide.

For the case considered, the slope obtained from Fig. 6.22 is between 45 and 48 V/µm, and the electric field presented in Fig. 6.23 is 14 V/µm. Using a value of 3 for the ratio of silicon to oxide dielectric constants, and assuming there is no interface charges, we obtain an electric field of 42 V/µm in the oxide, which is very close to the slopes of 45 and 48 V/µm found in Fig. 6.22.

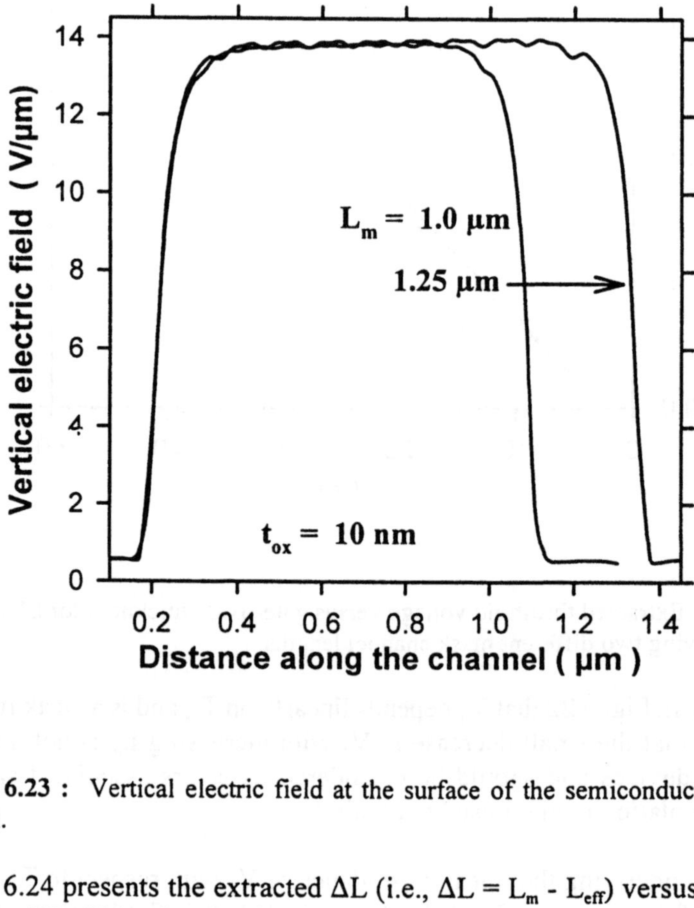

Figure 6.23 : Vertical electric field at the surface of the semiconductor along the channel.

Figure 6.24 presents the extracted ΔL (i.e., $\Delta L = L_m - L_{eff}$) versus T_{ox} at three different $(V_G - V_T)$ using the simulation results and extraction method developed recently by Takeuchi and co-workers [10]. It is shown that ΔL decreases with decreasing T_{ox}, which is in agreement with the finding reported in [19].

Note that the dependence of T_{ox} on V_T has been accounted for, and thus V_T used here varies with T_{ox}. Physically, the change of ΔL, and thus L_{eff}, with respect to T_{ox} is caused by the fact that the vertical electric field (i.e., from the oxide into semiconductor) in the channel region is altered by the change of T_{ox}. For a fixed

(V_G - V_T), the thinner the T_{ox}, the larger the vertical electric field, and the more accumulation of the holes in those lightly-doped p-type drain and source regions underneath the gate. These hole-accumulated regions become part of the effective channel length [20], and the effective channel length is increased (or ΔL is decreased).

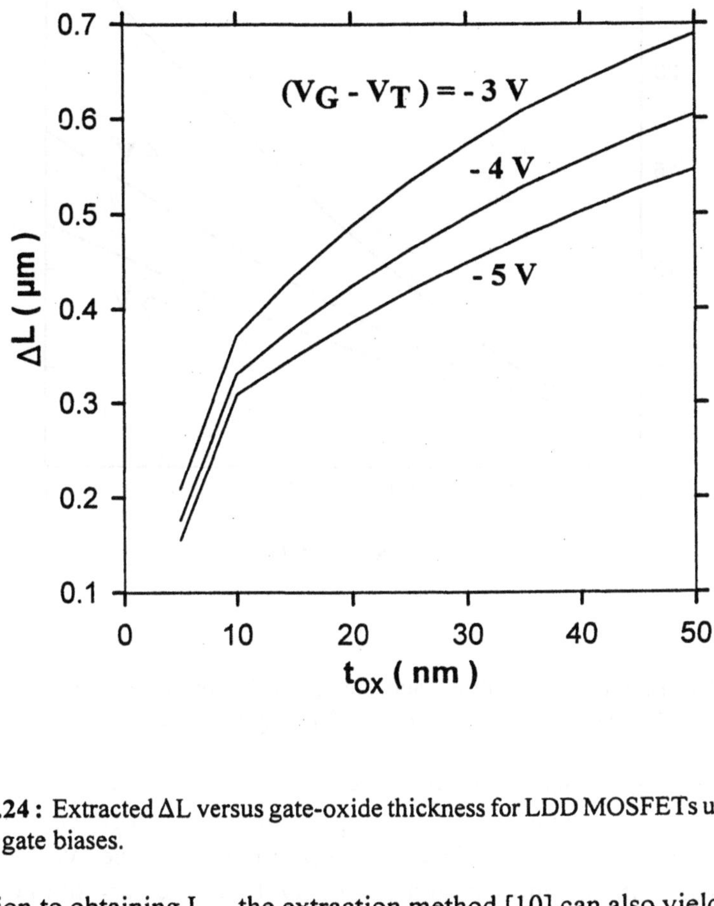

Figure 6.24 : Extracted ΔL versus gate-oxide thickness for LDD MOSFETs under three different gate biases.

In addition to obtaining L_{eff}, the extraction method [10] can also yield the total drain and source series resistance (R_d + R_s) of the LDD MOSFET. Figure 6.25 presents the extracted (R_d + R_s) versus T_{ox} for three different (V_G - V_T). We observe that (R_d + R_s) decreases as T_{ox} decreases. This is because, for a fixed (V_G - V_T), a thinner oxide thickness gives rise to a larger vertical electric field from the oxide into semiconductor, hence resulting in more hole accumulation in those N_A^- drain and source regions underneath the gate and thus lower series resistances in these regions.

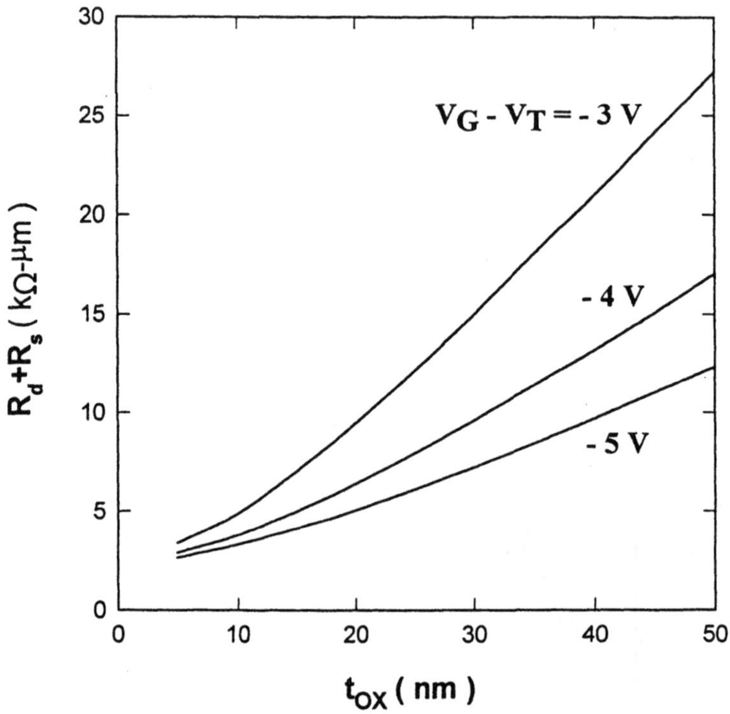

Figure 6.25: Extracted total drain and source series resistance versus gate-oxide thickness for LDD MOSFETs under three different gate biases.

We have also examined the oxide thickness dependence of the conventional MOSFET's parameters using the same approach described above for the LDD MOSFET, and have the following observations: 1) The V_T versus T_{ox} characteristics of conventional and LDD MOSFETs are very similar. This is due to the fact that V_T is determined mainly by the inversion mechanism in the n-type substrate, not the make-up of the drain and source; 2) The LDD MOSFET has a stronger oxide thickness dependence of ΔL than conventional MOSFET. This is caused by the presence of lightly-doped drain and source regions in the LDD MOSFET, the hole concentrations in which can be more easily modulated by the electric field change than those in the heavily-doped drain and source regions in the conventional MOSFET; and 3) $(R_d + R_s)$ of LDD MOSFET is also more sensitive to T_{ox} than that of MOSFET. This again arises from the lightly-doped drain and source regions in the LDD MOSFET, the conductivity of which depends more strongly on the vertical electric field than that of the heavily-doped drain and source regions in the conventional MOSFET.

REFERENCES

[1] J. J. Liou, *Advanced Semiconductor Device Physics and Modeling*, Boston, MA: Artech House, Inc., 1994.

[2] E. Takeda, H. Kume, T. Toyabe, and S. Asai, "Submicron MOSFET structure for minimizing hot-carrier generation," IEEE Trans. Electron Devices, vol. 29, 1982.

[3] T. Terada and H. Muta, "A new method to determine effective MOSFET channel length," Jap. J. Appl. Phys., vol. 18, p. 953, 1979.

[4] MEDICI Manual, Technology Modeling Associates, Inc., Palo Alto, CA, 1993.

[5] R. Narayanan, A. Ortiz-Conde, J. J. Liou, F. J. Garcia Sanchez, and A. Parthasarathy, "Two-dimensional numerical analysis for extracting the effective channel length of short-channel MOSFETs," Solid-St. Electron., vol. 38, p. 1155, 1995.

[6] B. J. Sheu, C. Hu, P. K. Ko, and F. C. Hsu, "Source-and-drain series resistance of LDD MOSFETs," IEEE Electron Device Lett., vol. EDL-5, p. 365, 1984.

[7] P. Antognetti, C. Lombardi, and D. Antoniadis, "Use of process and 2-D MOS simulation in the study of doping profile influence on S/D resistance in short-channel MOSFETs," Tech. Digest IEDM, 1981, p. 574.

[8] G. J. Hu, C. Chang, and Y. Chia, "Gate-voltage-dependent effective channel length and series resistance of LDD MOSFETs," IEEE Trans. Electron Devices, vol. ED-34, p. 2469, 1987.

[9] M. H. Seavey, "Source and drain resistance determination for MOSFETs," IEEE Electron Device Lett., vol. EDL-5, p. 479, 1984.

[10] K. Takeuchi, N. Kasai, T. Kunio, and K. Terada, "An effective channel length determination method for LDD MOSFETs," IEEE Trans. Electron Devices, vol. 43, p. 580, 1996.

[11] K. Takeuchi, N. Kaisi, and K. Terada, "A new effective channel length determination for LDD MOSFETs," Proc. 1991 Int. Conf. Microelectronic Test Structures, 1991, p. 215.

[12] S. -W. Lee, "A capacitance-based method for experimental determination of metallurgical channel length of submicron LDD MOSFET's," IEEE Trans. Electron Devices, vol. 41, p. 403, 1994.

[13] S. Ogura, P. J. Tsang, W. W. Walker, D. L. Critchklow, and J. F. Shepard, "Design and characteristics of the lightly doped drain-source (LDD) insulated gate field-effect transistor," IEEE J. Solid-State Circuits, vol. SC-15, p. 424, 1980.

[14] A. Raychaudhuri, J. Kolk, M. J. Deen, and M. I. H. King, "A simple method to extract the asymmetry in parasitic source and drain resistances from measurements on a MOS device," IEEE Trans. Electron Dev., vol. 42, p. 1388, 1995.

[15] A. Ortiz-Conde, J. J. Liou, W. Wong, and F. J. García Sánchez, "A simple method to extract the difference of the drain and source series resistances in MOSFETs," Electron. Lett., vol. 30, p. 1013, 1994.

[16] A. Ortiz-Conde, F. J. García Sánchez and J. J. Liou, "An improved method for extracting the difference between the drain and source resistances in MOSFETs," Solid-St. Electron., vol. 39, p. 419, 1996.

[17] A. Ortiz-Conde, J. J. Liou, R. Narayanan and F. J. García Sánchez, "Determination of physical mechanism contributing to the difference between drain and source resistances in short-channel MOSFETs," Solid-St. Electron., vol. 39, p. 211, 1996.

[18] D. K. Schroeder, *Semiconductor Material and Device Characterization*, Wiley, New York, 1990.

[19] J. Ida, A. Kita, and F. Ichikawa, "A new extraction method for effective channel length on lightly doped drain MOSFET's," Proc. IEEE Int. Conf. on Microelectronic Test Structures, vol. 3, p. 117, 1990.

[20] Z. Latif, A. Ortiz-Conde, J. J. Liou, F. J. Garcia Sanchez, W. Wong, and Y. G. Chen, "Analysis of the validity of methods for extracting the effective channel length of short-channel LDD MOSFETs," Solid-St. Electron., vol. 39, p. 1093, 1996.

Appendix A

Physical constants and unit conversions

Quantity	Symbol	Value	Unit
Boltzmann constant	k	$1.38x10^{-23}$	J/K
		$8.617x10^{-5}$	eV/K
Electron charge	q	$1.602x10^{-19}$	C
Electron rest mass	m_0	$9.109x10^{-31}$	kg
Permittivity of free space	ϵ_0	$8.8542x10^{-14}$	F/cm
Planck constant	h	$6.62607x10^{-34}$	J.s
		$4.13567x10^{-15}$	eV.s
Thermal voltage (T=300K)	kT/q	0.02585	V
Thermal energy (T=300K)	kT	0.02585	eV
Speed of light in vacuum	c	$2.998x10^{10}$	cm/s
Absolute temperature	T	273 + °C	K
Angstrom	Å	10^{-10}	m
Centimeter	cm	10^{-2}	m
Electron-volt	eV	$1.602x10^{-19}$	J
Micrometer	μm	10^{-6}	m
Nanometer	nm	10^{-9}	m
Picosecond	ps	10^{-12}	s
Wavelength of 1eV quantum	λ	1.2398	μm

Appendix B

Properties of germanium, silicon, and gallium arsenide (at 300 K)

Parameter	Unit	Ge	Si	GaAs
Atom density	#/cm^3	4.4×10^{22}	5.0×10^{22}	4.4×10^{22}
Atomic or molecular weight		72.6	28.08	144.63
Breakdown field	V/cm	10^5	3×10^5	4×10^5
Coefficient of linear				
thermal expansion	°C^{-1}	5.8×10^{-6}	2.6×10^{-6}	5.9×10^{-6}
Density	g/cm^3	5.323	2.32	5.316
Diffusion coefficient	cm^2/s			
(electron)		93	31	224
(hole)		44	12.5	10.4
Effective density of states	cm^{-3}			
conduction band		1.04×10^{19}	2.8×10^{19}	4.7×10^{17}
valence band		6.1×10^{18}	1.02×10^{19}	7.0×10^{18}
Electron affinity	eV	4.0	4.05	4.07
Energy bandgap (indirect)	eV	0.67	1.12	1.77
Energy bandgap (direct)	eV	0.804	2.0	1.424
Intrinsic Debye length	µm	0.68	24	2250
Intrinsic carrier concentration	cm^{-3}	2.37×10^{13}	1.04×10^{10}	5×10^6
Intrinsic mobility (drift)	cm^2/V-s			
(electron)		3900	1350	8600
(hole)		1900	480	400
Lattice constant	Å	5.6575	5.4308	5.6534
Relative dielectric constant		16	11.8	13.1
Thermal conductivity	W/cm-°C	0.606	1.5	0.455
Melting point	°C	937	1420	1238

Appendix C

Properties of SiO_2 and Si_3N_4 (at 300 K)

Parameter	Unit	Si_2O	Si_3N_4
Breakdown field	V/cm	10^7	10^7
Density	g/cm^3	2.27	3.44
Coefficient of linear thermal expansion	°C^{-1}	$5x10^{-7}$	$2.8x10^{-6}$
Electron affinity	eV	0.9	
Energy bandgap	eV	9	5
Molecular density	#/cm^3	$2.3x10^{22}$	$1.48x10^{22}$
Molecular weight		60.08	140.28
Relative dielectric constant		3.9	7.5
Thermal conductivity	W/cm-°C	0.014	0.185
Melting point	°C	1700	1900

Appendix D

Derivation of the integral function and its applications to parameter extraction

In this appendix we will derive the integral function used in a threshold-voltage extraction method developed in Chapter 3 (see Sec. 3.2). The applications of such a function to parameter extraction of two-terminal devices will also be discussed.

D.1 General theory

Consider two generalized two-terminal devices, d and r, which are connected in series. A terminal voltage V is applied to the series combination causing a current I to flow through the devices which produces voltage drops V_d and V_r across them, respectively (see Fig. D.1). Let the current I be described by the functions $f(V_d)$ and $h(V_r)$ of the devices d and r, as well as by the function g(V) of the series combination of the two devices. This gives

$$I = f(V_d) = h(V_r) = g(V) \quad . \tag{D.1}$$

The voltages can be expressed by their corresponding inverse functions of current as

$$V = g^{-1}(I) = f^{-1}(I) + h^{-1}(I) \tag{D.2}$$

where $f^{-1}(I)=V_d$ and $h^{-1}(I)=V_r$. Using geometric considerations, equivalent to integrating by parts, the following equalities can be written:

$$\int_0^{V_{d0}} I \, dV_d + \int_0^{I_0} V_d \, dI = I_0 \, V_{d0} \quad , \tag{D.3}$$

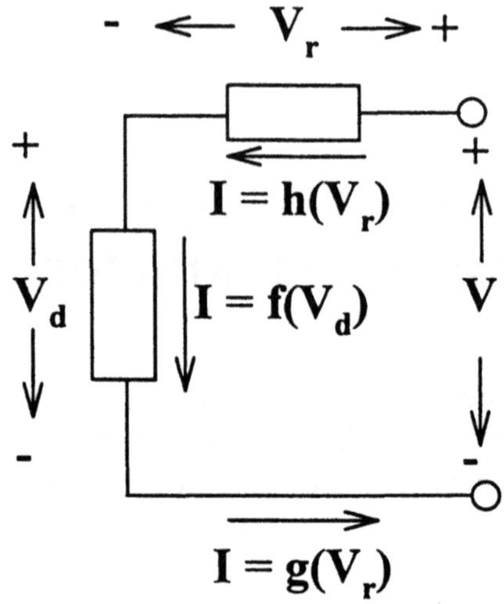

Figure D.1 : A generalized two-terminal device.

$$\int_0^{V_{r0}} I \, dV_r + \int_0^{I_0} V_r \, dI = I_0 \, V_{r0} \tag{D.4}$$

and

$$\int_0^{V_0} I \, dV + \int_0^{I_0} V \, dI = I_0 \, V_0 \tag{D.5}$$

where V_{d0}, V_{r0} and V_0 are upper voltage integration limits corresponding to an upper current integration limit I_0 at a certain point in the current-voltage characteristics of the devices and terminals. Integrating the sum of the two voltages of the devices with respect to current from zero to I_0 yields

$$\int_0^{I_0} V_d \, dI + \int_0^{I_0} V_r \, dI = \int_0^{I_0} V \, dI \tag{D.6}$$

Analogously, integrating the current gives:

$$\int_0^{V_{d0}} I\, dV_d + \int_0^{V_{r0}} I\, dV_r = \int_0^{V_0} I\, dV \tag{D.7}$$

We can define a "difference integral function" D as

$$D \equiv \int_0^{I_0} V\, dI - \int_0^{V_0} I\, dV \tag{D.8}$$

Finally, putting (D.6) and (D.7) into (D.8),

$$D = \left(\int_0^{I_0} V_d\, dI - \int_0^{V_{d0}} I\, dV_d \right) + \left(\int_0^{I_0} V_r\, dI - \int_0^{V_{r0}} I\, dV_r \right) \tag{D.9}$$

Each integral in (D.8) has units of power, and the addition of the two integrals represents the total power of the two devices in series. This integral function D is valid [1-2], not only for this case of two devices in series, but also in general for any number of generalized devices connected in series, provided the two-term addition on the right-hand side of (D.9) is replaced by a summation over all the devices involved. It could also be demonstrated that the integral function D holds for parallel or series-parallel mixed connections of generalized two-terminal devices. It should be mentioned that the D function has a form analogous to the Tellegen theorem of conservation of power [3], which has the same expression as (D.9) except the minus signs are replaced by plus signs.

In the following, we will prove that the summation of the nonlinearities, as defined by (D.8), of an arbitrary network with any number of generalized elements is zero. This will be carried out by first demonstrating that the summation of the integrations of voltage with respect to its corresponding current (i.e., first term on the right-hand side of (D.8)) over all the elements is zero. Analogously, it will be shown that the summation of the integrations of each current with respect to the corresponding voltage (i.e., second term on the right-hand side of (D.8)) is zero.

Following the notation of Chua et al. [4] for a generalized circuit with n nodes and b branches, we can write Kirchhoff's Current Law,

$$A\ I = 0 \tag{D.10}$$

and Kirchhoff's Voltage Law,

$$V = A^T E \tag{D.11}$$

where A is the reduced incidence matrix of dimension (n-1)b, which defines the topology of the network, $I = (i_1,i_2,....,i_b)^T$ is the branch current vector, $V = (v_1,v_2,....,v_b)^T$ is the branch voltage vector and $E = (e_1,e_2,....,e_b)^T$ is the node voltage vector.

We now proceed to prove that the summation of the integrations of each branch current with respect to the corresponding branch voltage over all the elements is zero:

$$\sum_{k=1}^{b} \int_{v_{ki}}^{v_{kf}} i_k\, dv_k = 0 \tag{D.12}$$

where v_{ki} and v_{kf} are the initial and final integration limits for the k branch, respectively. We start by changing the variable of integration,

$$v_k = (v_{kf} - v_{ki})x + v_{ki} \tag{D.13}$$

such that the integral can be factored out of the summation, since the integration limits become the same for all k:

$$\begin{aligned} \sum_{k=1}^{b} \int_{v_{ki}}^{v_{kf}} i_k\, dv_k &= \sum_{k=1}^{b} \int_{0}^{1} i_k\, (v_{kf} - v_{ki})\, dx \\ &= \int_{0}^{1} \sum_{k=1}^{b} i_k\, (v_{kf} - v_{ki})\, dx \quad . \end{aligned} \tag{D.14}$$

Second, we use the matrix notation,

$$\int_0^1 \sum_{k=1}^{b} i_k \left(v_{kf} - v_{ki} \right) dx = \int_0^1 I^T \Delta V \, dx \tag{D.15}$$

where $\Delta V = \{(v_{1f} - v_{1i}),(v_{2f} - v_{2i}) \ldots,(v_{bf} - v_{bi})\}^T$ is the region of integration vector of the branch voltages. Third, we use the relationship between ΔV and $\Delta E = \{(e_{1f} - e_{1i}),(e_{2f} - e_{2i}) \ldots,(e_{bf} - e_{bi})\}^T$ given by (D.11). Putting this into (D.15) yields

$$\int_0^1 I^T \Delta V \, dx = \int_0^1 I^T A^T \Delta E \, dx \tag{D.16}$$

Finally, using matrix identities and (D.10) we obtain the proof:

$$\int_0^1 I^T A^T \Delta E \, dx = \int_0^1 (A \, I)^T \Delta E \, dx = 0 \quad . \tag{D.17}$$

Next, we will prove that the summation of the integrations of each branch voltage with respect to the corresponding branch current over all the elements is zero:

$$\sum_{k=1}^{b} \int_{i_{ki}}^{i_{kf}} v_k \, di_k = 0 \tag{D.18}$$

We start by changing the variable of integration,

$$i_k = (i_{kf} - i_{ki}) \, y + i_{ki} \tag{D.19}$$

such that the limits of all the integrations are from 0 to 1:

$$\sum_{k=1}^{b} \int_{i_{ki}}^{i_{kf}} v_k \, di_k = \sum_{k=1}^{b} \int_0^1 v_k \left(i_{kf} - i_{ki} \right) dy = \int_0^1 \sum_{k=1}^{b} v_k \left(i_{kf} - i_{ki} \right) dy \, . \tag{D.20}$$

Second, we use the matrix notation,

$$\int_0^1 \sum_{k=1}^{b} v_k \left(i_{kf} - i_{ki} \right) dy = \int_0^1 V^T \, \Delta I \, dy \tag{D.21}$$

where $\Delta I = \{(i_{1f} - i_{1i}),(i_{2f} - i_{2i}),\ldots,(i_{bf} - i_{bi})\}^T$ is the region of integration vector of the branch currents. Finally, substituting (D.11) into (D.21), we obtain the proof

$$\int_0^1 V^T \, \Delta I \, dy = \int_0^1 (A^T E)^T \, \Delta I \, dy = \int_0^1 E^T A \, \Delta I \, dy = 0 \tag{D.22}$$

Note that adding (D.12) and (D.18) and using integration by parts, we obtain

$$\sum_{k=1}^{b} \left(\int_{i_{ki}}^{i_{kf}} v_k \, di_k + \int_{v_{ki}}^{v_{kf}} i_k \, dv_k \right) = \sum_{k=1}^{b} \left(i_{kf} \, v_{kf} - i_{ki} \, v_{ki} \right) = 0 \tag{D.23}$$

which is a generalized form of Tellegen's Theorem [3].

On the other hand, subtracting (D.12) from (D.18) yields that the summation of the function D, defined in (D.8), over all the branches is zero:

D.2 Case of a constant series resistance

We now analyze the case for which one of the two devices in series is linear (i.e., a resistor) and the other device is nonlinear. Let device d be the nonlinear device and device r be the resistor with a resistance R. The current is

$$I = h(V_r) = \frac{V_r}{R} \tag{D.24}$$

Substituting (D.24) into the integral function D in (D.8) yields an expression that no longer contains the series resistance:

$$D \equiv \left(\int_0^{I_0} V\, dI - \int_0^{V_0} I\, dV \right) = \left(\int_0^{I_0} V_d\, dI - \int_0^{V_{d0}} I\, dV_d \right) \tag{D.25}$$

Thus the nonlinear behavior of the device has been isolated. This equation can be expressed in a way that only one numerical integration of the measured I-V data is required:

$$D = I_0\, V_0 - 2 \int_0^{V_0} I\, dV = I_0\, V_0 - 2 \int_0^{V_{d0}} I\, dV_d \tag{D.26}$$

This expression relates the I-V data measured at the terminals of the series combination, shown on the left-hand side, and the model of the nonlinear device, shown on the right-hand side of (D.26).

D.3 Diode with constant series resistance

In this section, we will illustrate how to use the integral function to extract parameters of a real p/n junction diode, which consists of an ideal diode (nonlinear device) and a series resistance (linear device).

The I-V characteristics of a real p/n junction diode can be modeled by the following exponential function:

$$I = g(V) = f(V_d) = I_S \left(\exp\left(\frac{V_d}{n V_{th}} \right) - 1 \right) \tag{D.27}$$

where I_S is the saturation current of the diode, V is the voltage across the terminals of the series combination, V_d is the intrinsic voltage across the diode junction (i.e., excluding the voltage drop on the series resistor R), n is the diode ideality factor, and V_{th} is the thermal voltage. Note that

$$V_d = V - I\,R \tag{D.28}$$

The extraction of the parameters I_S and n is customarily performed either by direct numerical fitting, or by graphical analysis of ln(I) versus V plots. For the latter approach, however, the presence of R can significantly reduce the linear portion of those plots to such an extent that the determination of I_S and n becomes unreliable. Eliminating the effect of R is therefore very important for the extraction procedure, which is the essence of the integral function developed.

Substituting (D.27) into the right-hand side of (D.26) and considering only the region where $I_0 >> I_s$, we obtain

$$D \approx I_0 \; V_{th} \; n \left(\ln\left(\frac{I_0}{I_S}\right) - 2 \right) \tag{D.29}$$

Dividing this equation by the current I_0, we can define an auxiliary function $G(I_0, V_0)$ of the experimental terminal current and voltage:

$$G(I_0, V_0) = \frac{D}{I_0} = \frac{I_0 \; V_0 - 2 \int_0^{V_0} I \; dV}{I_0} = V_{th} \; n \left(\ln\left(\frac{I_0}{I_S}\right) - 2 \right) \tag{D.30}$$

When the function G, obtained by numerical calculation from the experimental data, is plotted against $\ln(I_0)$, it should produce a straight line, whose slope and intercept on the voltage axis allow the determination of n and I_S, respectively. Note that the parameters extracted by the integral function method are not obscured and affected by the value of the parasitic resistance R. This is not possible from the conventional extraction method based on the ln(I) versus V plots, particularly when R is large.

REFERENCES

[1] F. J. García Sánchez, A. Ortiz-Conde, G. Mercato, J. J. Liou, and L. Recht, "Eliminating parasitic resistances in parameter extraction of semiconductor device models", Int. Caracas Conf. on Cir. Dev. and Sys., (Caracas, Venezuela), pp. 298-302, Dec. 1995.

[2] F.J. García Sánchez, A. Ortiz-Conde, and J.J. Liou, "A parasitic series resistance-independent method for device-model parameter extraction", IEE Proc. Cir. Dev. and Sys., vol. 143, pp. 68-70, Feb. 1996.

[3] B. D. H. Tellegen, "A general network theorem, with applications," Philips Res. Rep., vol. 7, p. 259, 1952.

[4] L. O. Chua, C. A. Desoer, and E. S. Kuh, *Linear and nonlinear Circuits,* New York: McGraw-Hill, 1987.

Subject index

About the Authors

Juin J. Liou received the B.S. (with honors), M.S., and Ph.D. degrees in electrical engineering from the University of Florida, Gainesville, Florida, USA in 1982, 1983, and 1987, respectively.

From 1985 to 1986, he was an instructor in the Department of Electrical Engineering at the University of Florida. In 1987, he joined the Department of Electrical Engineering at the University of Central Florida, Orlando, Florida, where he is now a full professor. His current research interests are in semiconductor device physics, modeling, simulation, and reliability.

Dr. Liou has published three textbooks, Advanced Semiconductor Device Physics and Modeling (Artech House, 1994), Principles and Analysis of AlGaAs/GaAs Heterojunction Bipolar Transistors (Artech House, 1996) and Semiconductor Device Simulation and Characterization (Plenum Press, 1998). In addition, he has authored more than 250 technical papers in refereed journals and international and national conference proceedings, and has presented more then 30 invited seminars or conference papers in several countries. He has held consulting positions with research laboratories and companies in the United States, Taiwan and Japan. He also serves as a technical reviewer for several journals and conferences, and is an associate editor for the Simulation Journal in the area of VLSI and circuit simulation. He has worked on numerous semiconductor device modeling and integrated circuit simulation projects and has been awarded more than $1.5 million in research grants from industry, state, and federal agencies.

Dr. Liou's honors and awards include Distinguished Researcher Award, Electrical Engineering Department, University of Central Florida, 1989, 1990, 1992, 1995, 1997, and 1998; Distinguished Researcher Award, College of Engineering, University of Central Florida, 1992 and 1998; Senior College Research Award, University of Central Florida, 1993; Faculty Outstanding Award, Student Engineering Council, University of Central Florida, 1993; Engineer of the Year, IEEE Orlando Section, 1992; Eminent Engineer, Tau Beta Pi, 1992; and Teaching Incentive Award, University of Central Florida, 1995. He was listed in Who's Who Among Young American Professionals, Who's Who in the South and Southwest, Who's Who in Technology, and Who's Who in Science and Engineering. In the Summer of 1992, 1993, and 1994, Dr. Liou was selected as a Summer Research Faculty at Solid State Laboratory, Wright-Patterson Air Force Base, Ohio, where he conducted research on modeling the

AlGaAs/GaAs heterojunction bipolar transistor. In the Fall of 1997, he held the position of Visiting Senior Fellow during his sabbatical leave with the National University of Singapore, Singapore.

Dr. Liou is a Senior Member of the Institute of Electrical and Electronics Engineers.

Adelmo Ortiz-Conde was born in Caracas, Venezuela, on November 28, 1956. He received the B.S. degree in Electronics from the Universidad Simón Bolívar, Caracas, in 1979 and the M.E. and Ph.D. from the University of Florida, Gainesville, in 1982 and 1985, respectively. His doctoral research was in the area of semiconductor device modeling under the guidance of Professor J. G. Fossum.

From 1979 to 1980, he served as an instructor in the Department of Electronics at the Universidad Simón Bolívar. In 1985, he joined the technical Staff of Bell Laboratories, Reading, PA, where he was engaged in the development of high voltage integrated circuits. Since 1987 he has been with the Department of Electronics at the Universidad Simón Bolívar and he was promoted to Full Professor in 1995. During his sabbatical leave 1993-1994, he was with Florida International University from September to December 1993, and with University of Central Florida from January to August 1994. His present research interest includes the modeling and parameter extraction of semiconductor devices.

He has published more than 60 international technical papers in specialized journals and conferences. Dr. Ortiz-Conde is a member of the Editorial Advisory Board of Microelectronic and Reliability and he has served as reviewer for national and international journals and conferences. He was the technical chairperson of the Second IEEE International Caracas Conference on Devices, Circuits and Systems, held in March 1998, and he was the general chairperson of the first edition of this conference in 1995.

Dr. Ortiz-Conde is a Senior Member of the IEEE and a member of Eta Kappa Nu, Tau Beta Pi, Phi Kappa Phi and the Galilean Society.

Francisco J. García-Sánchez received the B.E.E., M.E.E. and Ph.D. degrees in electrical engineering from the Catholic University of America, Washington, DC., USA, in 1970, 1972 and 1976, respectively. He is a Full Professor of the Electronics Department at Simon Bolivar University, Caracas, Venezuela, where he has held several academic positions, such as: member of the University's Academic Council, Coordinator of Research and Development for Engineering and Applied Sciences, and Coordinator for Undergraduate and Graduate Studies in Electronics Engineering. Prof. García-Sánchez presently is the head of the Solid State Electronics Laboratory and a member of the University's Superior Council.

He made contributions to the area of photovoltaic devices and materials and to the development of thin and thick film low-cost fabrication techniques for compound semiconductors, such as spray pyrolysis and screen printing. Currently his research interests are in the areas of semiconductor device modeling and electrical characterization and modeling of biological tissues. He has authored numerous technical papers and presentations in specialized journals and conferences. He has written technical and general interest science and engineering invited papers and presented invited talks at different seminars, meetings and conferences. He has served as reviewer for national and international journals and as session chairman at technical conferences. He was the editor of a book in Spanish on semiconductor devices and integrated circuits.

Prof. García-Sánchez has directed international projects and advanced courses on solar energy conversion. He was the Technical Committee Chairman of the first (1995) IEEE International Caracas Conference on Devices, Circuits and Systems (ICCDCS) and was the Organization Chairman of the second edition (ICCDCS-1998). Prof. García-Sánchez participated in drawing up Venezuela's National Program for New Technologies, where he also acted as that Program's first national Coordinator for the area of Electronics. He has received several prizes for excellence in Applied Sciences research. He is listed in Who's Who in Science and Engineering.

Prof. García-Sánchez is a founding member of the Galilean Society and its present Vice-President. He is a Senior Member of the IEEE and chairs IEEE-Venezuela's EDS/CSS/PEL/EMB chapter.

MIX
Papier aus verantwortungsvollen Quellen
Paper from responsible sources
FSC® C105338

If you have any concerns about our products,
you can contact us on
ProductSafety@springernature.com

In case Publisher is established outside the EU,
the EU authorized representative is:
Springer Nature Customer Service Center GmbH
Europaplatz 3, 69115 Heidelberg, Germany

Printed by Libri Plureos GmbH
in Hamburg, Germany